Sustainable Agriculture Reviews

Volume 51

Other Publications by Dr. Eric Lichtfouse

Books
Scientific Writing for Impact Factor Journals
https://www.novapublishers.com/catalog/product_info.php?products_id=42242

Environmental Chemistry
http://www.springer.com/978-3-540-22860-8

Sustainable Agriculture
Volume 1: http://www.springer.com/978-90-481-2665-1
Volume 2: http://www.springer.com/978-94-007-0393-3

Book series
Environmental Chemistry for a Sustainable World
http://www.springer.com/series/11480

Sustainable Agriculture Reviews
http://www.springer.com/series/8380

Journal
Environmental Chemistry Letters
http://www.springer.com/10311

Sustainable agriculture is a rapidly growing field aiming at producing food and energy in a sustainable way for humans and their children. Sustainable agriculture is a discipline that addresses current issues such as climate change, increasing food and fuel prices, poor-nation starvation, rich-nation obesity, water pollution, soil erosion, fertility loss, pest control, and biodiversity depletion.

Novel, environmentally-friendly solutions are proposed based on integrated knowledge from sciences as diverse as agronomy, soil science, molecular biology, chemistry, toxicology, ecology, economy, and social sciences. Indeed, sustainable agriculture decipher mechanisms of processes that occur from the molecular level to the farming system to the global level at time scales ranging from seconds to centuries. For that, scientists use the system approach that involves studying components and interactions of a whole system to address scientific, economic and social issues. In that respect, sustainable agriculture is not a classical, narrow science. Instead of solving problems using the classical painkiller approach that treats only negative impacts, sustainable agriculture treats problem sources.

Because most actual society issues are now intertwined, global, and fast-developing, sustainable agriculture will bring solutions to build a safer world. This book series gathers review articles that analyze current agricultural issues and knowledge, then propose alternative solutions. It will therefore help all scientists, decision-makers, professors, farmers and politicians who wish to build a safe agriculture, energy and food system for future generations.

More information about this series at http://www.springer.com/series/8380

Praveen Guleria • Vineet Kumar
Eric Lichtfouse
Editors

Sustainable Agriculture Reviews 51

Legume Agriculture and Biotechnology Vol 2

Springer

Editors
Praveen Guleria
Plant Biotechnology and Genetic Engineering Lab, Department of Biotechnology
DAV University
Jalandhar, Punjab, India

Vineet Kumar
Department of Biotechnology, School of Bioengineering and Biosciences
Lovely Professional University
Phagwara, India

Eric Lichtfouse
CNRS, IRD, INRAE, Coll France, CEREGE
Aix-Marseille University
Aix-en-Provence, France

ISSN 2210-4410 ISSN 2210-4429 (electronic)
Sustainable Agriculture Reviews
ISBN 978-3-030-68827-1 ISBN 978-3-030-68828-8 (eBook)
https://doi.org/10.1007/978-3-030-68828-8

This Springer imprint is published by the registered company Springer Nature Switzerland AG
The registered company address is: Gewerbestrasse 11, 6330 Cham, Switzerland

Preface

If you want a healthy, nutrient dense, poor man's diet you can't go wrong with whole grains and legumes

Harken Headers

Legumes are major constituents of vegetarian diets. Grain and forage legumes are cultivated on more than 12% of the world's arable land. Grain legumes alone contribute more than 33% of the total human protein, and 35% of the total needs in vegetable oil. The protein-rich nature of forage legumes is adapted as a feed source in the chicken and poultry industry. Legumes are also rich in polyhenols, antioxidants, and carbohydrates. In the context of climate change, pollution, and food safety, the current challenge is to enhance legume production to sustain the growing population needs by 2050. This is a daunting task because abiotic and biotic stresses are threatening the growth, survival, and productivity of legumes. For instance, the productivity of legumes is documented to be reduced by 14–88% by abiotic stresses (Figure). The co-occurrence of abiotic and biotic stresses under field conditions leads to interactive stress types, thus yielding positive or negative outcomes. Legumes react using antioxidant defense, osmoregulatory adjustments, hormonal regulations, and molecular mechanisms to tolerate stress. Hence, improving legume productivity requires knowledge on the sensitivity, mechanisms and approaches of stress tolerance in legumes, in order to design new crops and alternative management systems. This book presents advances on bioactive compounds, applications, effect of various stresses, and biotechnology-based stress tolerance mechanisms of legumes. This is our second volume on legume agriculture and biotechnology, published in the series Sustainable Agriculture Reviews.

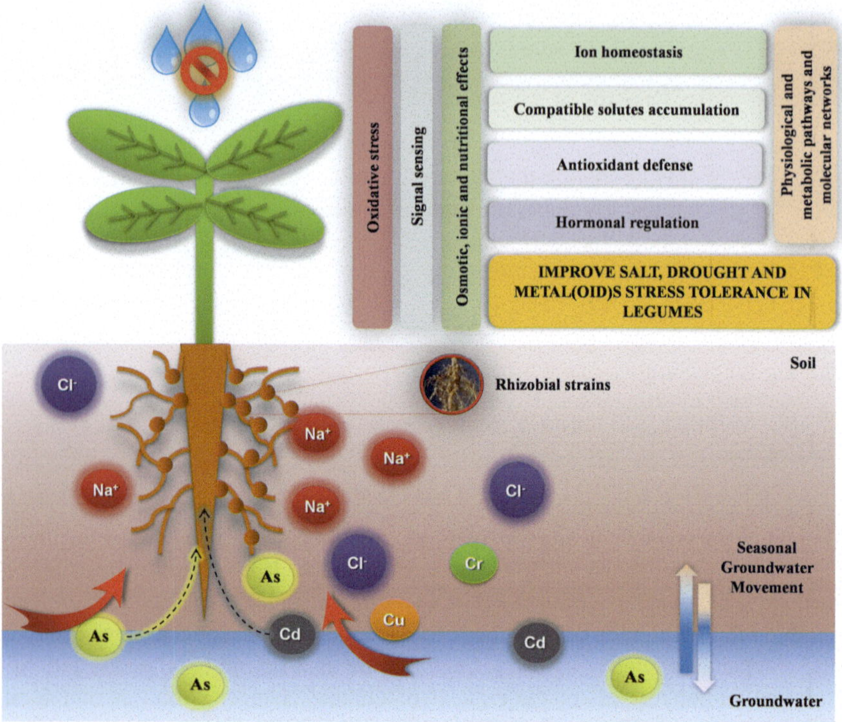

Abiotic stresses on legume crops. Legumes can be stressed by salinity, e.g., high concentration of Na^+ and Cl^-; drought; and metals and metaloids, e.g., high concentrations of Cu, Cd, Cr, and As in groundwater and soil. Plant stress typically generates the production of reactive oxygen and nitrogen species (ROS, RNS), which act both as cell signaling and damaging agents to biomolecules, thus inducing oxidative stress. Plant sensing of stresses generate osmotic, ionic, and nutritional effects that trigger ionic homeostasis maintenance, compatible solutes accumulation, antioxidant defense, and hormonal regulation. Therefore, understanding physiological and metabolic pathways helps to design strategies to improve legume tolerance under unfavorable conditions. From Furlan et al. in Chap. 6

The first chapter by Vijayakumar and Haridas introduces the various aspects of legumes such as nutritional composition and pharmaceutical applications. The second chapter by Agnihotri and Rana focuses on the nutraceutical potential of the underutilized horse gram. The nutritional profile and the application of grain legumes in the poultry industry is reviewed in Chap. 3 bay Nalluri and Karri. Avci discusses annual forage legumes in crop rotations and sustainable agriculture production in Chap. 4. The alternative splicing-based strategy for gene regulation to enhance legume productivity is presented by Dass et al. in Chap. 5; this strategy improves pathogenic immunity and prevents cell aging. Furlan et al. discuss abiotic stresses affecting legume yield and tolerance mechanisms in Chap. 6. Biotic stress inducers with focus on fungal diseases and their management are reviewed by Mahmoud in Chap. 7. Suthar et al. decipher the molecular interactions of chickpea with fungus *Fusaruim oxysporum* f. sp. *ciceris* in Chap. 8. In Chap. 9, Chandra et al. discuss biotechnological tools for stress tolerance in leguminous crops to improve productivity. The final chapter by Chakraborty et al. summarizes molecular and physiological mechanisms regulating legume–pathogen interaction and various plant defense approaches.

The editors are thankful to all the authors who contributed to this book for their efforts in producing timely and high-quality chapters. The creation of this book would not have been possible without the assistance of several colleagues and friends deserving acknowledgment. They have helped by choosing contributors, reviewing chapters, and in many other ways. Finally, we would like to thank the staff at Springer Nature for their highly professional editing of the publication.

Jalandhar, Punjab, India Praveen Guleria

Phagwara, Punjab, India Vineet Kumar

Aix-en-Provence, France Eric Lichtfouse

Contents

About the Editors

Praveen Guleria is presently working as assistant professor in the Department of Biotechnology at DAV University, Jalandhar, Punjab, India. She has worked in the areas of plant biotechnology, plant metabolic engineering, and plant stress biology at CSIR–Institute of Himalayan Bioresource Technology, Palampur, H.P., India.

Her research interests include plant stress biology, plant small RNA biology, plant epigenomics, and nanotoxicity. Dr. Guleria has published several research articles in various peer-reviewed journals. She is also serving as the editorial board member and reviewer for certain international peer-reviewed journals. Dr. Guleria has been awarded the SERB – Start Up Grant by DST, GOI. She has also been awarded the prestigious "Bharat Gaurav Award" by the India International Friendship Society, New Delhi. She has also received various awards like CSIR/ ICMR – Junior research Fellowship, CSIR – Senior research fellowship, and state-level merit scholarship awards.

Vineet Kumar is currently working as assistant professor in the Department of Biotechnology, Lovely Professional University, Jalandhar, Punjab, India. He has worked in different areas of biotechnology and nanotechnology at various institutes and universities in India, namely Panjab University Chandigarh; CSIR-Institute of Microbial Technology, Chandigarh, India; CSIR–Institute of Himalayan Bioresource Technology; and Himachal Pradesh University. He has published many articles in these areas featuring in peer-reviewed journals. Dr. Kumar is also serving as editorial board member and reviewer for international peer-reviewed journals. He has received various awards like Dr DSK-postdoctoral fellowship, senior research fellowship, and best poster awards.

Eric Lichtfouse is a biogeochemist at Aix Marseille University who has invented carbon-13 dating, a molecular-level method allowing to study the dynamics of organic compounds in temporal pools of complex environmental media. He is chief editor of the journal *Environmental Chemistry Letters* and the book series Sustainable Agriculture Reviews and Environmental Chemistry for a Sustainable World. He is the author of the book *Scientific Writing for Impact Factor Journals*, which includes an innovative writing tool: the Micro-Article.

Contributors

Vasudha Agnihotri Centre for Land and Water Resource Management, G.B. Pant National Institute of Himalayan Environment, Almora, Uttarakhand, India

Süleyman AVCI Faculty of Agriculture, Department of Field Crops, Eskişehir Osmangazi University, Eskişehir, Turkey

Rajib Bandopadhyay UGC-Center of Advanced study, Department of Botany, The University of Burdwan, Bardhaman, West Bengal, India

Jolly Basak Department of Biotechnology, Sikhsha Bhavan, Visva-Bharati, Santiniketan, West Bengal, India

Eliana Bianucci Instituto de Investigaciones Agrobiotecnológicas - Consejo Nacional de Investigaciones Científicas y Técnicas (INIAB- CONICET), Departamento de Ciencias Naturales, Facultad de Ciencias Exactas, Físico-Químicas y Naturales, Universidad Nacional de Río Cuarto (UNRC), Río Cuarto, Córdoba, Argentina

Stella Castro Instituto de Investigaciones Agrobiotecnológicas - Consejo Nacional de Investigaciones Científicas y Técnicas (INIAB- CONICET), Departamento de Ciencias Naturales, Facultad de Ciencias Exactas, Físico-Químicas y Naturales, Universidad Nacional de Río Cuarto (UNRC), Río Cuarto, Córdoba, Argentina

Nibedita Chakraborty Department of Biotechnology, National Institute of Technology, Durgapur, West Bengal, India

Priyanka Chakraborty UGC-Center of Advanced study, Department of Botany, The University of Burdwan, Bardhaman, West Bengal, India

P. S. Chandrashekharaiah Reliance Industries Ltd., Jamnagar, Gujarat, India

Santanu Dasgupta Reliance Industries Ltd., Navi Mumbai, Maharashtra, India

Regina Sharmila Dass Fungal Genetics and Mycotoxicology Laboratory, Department of Microbiology, School of Life Sciences, Pondicherry University, Pondicherry, India

Ana Furlan Instituto de Investigaciones Agrobiotecnológicas - Consejo Nacional de Investigaciones Científicas y Técnicas (INIAB- CONICET), Departamento de Ciencias Naturales, Facultad de Ciencias Exactas, Físico-Químicas y Naturales, Universidad Nacional de Río Cuarto (UNRC), Río Cuarto, Córdoba, Argentina

Haridas M Inter University Centre for Biosciences, Department of Biotechnology and Microbiology, Kannur University, Kannur, Kerala, India

Vasavi Rama Karri Department of Biotechnology, GITAM Institute of Technology, GITAM (Deemed to be University), Visakhapatnam, Andhra Pradesh, India

Shivbachan Kushwaha Reliance Industries Ltd., Jamnagar, Gujarat, India

Analía Llanes Instituto de Investigaciones Agrobiotecnológicas - Consejo Nacional de Investigaciones Científicas y Técnicas (INIAB- CONICET), Departamento de Ciencias Naturales, Facultad de Ciencias Exactas, Físico-Químicas y Naturales, Universidad Nacional de Río Cuarto (UNRC), Río Cuarto, Córdoba, Argentina

Ghada Abd-Elmonsef Mahmoud Botany and Microbiology Department, Faculty of Science, Assiut University, Assiut, Egypt

Rathijit Mallick Fungal Genetics and Mycotoxicology Laboratory, Department of Microbiology, School of Life Sciences, Pondicherry University, Pondicherry, India

Nirmala Nalluri Department of Biotechnology, GITAM Institute of Technology, GITAM (Deemed to be University), Visakhapatnam, Andhra Pradesh, India

Preeti R. Parmar Main Cotton Research Station, Navsari Agricultural University, Surat, Gujarat, India

Vishal Paul Reliance Industries Ltd., Jamnagar, Gujarat, India

Juan Manuel Peralta Instituto de Investigaciones Agrobiotecnológicas - Consejo Nacional de Investigaciones Científicas y Técnicas (INIAB- CONICET), Departamento de Ciencias Naturales, Facultad de Ciencias Exactas, Físico-Químicas y Naturales, Universidad Nacional de Río Cuarto (UNRC), Río Cuarto, Córdoba, Argentina

B. K. Rajkumar Main Cotton Research Station, Navsari Agricultural University, Surat, Gujarat, India

Smita Rana Centre for Land and Water Resource Management, G.B. Pant National Institute of Himalayan Environment, Almora, Uttarakhand, India

Debanjan Sanyal Reliance Industries Ltd., Jamnagar, Gujarat, India

Diwakar Singh Department of Plant Molecular Biology and Biotechnology, ASPEE College of Horticulture and Forestry, Navsari Agricultural University, Navsari, Gujarat, India

K. P. Suthar Department of Plant Molecular Biology and Biotechnology, ASPEE College of Horticulture and Forestry, Navsari Agricultural University, Navsari, Gujarat, India

Pooja Thorat Fungal Genetics and Mycotoxicology Laboratory, Department of Microbiology, School of Life Sciences, Pondicherry University, Pondicherry, India

Vijaytha Vijayakumar Department of Biotechnology and Microbiology, Kannur University, Kannur, Kerala, India

Chapter 1
Nutraceutical Legumes: A Brief Review on the Nutritional and Medicinal Values of Legumes

Vijaytha Vijayakumar and Haridas M

Abstract The term 'Nutraceuticals' was coined to project the food or modified food items that can ameliorate diseases together with meeting the nutrients requirements. Legumes are a group of plants implicated with high nutritional quality, cheap cultivation needs and adaptation to climatic conditions which made them an important and essential food among the common men. The safe levels of anti-nutritional factors of legumes are significant in eliciting some bioactivities. Anti-nutritional factors of legume are mainly contributing to their richness of ingredients. They include phytochemicals like phytoestrogens, phytosterols, phytates, various enzyme inhibitors, lectins, polyphenols, alkaloids, saponins, flavonoids, oligosaccharides etc.

Here we review the nutritional and anti-nutritional profiles of legumes and how they provide the medicinal properties alleged on them. The anti-nutritionals, though limit bioavailability of nutrients, have very high bioactivities like anti-oxidant, anti-inflammatory, anti-cancerous, anti-diabetic, hepato-protective etc. Essaying the nutritional and pharmaceutical effects of the legumes would highlight them as better nutraceuticals of future.

Keywords Nutraceutical-legumes · Anti-nutrients · Phytochemical · Bioactivities

---

V. Vijayakumar (✉)
Department of Biotechnology and Microbiology, Kannur University, Kannur, Kerala, India

Haridas M
Inter University Centre for Biosciences, Department of Biotechnology and Microbiology, Kannur University, Kannur, Kerala, India

P. Guleria et al. (eds.), *Sustainable Agriculture Reviews 51*, Sustainable Agriculture Reviews 51, https://doi.org/10.1007/978-3-030-68828-8_1

1.1 Introduction

The famous Hippocratic prayer, "let food be thy medicine and medicine be thy food", explains profoundly the medicinal values of our food. Nowadays, food is considered just only for nutrition that satisfies our hunger too. Medicine is something that cures diseases and saves our life. Modern science appropriated the words of Hippocrates in to "Nutraceuticals". Great amounts of researches are going on in this area to find out better nutraceuticals that can meet the need of present day world. The global population has been projected to be in between 9.4 and 10.1 billion (United Nations 2019), it is also projected that the people who would be below poverty line then is still going to be unappealingly high. Rosa (2017) reported that several model programs of UN have been running successful only to a limited extent, and required to go further to make the world free from poverty by 2030 at least. Major challenges for alleviating poverty are the population growth and the climatic changes. Climatic changes can particularly bring down the world food production which in turn widen the gulf between the demand and supply; also between the demand and purchase power (Ray et al. 2019). As a consequence morbidity and mortality rate would increase. This disaster can be overcome to an extent by sustainable improvement in the production of quality food. Quality means of both nutritional and medicinal. In this context, the nutritional and medicinal values that legumes possess must be seen with novel ideas. In general, legumes are a multifunctional food crop with high nutritional and medicinal properties. They also have a tremendous capacity to withstand adverse environmental conditions (Yarlagadda 2013).

"Legume" is a word derived from the Latin word *legere* which means 'to gather' (Albala 2007). The term legume is referred to a group of plants with its seeds arranged inside a pod. Technically the word is used to refer to plants coming under the family Leguminosae or Fabaceae or the seeds produced by such plants. Family Leguminosae is one of the largest families in the plant kingdom with around 20,000 species in 800 genera (Smýkal et al. 2015). This includes pasture, ornamentals, medicinal and agro-forestry species (Sutjaritjai et al. 2019). They have characteristic flower structure, seed pods and ability to form symbiotic relationship with soil rhizobia. Some of the important beneficiary roles of legumes include (i) food for both human and animal consumption, (ii) major plant protein source for improving human health, (iii) increasing soil fertility by fixing atmospheric nitrogen, (iv) to alleviate the effects of green house gases and (v) also used as diversification crop in certain agro ecosystem (Stagnari et al. 2017). Legumes commonly included in our foods are alfalfa, chick peas, kidney beans, lentils, mung beans, peanuts, peas, soya beans, pigeon peas etc. (Ahmed and Hasan 2014). Nutritionally and in health benefits Leguminosae comes only second to Gramineae (Graham and Vance 2003).

1.2 Legumes in History

Extensive molecular analysis carried out by Lavin et al. (2005) concluded that legumes evolved around 60 million years ago. According to older hypothesis legumes were evolved from western Gondwana whereas recent hypothesis points towards its Laurasian origin. Boreal hypothesis described the spread of legumes from America directly to northwards and then to southwards. This hypothesis was later tested by Schrire et al. (2005) and concluded that legumes were first originated in the northern region of Tethys Sea and spread to the other regions later (Sprent 2007). Man used legumes for enriching soil years before. He knew about its food value. Since then they played important role in the cropping system. Romans introduced the crop rotation system with special emphasis on legumes. Greek philosopher and botanist Theophrastus described legumes as "reinvigorating" the soil; beans are not a burden to the soil but manure. Grain legumes were constant companions of cereals in the beginning of crop domestication. Clear evidences of lentil cultivation (10,100 to 9700 cal BP), pea, chick pea and vetch cultivation (9900 to 9500 cal BP) support the understanding (Weiss et al. 2012). Many millennia old Indian practices of mixed farming and crop rotation of cereals, legumes and other plants of human uses have been well documented. Legumes are held equally important as cereals even in that period of civilization (Roy 2009).

Uses of various legumes were found in the data from the oldest civilizations of Egypt and Eastern Asia. They used soya beans, beans, alfalfa, peas and vetches (Smýkal et al. 2020). Evidences of wild lentils eaten naturally at about 11,000 BC in Greece before their agrarian production, evidences of carbonized seeds of some legumes in the Neolithic age (7000–8000 BC) were found from Turkey, pea cultivation among lake dwellers in Switzerland (4000–5000 BC), soya bean cultivation in China (2000–3000 BC), beans and soya bean cultivation in America and Staple crops cultivation in Asia (1000–2000 BC) adds to legume history (Ahmed and Hasan 2014). Some Cuniform recipes of 1500 BC are found with lentils (Albala 2007). Meditteranian regions were in surplus, with different varieties of legumes under different climatic conditions (Flint-Hamilton 1999). Soya beans, though originated in China, spread worldwide and gained its name "miracle crop" due to its versatile growth manifestation (Hymowitz 2008). Every crop in the legume family has its own history with distinct origin, distribution in agro ecosystem, food, forage and medicinal areas.

1.3 Nutritional Values of Legumes

Food legumes are of two types, oil seeds and pulses. Oil seeds, as the name suggests, are high oil and high protein foods like soya bean, ground nut, peanut, lupine etc. Pulses are moderate protein and low oil foods like peas, lentils, cowpeas, grams etc. They are the dry seeds of cultivated legumes (Yarlagadda 2013; Maphosa and

Jideani 2017). Food and Agriculture Organization (FAO) defines pulses as the crops harvested solely for the dry seeds, excluding those legumes which are consumed green as vegetables, oil seeds and the ones used for sowing purposes. Word legume and pulses may be used interchangeably (Kouris-Blazos and Belski 2016). Legumes, as they are rich in proteins, unsaturated fats, dietary fiber, essential minerals, vitamins and complex carbohydrates are rich in their nutritional constituents as may be learnt from the general description (Maphosa and Jideani 2017). Along with high nutritional quality, cheap cultivation needs and adaptation to climatic conditions made legumes an essential food among common man. Hence they have been called as "poor man's meat" (Messina 1999; Yarlagadda 2013). For the less earning groups of population, legumes are a good protein source which are inexpensive to animal proteins (Iqbal et al. 2006).

Traditional diets of many countries like India, South America, Mexico, Middle East etc. are surplus with legumes but their consumption in western countries were low and somewhat declined in European countries (Zander et al. 2016). The main reasons behind this rejection are their bloating or flatulent effect. Their rustic image and impression of them as animal fodder were other reasons of reduced acceptance among Europeans. Now a day there is improvement in this perception as there is increase in awareness and evidence of the health benefits of legumes. There is a change in man's attitude towards consumption of legumes as a worthy food stuff. As a consequence legume cultivation is being shifted from household level to large areas of farmable land (Erbersdobler et al. 2017). 2016 was declared as the International Year of Pulses by FAO of the United Nations and they mainly covered 11 primary classes of legumes.

Since legumes are found as an economical dietary source of nutrients there is an increase in its consumption rate. In many countries foods complemented with legumes are used to get rid of malnutrition and this reflects how promising legumes could be and are useful for the undernourished (Mudryj et al. 2014). Protein Energy Malnutrition (PEM) is one of most common nutrient related deficiency condition in the developing countries. Out of 852 million nutrient related deficiency condition affected people in the world 815 million are of the developing countries (Müller and Krawinkel 2005). Legumes, as they can provide dietary proteins and micronutrients, furnish possible benefits in developing countries to meet their growing needs of nutrition (de Jager et al. 2019). *Dietary guideline for Indians* by Indian Council of Medical Research identified pulses as one of the most recommended food for daily consumption. It is specifically mentioned that 30 g and 60 g of uncooked pulses should be consumed respectively by non-vegetarians and vegetarians daily (Marinangeli et al. 2017). Dietary guidelines and recommended intake in many other countries are also available (Marinangeli et al. 2017) (Table 1.1).

Table 1.1 Dietary guidelines for the use of pulses in different countries

Name of the Country	Food categories and legumes included	Required quantity of cooked pulses	Frequency of consumption
Australia	Vegetables and legumes (beans, lentils, chickpeas) Meat, poultry, fish, eggs and legumes (beans, lentils, chickpeas)	75 g 150 g	5 times/ day Minimum 2 times/ week
Bulgaria	Meat and legumes (beans, chickpeas, peas, lentils)	200 to 300 g	Minimum 2 times/ week
Canada	Meat and legumes (kidney beans, lentils, chickpeas)	0.75 of a cup	Couple of times/ week
Greece	Olives, nuts and pulses	100 g	3–4 times/ week
India	Cereals, millets, and pulses	30 g	1 time/day (non-vegetarians) 2 times/day (vegetarians)
Ireland	Meat, fish, poultry, eggs, nuts and legumes (beans, lentils, peas)	0.75 of a 200 mL cup	2 times/ day
Spain	Beans, lentils, and chickpeas	150 to 200 g	Minimum 2 times/ week
South Africa	Dry beans, peas, lentils, and soya	0.5 of 200 mL cup	Included in food each week
United Kingdom	Vegetables, fruit and legumes (beans, peas, lentils) Meat, fish, eggs, beans, and other protein sources	80 g	5 times/ day Included in every day meals
United States	Vegetables (beans, peas, lentils, and chickpeas)	0.5 of 250 mL cup	1–3 times/ week

1.4 Nutrient Profile of Legumes

Legumes are mainly known for their rich protein content. They are the main protein source for vegetarians. They can also supply different types of essential nutrients needed for human body like carbohydrates, dietary fiber, fats, vitamins and minerals. Pulses are high in protein, folic acid and minerals like iron, magnesium, potassium and zinc (Singh 2017). Legumes are consumed along with cereals by majority of the people. The only exemption is soya bean, which has low carbohydrate, high fat and sulfur containing amino acids, when compared to other pulses (Erbersdobler et al. 2017; Kouris-Blazos and Belski 2016).

1.4.1 Proteins

Legumes fix atmospheric nitrogen, which in turn is used for the protein biosynthesis or metabolism. Most other plants do not have this ability, thus higher protein content of legumes is not just a coincidence (Mudryj et al. 2014). Protein constitutes about 20–30% of nutrients in legumes with a higher concentration of essential amino acid lysine and limiting concentration of sulfur containing amino acids (SCA) (Maphosa and Jideani 2017). Dry weight protein varies in different legumes such as peas and beans (20%), and soybean and lupin (38–40%) (Awad et al. 2014). This profile really contrast with that of cereals which contain only 7–13% of proteins (Temba et al. 2016). Among the entire legumes, whole seed of sweet lupin has the highest protein content (about 35–44% dry weight). It is nearly double that of the legumes commonly found in human diet (Uauy et al. 1995). All the above given data may vary with the variety, germination and the environment of the plant. Low SCA in legumes can be related to their calcium retention capacity, a hypercalciuric affect developed at least partially by the metabolism of SCAs. The H ions from the SCA metabolism can demineralize bones and excrete through urine. As legume SCA composition is less their hypercalciuric effect is also less (Allen et al. 1979).

Legume proteins are mainly storage proteins. They comprise mainly of two classes viz. globulins and albumins. Globulins are the structural proteins that constitute about 70% of the total protein content and are rich in aspartic acid, glutamine, lysine and arginine (Dahl et al. 2012). Albumins on the other hand are present in lower concentration and are mainly involved in physiological activities. Cystine, methionine and lysine are the important amino acids present in albumins (Agarwal 2017). A major constraint of legume protein is their poor digestibility, mainly due to the presence of inter molecular disulphide bonds found among the storage proteins, and their poor solubility. Digestibility of the protein can be increased with starch content; an opening will be formed in the protein structure when it is bound with the starch molecules. Thus, increases its access to proteolytic enzymes (Ghumman et al. 2016).

1.4.2 Carbohydrates

Carbohydrates are the main source of energy in all food items. Legumes also contain a high concentration of carbohydrates numerically about 60–65% (Singh 2017). Low glycemic index (GI) carbohydrates, resistant starch (RS), oligosaccharides (OS) like raffinose and dietary fiber constitutes the major portion (Herath et al. 2018; Singh 2017). Starches from legumes are highly different in their physicochemical and technological properties and their pastes have high retrogradation tendency. These properties make them resistant towards the digestive enzymes and thus reduce GI (Singh 2011). Even though there is a controversy in the clinical utility of GI, a number of research publications made strong arguments to feature the importance of GI factor in a balanced diet and health (Vega-López et al. 2018). Low GI food have low blood sugar raising potential, made legumes important for diabetes patients (Jenkins et al. 2012). The RS, OS and dietary fibers are difficult to get digested in the stomach and small intestine and in the colon they act as food for commensal bacteria. Fermentation of the RS and dietary fiber favors the formation of butyrate, which lowers the risk of colon cancer (Bird et al. 2010). For patients with celiac disease or individuals who are sensitive to gliadin and glutenin, legumes are a suitable component of diet as they lack glutens (Hosseini et al. 2018; Mlyneková et al. 2014). They are also important in weight management due to the satiating effect produced by them which reduces the food intake level in individuals (Li et al. 2014).

Dietary fibers are defined as macromolecules made up of plant components which are resistant to human endogenous enzymes (Tharanathan and Mahadevamma 2003). In legumes they are of two types, water insoluble fibers in seed coat and soluble fibers in cotyledons and a total concentration of about 14–32%. They are made up of heterogeneous molecules like cellulose, hemicellulose, pectin, lignin etc. (Singh et al. 2017). Presence of dietary fiber in the food provide many health benefits like prevention of constipation, diabetes, obesity, heart diseases, cancer, piles etc. and can also prevent the accumulation of cholesterol and sugar in blood streams (Maphosa and Jideani 2016). They are also widely used in industries and bakeries as fortifiers, stabilizers, texturing agents, fat replacers, bulking agents etc.

1.4.3 Fats

Fat content in legumes are found to be comparatively less, which contribute only ≈ 5% of the total energy supply. Cholesterol is absent in most of the legumes (Messina 1999). Some of the exceptions in percentage of energy supply are soya beans, peanuts and chick peas, which have an energy supply of ≈ 47%, ≈ 45% and ≈ 15% respectively (Maphosa and Jideani 2017). No saturated fatty acids are present in legumes, fat in them is mainly suppliers of essential fatty acids like mono unsaturated (MUFA) and poly unsaturated fatty acids (PUFA) (Maphosa and Jideani 2017; Ofuya and Akhidue 2006). The major fatty acid in legume fat is linoleic acid which is a poly unsaturated essential fatty acid. α Linoleic acid is the another important fatty acid present in legumes. In soy bean, linoleic acid and α linoleic acid are present in the ratio 7.5:1 (Hepburn et al. 1986). Anti-cancerous and cardiovascular benefits of these fatty acids and their derivatives are progressively in study (Nair et al. 1997; Ofuya and Akhidue 2006). Lopes et al. (2018) reported that legume consumption can reduce cardiovascular diseases by creating hypocholesterolemic and hepatoprotective effects.

1.4.4 Vitamins and Minerals

Legumes are good suppliers of vitamins and minerals. B group vitamins like thiamine, riboflavin and folic acid are present in larger amount than others (Kouris-Blazos and Belski 2016; Ofuya and Akhidue 2006). Vitamin E and K are also found to be present in minor quantities. Folate can reduce the risk of neural tube defects in addition to its nutritional benefits (Imbard et al. 2013). Essential minerals like iron, zinc, phosphorous, calcium, magnesium, potassium, selenium, copper, chromium etc. are also present in legumes (Kouris-Blazos and Belski 2016; Brigide et al. 2014). All these vitamins and minerals are essential for the metabolic and physiological functioning of body. Bioavailability of iron from legumes is found to be less when compared to other sources but acute studies proved that this can be overcome by consuming along with vitamin C (Hurrell and Egli 2010). Zinc and calcium has comparatively better bioavailability than iron. It is estimated that $\approx$ 25% zinc and $\approx$ 20% calcium can be absorbed from legumes (Sandström et al. 1989; Weaver et al. 2006). Soy beans showed a high calcium bio-availability which is equivalent to that of milk (Reinwald and Weaver 2010).

1.4.5 Anti-nutrients

Legumes are not only rich in nutritional qualities but also in anti-nutritional qualities. The anti-nutritional quality of the legume is mainly by the richness of bioactive compounds. This includes phytochemicals like phytoestrogens, phytosterols, phytates, various enzyme inhibitors, lectins, polyphenols, alkaloids, saponins, flavonoids, oligosaccharides etc. (Messina 1999; Maphosa and Jideani 2017; Kouris-Blazos and Belski 2016; Singh 2017). They are called anti-nutrients since they affect the digestibility and bioavailability of nutrients. There are no toxic effects reported by the anti-nutrients from legumes. All these anti-nutritional effects can be nullified by cooking as most of them are heat labile. Various other processing methods can also remove considerable amount of anti-nutrients from the legumes (Ndidi et al. 2014).

Enzyme inhibitors in legumes can inhibit principal enzymes in the protein digestion and other physiological pathways. Main examples of proteinaceous anti-nutritional factors are trypsin and chymotrypsin inhibitors, amylase inhibitors, lectins, lipoxygenase and cycloxygenase inhibitors etc. The effect of these proteinaceous anti-nutritional factors on pancreas can create major problems like enlargement of pancreas and development of pancreatic tumors. Cellular hyperplasia, increase in organ sizes, impairment in active glucose transport and expression of tight junction proteins were seen in animals which are fed with soya meal and pea meal (Grant 1989; Röhe, Boroojeni, and Zentek 2017). Apart from their anti-nutritional roles many of these enzyme inhibitors are studied for their roles in anti-cancerous and anti-inflammatory activities (Clemente and Arques 2014; Kennedy and Manzone 1995).

In anti-nutritional factors, polyphenols can bind with the proteins and make them unavailable for digestion. Tannins, which are the major defense tool in plants, can chelate metal ions and bind to proteins by forming hydrogen bonds thus make them unavailable (Beebe et al. 2000; Carbonaro et al. 1996). Along with this the anti-oxidant property of polyphenols is a topic which is widely studied especially of seeds with colored coat. The potential benefits of polyphenols in human health includes treatment for many serious diseases like cardiovascular, cerebrovascular, neurodegenerative diseases, cancer, diabetes, etc. (Ganesan and Xu 2017).

Phytate present in legumes can reduce mineral bioavailability and digestibility still they are gaining much importance for their action against diseases like cancer, diabetes mellitus, coronary heart diseases, renal lithiasis, caries etc. (Nissar et al. 2017). Legumes are also a major source of dietary saponins, which are absorbed very poorly. They can form insoluble complexes with 3-β-hydroxysteroids that interact with cholesterol and bile acids to form large complexes; this will make them poorly absorbed. Hypocholesterolemic, hypoglycemic, anti-fungal, cytotoxic, immunostimulant, anti-cancerous, anti-obesity effects of saponins are also discussed widely along with their anti-nutritional effects (Shi et al. 2004). Isoflavones are another group of phytochemicals which are found in beans especially in soya beans. They are becoming relevant due of their potential effects in treating cancer, osteoporosis, and heart diseases. Estrogenic effects of the compounds are useful in the treatment of menopausal symptoms. Diets with beans are found to cause flatulence more than 40 years ago which is mainly due to the oligosaccharides present in them. As human intestinal mucosa lack α – galactosidase, α (1–6) linkage of galactose containing oligosaccharides are difficult to break and which pass to the large intestine where it is metabolized by bacteria and produce carbon dioxide, hydrogen and methane (Steggerda and Dimmick 1966).

1.5 Medicinal Values of Legumes

Plants are the source of different types of medicines all over the world. Their rich components can be either directly consumed as drug or can be used for drug development. Most of the medicines in use today are either extracted from plants or derived from any plant based compounds. Morphine, vincristine, taxol, artimisinin, forkolin etc. are a few among them. Traditional medicines in most part of the world rely completely on plants for their medicines. About 3.3 billion people in less developed and developing countries use this in regular basis (Singh 2015). Countries like India, China, Japan, Thailand, Africa, etc. had their on traditional medicinal practices. Legumes being a big and widespread family with rich nutrients and phytochemicals, is an indispensible part of this medicinal practices.

Leguminosae family ranked fourth among the oriental herbal medicines following Gramineae, Compositae, and Orchidacea. Papilionoideae, one of the subfamilies of Leguminoseae is the largest source of medicinal plants. Items derived from this family can be used interchangeably; 7 species of *Astragalus* are used as source of astragalus root, 6 species of *Glycyrrhiza* are used for licorice root etc. Also the Chinese herb *jixueteng* is derived from three genera-*Spatholobus*, *Mucuna*, and *Millettia* of the Leguminosae family. This interchangeability of the plant source made them essential among the medicinal preparations ("Legumes" n.d.). Phytoestrogens, flavonoids, isoflavonoids, phytates, polyphenols, alkaloids etc. in the plant are responsible for their medicinal property and can serve as the source of various drugs. As mentioned earlier all these phytochemicals possess various bioactivities. Health benefits of consuming legumes include decrease in the incidence of chronic diseases like cancer, cardiovascular disease, diabetes, ageing etc. (Kris-Etherton et al. 2002).

1.5.1 Anti-oxidant Activities

Legumes have natural endogenous anti-oxidants like phenolic compounds, vitamin C, tannins and tocopherols. They can act as radical scavengers, reducing agents and metal chelators which add to the anti-oxidant property of the plant (Zhao et al. 2014). A number of studies reported the involvement of phenolic compounds in the anti-oxidant property (Djordjevic et al. 2011; Gujral et al. 2013). Germination of legume seeds increases the concentration of phenolics and thus its anti-oxidant activity (Saleh et al. 2019). So consuming germinated seeds can reduce oxidative stress by increasing the activity of anti-oxidant enzymes in the tissues and in blood plasma (Me et al. 2013). Total phenolic content and anti-oxidant property get varied with the plant variety, the morphological part used and the processing methods employed (Gujral et al. 2013). Legume seeds contain some of the major phenolics like phenolic acid, procyanidins, flavonoids and anthocyanidins (seeds with colored coat) (Amarowicz and Pegg 2008; Barman et al. 2018). A number of proteins and peptides with anti-oxidant properties were also reported from different varieties of legumes (Chen et al. 2017).

1.5.2 Anti-inflammatory Activities

Inflammation inhibition potential of the legumes made them an important drug source for various diseases. Many of the constituents that reduce oxidative stress can also reduce pro-inflammatory mechanisms by inhibiting various enzymes in the pathway. Cycloxygenase inhibitory activity of the selected legumes showed their potential to prevent the level of thromboxanes and prostaglandins (Ziaul Haq et al. 2013) which affects the platelet activity, cytoprotection and causes fever and pain during inflammation. Akash, Singh, and Singh (2018) reported the protein denaturation, anti-protease, and lipoxygenase inhibitory potential of the bean extracts. They can also inhibit pro-inflammatory cytokines and regulate the action of interleukin-10 for reducing the effect of inflammation (Moreno-Jiménez et al. 2015). Lipopolysaccharide induced inflammation in RAW 264.7 macrophages of mouse showed considerable reduction when treated with phenolic extracts of mung bean (Zhang et al. 2013). A number of proteins, peptides, amino acids and non protein amino acids with anti-inflammatory potential were isolated from legumes. Germination of seeds and processing of plant material can make considerable changes in these activities. As there is an increase in the phytochemical and protein content during germination bioactivities are found to increase accordingly but certain processing protocols may destroy the plant constituents, in turn reduces bioactivities.

1.5.3 Anti-cancerous Activities

Along with anti-oxidant and anti-inflammatory compounds, phytochemicals that exhibit anti-cancer activity are also present in legumes. These includes enzyme inhibitors (protease inhibitors), phytates, phytosterols, saponins, isoflavonones, dietary fibers etc. (Messina and Barnes 1991; Barman et al. 2018). All these compounds exert their anti-cancer activity by anti-proliferative, pro-apoptotic, anti-metastatic, anti-inflammatory or anti-oxidant actions. Compounds which reduce oxidative stress and inhibit inflammatory pathways can also reduce the incidence of cancer and other chronic diseases (Amarowicz and Pegg 2008; Rao et al. 2018).

Dietary fibers, non digestible carbohydrates and some phenolic compounds reach colon during the progress of digestion, where they get fermented by colon bacteria to form many short chain fatty acids, can inhibit tumor cell proliferation, induce apoptosis and can reduce the colorectal cancer incidence (Campos-Vega et al. 2013). Inhibitory effects of bean based diets on the development of azoxymethane induced colon cancer have been reported (Haydé et al. 2012; Vergara-Castañeda et al. 2010; Rondini and Bennink 2012).

Pulse phenolics (Fig. 1.1) which were reported to have anti-inflammatory and anti-oxidant activities were also reported with anti-cancer activity. There is a correlation exist between these activities with tumor cell proliferation (Luo et al. 2016; Ombra et al. 2016; Moreno-Jiménez et al. 2015). Phenolic extracts of various pulses showed considerable inhibition in the proliferation of cancer cell lines (Ombra et al. 2016). They not only inhibited proliferation of cancer cells but also induced apoptosis. Jamapa bean (Aparicio-Fernández et al. 2008) and mung sprout phenolic extracts could induce apoptosis in HeLa cells by up-regulating pro-apoptotic proteins whereas Korean kidney bean extracts (Lee et al. 2009) by an activated protein kinase pathway (Rao et al. 2018). The metastatic effect on the cancer cell lines were also found to be greatly influenced by the extracts of various pulses (Lima et al. 2016).

Isoflavones (Fig. 1.1) of soya beans were found to be very effective in inhibiting cancer. Daidzein can exert anti-cancer activity by arresting cell proliferation,

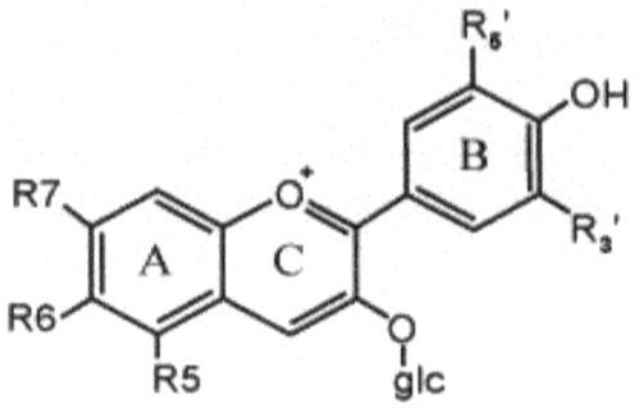

(g)

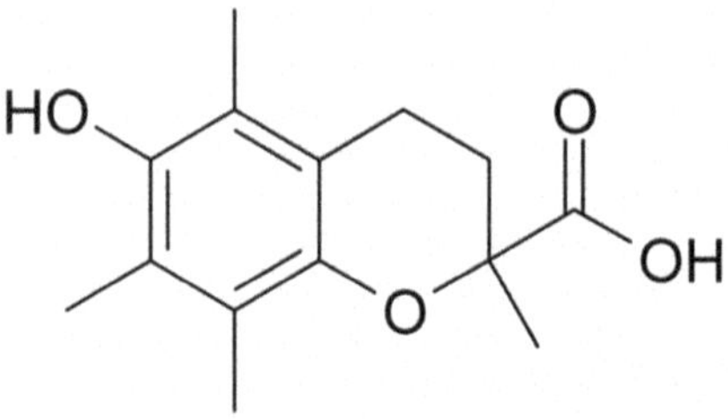

(h)

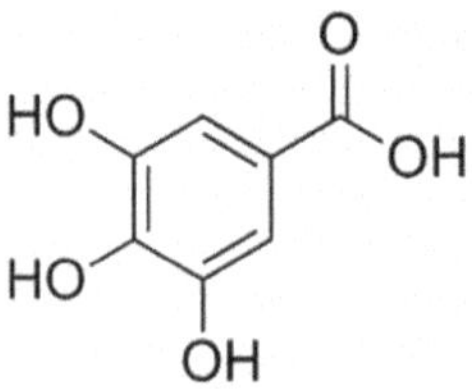

(i)

Fig. 1.1 (continued)

survivin, BIRC7 and cytC (Luna Vital et al. 2014). Some anti-cancer peptides from legumes are listed in Table 1.2 (Marcela 2017).

1.5.4 Cardiovascular Effects

Cardiovascular disease (CVD) being the leading cause of death worldwide (Townsend et al. 2016), public awareness on reducing CVD risk is necessary. A Mediterranean diet pattern to lower the CVD mortality and morbidity has been reported by (Panagiotakos et al. 2007). Legumes were considered as one among the frequently consumed foods according to this diet pattern. Grains and Legumes Nutrition Council of Australia indicates that consuming half to two cups of cooked legumes per day can reduce the risk of CVD by reducing blood pressure, LDL cholesterol level, increasing blood HDL cholesterol, maintaining blood glucose level and weight management. There are experimental evidences that rich sources of antioxidants could prevent CVD and legumes are rich in them. Chick pea consumption

Table 1.2 Bioactive peptides from legumes and their mode of action

Peptide source	Name or sequence of the peptide	Models used (*in vivo* and *in vitro*)	Action
Glycine max (soybean)	Bowman Birk iso inhibitors – IBB1, IBB2	Human colon cancer cell line (HT-29)	Cells are blocked in G_0 and G_1 phase and can inhibit serine protease.
Glycine max	Lunasin	C3H/10 T1/2	Cell transformation by chemical carcinogen is inhibited and can induce apoptosis in transformed cells selectively by avoiding non-transformed cells.
Glycine max	Lunasin	SENCAR	Chance of skin tumor was reduced by 70% and delayed the display of tumors by 2 weeks.
Glycine max	Lunasin-like peptide, ASKWQHQQDS, IQGRGDDDDD, TPCEKHIMEK, CRKQLQGVNL	Raw cell lines (264.7)	Pro-inflammatory markers can be inhibited.
Glycine max	XMLPSYSPY	Macrophage cell line P388D1	Arrest at G2/M phase.
Phaseolus vulgaris	ANDISFNFVRFNETNLILGG	Breast cancer cell line MCF-7	Cell cycle arrest at G2/M phase, externalization of phosphatidyl serine and depolarization of mitochondrial membrane.
Azufrado higuera and *Bayo madero* cultivars	GLTSK, MPACGSS, LSGNK, MTEEY, GEGSGA	Human colon cancer cell line HCT-116	p53 expression was increased by both cultivars by 76% and 68%.
Vigna unguiculata	ANEIYFSFQRFNETNLILQR	Cellosaurus cell lines, breast cancer cell lines, human liver cancer cell lines	Induce the production of apoptotic bodies and nitric oxide which in turn can inhibit the tumor cells
Cicer arietinum	ARQSHFANAQP	Breast cancer cell lines MCF-7 and MDAMB-231	Breast cancer cell proliferation can be inhibited and can increase the level of p53 by molecular docking.
Pisum sativum	Bowman-Birk iso inhibitor – TI1B	Human colon cancer cell line HT-29	Cancer cells viability reduction

(continued)

Table 1.2 (continued)

Peptide source	Name or sequence of the peptide	Models used (*in vivo* and *in vitro*)	Action
Vicia faba	Bowman-Birk type trypsin inhibitor - VFTI-G1	Human liver cancer cell line HepG2	Nuclear morphological changes can be induced; reduction in cell viability, increased condensation of chromatin & 60% cells were apoptosized.
Lens culinaris	Bowman-Birk iso inhibitors	Human colon cancer cell line HT-29	Dose dependent inhibition of cell proliferation in colon cancer cells.

Table 1.3 Cardiovascular effects of compounds from legumes

Legume phytochemicals	Mechanism of action
Isoflavones	Treatment of metabolic disorders and cardiovascular disease
Isoflavonoids of soy products	Beneficial effects on cardiovascular system; protection against atherosclerosis; tumor biomarkers in angiogenesis were altered.
Flavonoids, Isoflavones	Possess estrogenic and anti-estrogenic effects and has the ability to protect heart diseases
Phyto sterols	Protection against atherosclerosis, oxidative stress, cardiovascular diseases, metabolic disorders and DNA damage
Phenolic compounds (lentil, chick pea, peas green, soya bean)	Block angiotensin converting enzyme; inhibit cycloxygenase; block enzymes in estrogen production pathway.
Polyphenols in azuki bean seed coats (ABSC)	Anti-inflammatory responses, regulate cycloxygenase-2 (COX-2)
Flavonoids, isoflavonoids (soy bean)	Natural anti-oxidants; harmonious action in scavenging free radicals
Phytostrogens; soy isoflavone metabolite	Boost up bioavailability of NO
Isoflavonoides (genistein and daidzein) (soybean)	Potentially preventing and treating osteoporosis; protection against cardiovascular diseases; inhibit angiogenesis; reduction in cell cycle progression, aromatase enzyme inhibition, stimulation of sex hormone binding globulins.

can normalize the triacylglycerol levels in hyper cholesterolemic rats (Zulet et al. 1999) and they can create a hypocholesterolemic effect by reducing cholesterol, triglycerides and low density lipoproteins (Mughal 2019). Legume consumption can reduce the factors leading to coronary heart disease (CHD). Anti-hyperlipidemic effects of the legume phytochemicals can also account for its CHD and CVD effects (Bouchenak and Lamri-Senhadji 2013). Systemic reviews and meta-analysis of clinical incidences proved that legume consumption can reduce CVD risks (Li et al. 2017; Afshin et al. 2014). CVD effect of some of the legume phytochemicals are explained in Table 1.3 (Bouchenak and Lamri-Senhadji 2013).

1.5.5 Anti-diabetic Properties

Diabetes Mellitus (DM) is an emerging epidemic among all the age groups in the world. A grim situation exists in India, world's largest diabetic populated country. Diet control and physical exercise are the basic necessity for controlling DM epidemic. American Diabetes Association developed a food pyramid in which low glycemic index (GI) foods are recommended for daily consumption. Legumes and cereals are the major groups of low GI foods (Singhal et al. 2014). In the early twentieth century clinical trials with bean pod tea on diabetic patients were found successful by many researchers and also alcoholic extracts of the beans were found to be effective against hyperglycemic condition in experimental animals. Recent years also witnessed a number of studies on the anti-diabetic effect of beans and other legumes (Helmstädter 2010). Some of the processing conditions can remove the anti-hyperglycemic principles of the legumes. Roasted pigeon peas at high temperature for 30 minutes lost its ability to reduce the blood glucose level (Amalraj and Ignacimuthu 1998). A reduction in the postprandial glycemic action was observed in volunteers who were fed with sphagetti made of wheat and chickpea flour when compared to those who ate wheat flour sphagetti alone. GI of wheat and chickpea sphagetti is 58 ± 6 and wheat sphagetti is 73 ± 5 (Goñi and Valentín-Gamazo 2003). Soya bean isoflavones are reported to have a very high inhibition on carbohydrate digesting enzyme α-glucosidase. Genistein showed a high percentage of dose dependent enzyme inhibition and it reduced the urinary excretion of glucose (Gilbert and Liu 2013; Jin et al. 2018). Soya saponins and isoflavone aglycones were also studied for their anti-diabetic abilities. Horse gram, field bean, lentils, moth bean, adzuki bean, green gram, lima bean, black gram, kidney bean, garden pea, field bean etc. were also reported to possess anti-hyperglycemic properties (Singhal et al. 2014). Germination of seeds can increase the effects in many folds due to increase in phytochemical content.

1.5.6 Hepato-Protective Activities

Being a pivotal organ in inflammation liver plays major role in the metabolism and excretion of toxic substances. Liver cells secrete many proinflammatory cytokines upon activation by the hepatotoxins and which will create reactive oxygen species that will cause liver damage and inflammation. Hence anti-oxidant and anti-inflammatory compounds that can suppress these reactive pathways can protect liver cells from damages (Mohd Ali et al. 2013). Mung bean, adzuki bean, black bean, rice bean and chick pea extracts were reported to possess this hepato-protective activity. Histopathological examination of liver cells from affected animals and the extract treated animals showed considerable reduction in the pathological changes of liver cell architecture in the treated ones (Mekky et al. 2016).

1.5.7 *Other Important Activities of Legumes*

39% of women above 50 years are getting affected with osteoporosis due to the hormonal changes in menopause stage. Deficiency of hormones increases the bone turnover rate and an imbalance occurs in between the formation and absorption. This will finally result in bone loss. Estrogen therapies are found to be successful but side effects prevented the continuous usage. Estrogenic effects of the isoflavones from soyabeans and other legumes can act under this condition to prevent the loss of bone mineral density. Also the rich mineral source of legumes will add up on their activity (Park et al. 2013). Slowly digestible carbohydrates, proteins and fiber contents of legumes can reduce the satiety and thereby reducing the risk of obesity. Observational studies showed an inverse relationship between BMI and legume consumption (McCrory et al. 2010; Li et al. 2014). Impacts of legumes on ageing and stress management have also been carried out by many researchers with positive results (Kapoor 2015).

1.6 Changes During Germination in Legumes

Sprouting or germination of seeds is a practice followed since thousands of years ago. It has been recommended as an effective and inexpensive method for improving the nutritional quality of the food. Germination is a natural process of sprouting where all the reserve materials are used for the growth and development of new plant. In a general procedure, seeds have to be soaked in water preceding germination. Metabolic enzymes get activated and start utilizing the wide chemical source of the plant. Reserved components are utilized for respiration and developing the embryo. Thus a number of changes in the sensory, nutritional and biochemical characteristics occur during germination.

Carbohydrates are been acted upon by hydrolytic and amylolytic enzymes and degraded to simple sugars thereby increasing its digestibility (Oghbaei and Prakash 2016), thus can be a good ingredient in complementary and infant foods. Duration of the process is also important. α-Amylase, major enzyme in starch metabolism has its maximum activity only after 48 hours of germination, thus hydrolysis of starch reaches its peak at this stage. Before and after this period enzyme activity was found to be decreased (Nkhata et al. 2018). The amount of reducing sugars in legumes was not significantly affected at the initial 12 hours of germination and later it was found to be increased 20 fold by the increasing action of amylases (Zhang et al. 2015). Breakdown of macromolecules and synthesis of structural components are the two main processes during germination. Total fiber content was also affected by the process of germination. As dietary fiber can slow down glucose release germinated foods are beneficial for diabetes patients.

Protein content in legumes during germination is affected both by the action of hydrolytic enzymes and the protein synthesized for the newly growing embryo. 72 h germination of buckwheat showed an increased protein content due to increased rate of protein synthesis than hydrolysis (Zhang et al. 2015). Digestibility of protein was also found to get increased by germination by the action of proteolytic enzymes. 14–18% increase in protein was observed in cowpea, chickpea, green gram and lentils and 68% increase was observed in finger millet (Nkhata et al. 2018) during germination.

Mineral availability generally depends on the type of legume and the presence of phytate content. Phytic acid, inhibits the bioavailability of many minerals, gets inactivated during germination by phytase enzyme which in turn make the minerals readily bioavailable (Liang et al. 2008). Mineral availability also depends on the methods and duration employed for germination process. A number of vitamins are synthesized during germination for the new sprout; this includes ascorbic acid, tocopherols, niacin, riboflavin etc. In the case of fat content during germination it was partially used for respiration and partially got hydrolyzed by the activated hydrolases. Thus reduces the total fat content (Nkhata et al. 2018).

A number of studies reported that germination will increase the nutrient contents and also their bioavailability. There was a considerable decrease in tannin, phytic acid and flatulence causing oligosaccharide content during germination (Megat Rusydi et al. 2011). This reduction was mainly due to increase in the level of proteinases like phytases, α-galactosidases, etc. Fat and ash content was also found to get decreased during germination (Camacho et al. 1992). Bioactivities of the legumes also have considerable differences before and after germination. Studies on different legume varieties showed considerable increase in the quality and quantity of phytochemicals like polyphenols, flavonoids, isoflavones etc. and the activities related to them. All these compounds and their germinated products are better candidates for drug development (Lin and Lai 2006; López-Amorós et al. 2006).

1.7 Beans in Ayurvedic Medicine

Ayurveda, the traditional Indian Medicine, is a 5000 years old system of curing from India. Herbal medicines are the core of *Ayurveda*. *Caraka samhita* and *Shushrutha samhita*, two principal books of *Ayurveda* describes about a number of herbal remedies for many diseases. It suggests a vegetarian diet which can nourish our body along with balancing the metabolic activities, the most essential aspect for maintaining health. Ayurvedic science explains that a right diet is the root of healing. *Ahara* (diet) is one among the three upasthambhas needed for a smooth life (Nishteswar 2016). According to *Caraka* rice ripened in 60 days, green gram, Indian goose berry, rock salt, meat of animals in arid areas, ghee, honey and rainwater are the foods that have to be consumed regularly. *Caraka samhita* describes pulses as *samidhanyavarga* that supplies proteins.

Balanced food or the right diet described by *Ayurveda* for many diseases is generally traditional foods of India. Indian traditional food is also described as "functional foods" as it has many body healing functional molecules, and this gets further

Kapoor S (2015) Bioactivities and therapeutic potential of legumes: a review. Int J Pharm Biol Sci 5(2):65–74

Kennedy AR, Manzone H (1995) Effects of protease inhibitors on levels of proteolytic activity in normal and premalignant cells and tissues. J Cell Biochem Suppl 22:188–194

Kouris-Blazos A, Belski R (2016) Health benefits of legumes and pulses with a focus on Australian sweet lupins. Asia Pac J Clin Nutr 25:1–17. https://doi.org/10.6133/apjcn.2016.25.1.23

Kris-Etherton PM, Hecker KD, Bonanome A, Coval SM, Binkoski AE, Hilpert KF, Griel AE, Etherton TD (2002) Bioactive compounds in foods: their role in the prevention of cardiovascular disease and cancer. Am J Med 113(Suppl 9B):71S–88S. https://doi.org/10.1016/s0002-9343(01)00995-0

Lad V (1987) Ayurveda, the science of self-healing: a practical guide, 2nd edn. Lotus Press, Twin Lakes

Lavin M, Herendeen PS, Wojciechowski MF (2005) Evolutionary rates analysis of Leguminosae implicates a rapid diversification of lineages during the tertiary. Syst Biol 54(4):575–594. https://doi.org/10.1080/10635150590947131

Lee Y-K, Hwang J-T, Lee M-S, Kim Y-M, Park OJ (2009) Kidney bean husk extracts exert antitumor effect by inducing apoptosis involving AMP-activated protein kinase signaling pathway. Ann N Y Acad Sci 1171:484–488. https://doi.org/10.1111/j.1749-6632.2009.04697.x

Legumes (n.d.) Retrieved August 19, 2019. from http://www.itmonline.org/arts/legumes.htm

Li SS, Kendall CWC, de Souza RJ, Jayalath VH, Cozma AI, Ha V, Mirrahimi A, Chiavaroli L, Augustin LSA, Blanco Mejia S, Leiter LA, Beyene J, Jenkins DJA, Sievenpiper JL (2014) Dietary pulses, satiety and food intake: a systematic review and meta-analysis of acute feeding trials. Obesity (Silver Spring, Md) 22(8):1773–1780. https://doi.org/10.1002/oby.20782

Li H, Li J, Shen Y, Wang J, Zhou D (2017) Legume consumption and all-cause and cardiovascular disease mortality. Biomed Res Int 2017:8450618. https://doi.org/10.1155/2017/8450618

Liang J, Han B-Z, Nout MJR, Hamer RJ (2008) Effects of soaking, germination and fermentation on phytic acid, total and in vitro soluble zinc in brown rice. Food Chem 110(4):821–828. https://doi.org/10.1016/j.foodchem.2008.02.064

Lima AIG, Mota J, Monteiro SA, Ferreira RM (2016) Legume seeds and colorectal cancer revisited: protease inhibitors reduce MMP-9 activity and colon cancer cell migration. Food Chem 197(Pt A):30–38. https://doi.org/10.1016/j.foodchem.2015.10.063

Lin P-Y, Lai H-M (2006) Bioactive compounds in legumes and their germinated products. J Agric Food Chem 54(11):3807–3814. https://doi.org/10.1021/jf060002o

Lopes LAR, do, de Martins M, C, CE, de Farias LM, da Brito AK, S, de Lima G, M, de Carvalho VBL, de Pereira CF, C, Conde Júnior AM, Saldanha T, Arêas JAG, Silva KJDE, de Frota K, MG (2018) Cholesterol-lowering and liver-protective effects of cooked and germinated mung beans (*Vigna radiata* L.). Nutrients 10(7). https://doi.org/10.3390/nu10070821

López-Amorós ML, Hernández T, Estrella I (2006) Effect of germination on legume phenolic compounds and their antioxidant activity. J Food Compos Anal 19:277–283. https://doi.org/10.1016/j.jfca.2004.06.012

Luna Vital DA, González de Mejía E, Dia VP, Loarca-Piña G (2014) Peptides in common bean fractions inhibit human colorectal cancer cells. Food Chem 157:347–355. https://doi.org/10.1016/j.foodchem.2014.02.050

Luo J, Cai W, Wu T, Xu B (2016) Phytochemical distribution in hull and cotyledon of adzuki bean (*Vigna angularis* L.) and mung bean (*Vigna radiate* L.) and their contribution to antioxidant, anti-inflammatory and anti-diabetic activities. Food Chem 201:350–360. https://doi.org/10.1016/j.foodchem.2016.01.101

Ma DZ, Wang HX, Ng TB (2009) A peptide with potent antifungal and antiproliferative activities from Nepalese large red beans. Peptides 30(12):2089–2094. https://doi.org/10.1016/j.peptides.2009.08.017

Maphosa Y, Jideani VA (2016) Physicochemical characteristics of Bambara groundnut dietary fibres extracted using wet milling. S Afr J Sci 112(1–2):01–08. https://doi.org/10.17159/sajs.2016/20150126

Maphosa Y, Jideani VA (2017) The role of legumes in human nutrition. Functional Food – Improve Health Through Adequate Food. https://doi.org/10.5772/intechopen.69127
Marcela G-M (2017) Bioactive peptides from legumes as anticancer therapeutic agents. Int J Cancer Clin Res 4:10.23937/2378-3419/1410081
Marinangeli CPF, Curran J, Barr SI, Slavin J, Puri S, Swaminathan S, Tapsell L, Patterson CA (2017) Enhancing nutrition with pulses: defining a recommended serving size for adults. Nutr Rev 75(12):990–1006. https://doi.org/10.1093/nutrit/nux058
McCrory MA, Hamaker BR, Lovejoy JC, Eichelsdoerfer PE (2010) Pulse consumption, satiety, and weight management. Adv Nutr 1(1):17–30. https://doi.org/10.3945/an.110.1006
Me N, Kadir K, Amom Z, Azlan A (2013) Antioxidant activity of white rice, brown rice and germinated brown rice (in vivo and in vitro) and the effects on lipid peroxidation and liver enzymes in hyperlipidaemic rabbits. Food Chem 141:1306–1312. https://doi.org/10.1016/j.foodchem.2013.03.086
Megat Rusydi MR, Noraliza CW, Azlan A, Amom Z (2011) Nutritional changes in germinated legumes and rice varieties. Int Food Res J:18
Mekky RH, Fayed MR, El-Gindi MR, Abdel-Monem AR, Contreras M d M, Segura-Carretero A, Abdel-Sattar E (2016) Hepatoprotective effect and chemical assessment of a selected egyptian Chickpea cultivar. Front Pharmacol:7. https://doi.org/10.3389/fphar.2016.00344
Messina MJ (1999) Legumes and soybeans: overview of their nutritional profiles and health effects. Am J Clin Nutr 70(3 Suppl):439S–450S. https://doi.org/10.1093/ajcn/70.3.439s
Messina M, Barnes S (1991) The role of soy products in reducing risk of cancer. JNCI: J Natl Cancer Inst 83(8):541–546. https://doi.org/10.1093/jnci/83.8.541
Mlyneková Z, Chrenková M, Formelová Z (2014) Cereals and legumes in nutrition of people with celiac disease. Int J Celiac Dis 2(3):105–109. https://doi.org/10.12691/ijcd-2-3-3
Mohd Ali N, Mohd Yusof H, Long K, Yeap SK, Ho WY, Beh BK, Koh SP, Abdullah MP, Alitheen NB (2013) Antioxidant and hepatoprotective effect of aqueous extract of germinated and fermented mung bean on ethanol-mediated liver damage. Biomed Res Int. https://doi.org/10.1155/2013/693613
Moreno-Jiménez MR, Cervantes-Cardoza V, Gallegos-Infante JA, González-Laredo RF, Estrella I, García-Gasca T, Herrera-Carrera E, Díaz-Rivas JO, Rocha-Guzmán NE (2015) Phenolic composition changes of processed common beans: their antioxidant and anti-inflammatory effects in intestinal cancer cells. Food Res Int 76:79–85. https://doi.org/10.1016/j.foodres.2014.12.003
Mudryj AN, Yu N, Aukema HM (2014) Nutritional and health benefits of pulses. Appl Physiol Nutr Me = Physiologie Appliquee, Nutrition Et Metabolisme 39(11):1197–1204. https://doi.org/10.1139/apnm-2013-0557
Mughal MH (2019) Determining the effective role of chapattis prepared from composite flour against lipid profile of experimental subjects. Biomed J Sci Tech Res 19(2):14164–14169. https://doi.org/10.26717/BJSTR.2019.19.003265
Müller O, Krawinkel M (2005) Malnutrition and health in developing countries. CMAJ: Can Med Assoc J 173(3):279–286. https://doi.org/10.1503/cmaj.050342
Nair SSD, Leitch JW, Falconer J, Garg ML (1997) Prevention of cardiac arrhythmia by dietary (n-3) polyunsaturated fatty acids and their mechanism of action. J Nutr 127(3):383–393. https://doi.org/10.1093/jn/127.3.383
Ndidi US, Ndidi CU, Aimola IA, Bassa OY, Mankilik M, Adamu Z (2014) Effects of processing (boiling and roasting) on the nutritional and antinutritional properties of bambara groundnuts (*Vigna subterranea* [L.] Verdc.) from Southern Kaduna, Nigeria. J Food Process. https://doi.org/10.1155/2014/472129
Nishteswar K (2016) Ayurvedic concept of food and nutrition. J Nutr Food Sci 6(4):1–8. https://doi.org/10.4172/2155-9600.1000530
Nissar J, Ahad T, Naik HR, Hussain SZ (2017) A review phytic acid: as antinutrient or nutraceutical. J Pharmacogn Phytochem 6(6):1554–1560

Nkhata SG, Ayua E, Kamau EH, Shingiro J (2018) Fermentation and germination improve nutritional value of cereals and legumes through activation of endogenous enzymes. Food Sci Nutr 6(8):2446–2458. https://doi.org/10.1002/fsn3.846

Ofuya Z, Akhidue V (2006) The role of pulses in human nutrition: a review. J Appl Sci Environ Manag 9(3). https://doi.org/10.4314/jasem.v9i3.17361

Oghbaei M, Prakash J (2016) Effect of primary processing of cereals and legumes on its nutritional quality: a comprehensive review. Cogent Food Agric 2(1):1136015. https://doi.org/10.1080/23311932.2015.1136015

Ombra MN, d'Acierno A, Nazzaro F, Riccardi R, Spigno P, Zaccardelli M, Pane C, Maione M, Fratianni F (2016) Phenolic composition and antioxidant and antiproliferative activities of the extracts of twelve common bean (*Phaseolus vulgaris* L) Endemic ecotypes of southern italy before and after cooking. Oxid Med Cell Longev:2016. https://doi.org/10.1155/2016/1398298

Panagiotakos DB, Pitsavos C, Arvaniti F, Stefanadis C (2007) Adherence to the Mediterranean food pattern predicts the prevalence of hypertension, hypercholesterolemia, diabetes and obesity, among healthy adults; the accuracy of the MedDietScore. Prev Med 44(4):335–340. https://doi.org/10.1016/j.ypmed.2006.12.009

Park Y, Moon H-J, Paik D-J, Kim D-Y (2013) Effect of dietary legumes on bone-specific gene expression in ovariectomized rats. Nutr Res Pract 7(3):185–191. https://doi.org/10.4162/nrp.2013.7.3.185

Pavese JM, Farmer RL, Bergan RC (2010) Inhibition of cancer cell invasion and metastasis by genistein. Cancer Metastasis Rev 29(3):465–482. https://doi.org/10.1007/s10555-010-9238-z

Pavese JM, Krishna SN, Bergan RC (2014) Genistein inhibits human prostate cancer cell detachment, invasion, and metastasis1234. Am J Clin Nutr 100(1):431S–436S. https://doi.org/10.3945/ajcn.113.071290

Poschner S, Maier-Salamon A, Zehl M, Wackerlig J, Dobusch D, Pachmann B, Sterlini KL, Jäger W (2017) The impacts of genistein and daidzein on estrogen conjugations in human breast cancer cells: a targeted metabolomics approach. Front Pharmacol:8. https://doi.org/10.3389/fphar.2017.00699

Rao S, Chinkwo KA, Santhakumar AB, Blanchard CL (2018) Inhibitory effects of pulse bioactive compounds on cancer development pathways. Diseases (Basel, Switzerland) 6(3). https://doi.org/10.3390/diseases6030072

Ray DK, West PC, Clark M, Gerber JS, Prishchepov AV, Chatterjee S (2019) Climate change has likely already affected global food production. PLoS One 14(5):e0217148. https://doi.org/10.1371/journal.pone.0217148

Reinwald S, Weaver CM (2010) Soy components vs. whole soy: are we betting our bones on a long shot?123. J Nutr 140(12):2312S–2317S. https://doi.org/10.3945/jn.110.124008

Röhe I, Boroojeni FG, Zentek J (2017) Effect of feeding soybean meal and differently processed peas on intestinal morphology and functional glucose transport in the small intestine of broilers. Poult Sci 96(11):4075–4084. https://doi.org/10.3382/ps/pex199

Rondini EA, Bennink MR (2012) Microarray analyses of genes differentially expressed by diet (black beans and soy flour) during azoxymethane-induced colon carcinogenesis in rats. J Nutr Metab. https://doi.org/10.1155/2012/351796

Rosa W (Ed.) (2017) Transforming our world: the 2030 agenda for sustainable development. In a new era in global health. Springer Publishing Company. https://doi.org/10.1891/9780826190123.ap02

Roy M (2009) Agriculture in the vedic period. Indian J Hist Sci 44(4):497–520

Saleh HM, Hassan AA, Mansour EH, Fahmy HA, El-Bedawey AE-FA (2019) Melatonin, phenolics content and antioxidant activity of germinated selected legumes and their fractions. J Saudi Soc Agric Sci 18(3):294–301. https://doi.org/10.1016/j.jssas.2017.09.001

Sanaei M, Kavoosi F, Atashpour S, Haghighat S (2017) Effects of genistein and synergistic action in combination with tamoxifen on the hepg2 human hepatocellular carcinoma cell line. Asian Pac J Cancer Prev: APJCP 18(9):2381–2385. https://doi.org/10.22034/APJCP.2017.18.9.2381

Sandström B, Almgren A, Kivistö B, Cederblad Å (1989) Effect of protein and protein source on zinc absorption in humans. J Nutr 119:48–53. https://doi.org/10.1093/jn/119.1.48
Sarkar P, Lohith Kumar DH, Dhumal C, Panigrahi SS, Choudhary R (2015) Traditional and ayurvedic foods of Indian origin. J Ethn Foods 2(3):97–109. https://doi.org/10.1016/j.jef.2015.08.003
Schrire BD, Lavin M, Lewis GP, Friis IB, Balslev H (2005) Global distribution patterns of the Leguminosae: insights from recent phylogenies. /paper/Global-distribution-patterns-of-the-Leguminosae%3A-Schrire-Lavin/1e6e799d11e0d9640e42b1083c8b2581fd17cbbe
Sen CT (2004) Food culture in India. Greenwood Publishing Group, Westport
Sharma P, Rao MV (2014) Shimbhi dhanya varga (group of legumes and pulses): a preventive and curative perspective. 2(6): 11
Shi J, Arunasalam K, Yeung D, Kakuda Y, Mittal G, Jiang Y (2004) Saponins from edible legumes: chemistry, processing, and health benefits. J Med Food 7(1):67–78. https://doi.org/10.1089/109662004322984734
Singh N (2011) Functional and physicochemical properties of pulse starch. Pulse Foods:91–119. https://doi.org/10.1016/B978-0-12-382018-1.00004-6
Singh R (2015) Medicinal plants: a review. https://doi.org/10.11648/j.jps.s.2015030101.18
Singh N (2017) Pulses: an overview. J Food Sci Tech 54(4):853–857. https://doi.org/10.1007/s13197-017-2537-4
Singh B, Singh JP, Shevkani K, Singh N, Kaur A (2017) Bioactive constituents in pulses and their health benefits. J Food Sci Tech 54(4):858–870. https://doi.org/10.1007/s13197-016-2391-9
Singhal P, Kaushik G, Mathur P (2014) Antidiabetic potential of commonly consumed legumes: a review. Crit Rev Food Sci Nutr 54:655–672. https://doi.org/10.1080/10408398.2011.604141
Smýkal P, Coyne CJ, Ambrose MJ, Maxted N, Schaefer H, Blair MW, Berger J, Greene SL, Nelson MN, Besharat N, Vymyslický T, Toker C, Saxena RK, Roorkiwal M, Pandey MK, Hu J, Li YH, Wang LX, Guo Y et al (2015) Legume crops phylogeny and genetic diversity for science and breeding. Crit Rev Plant Sci 34(1–3):43–104. https://doi.org/10.1080/07352689.2014.897904
Smýkal P, von Wettberg EJB, McPhee K (2020) Legume genetics and biology: from Mendel's pea to legume genomics. Int J Mol Sci 21(9). https://doi.org/10.3390/ijms21093336
Sprent JI (2007) Evolving ideas of legume evolution and diversity: a taxonomic perspective on the occurrence of nodulation. New Phytol 174(1):11–25. https://doi.org/10.1111/j.1469-8137.2007.02015.x
Stagnari F, Maggio A, Galieni A, Pisante M (2017) Multiple benefits of legumes for agriculture sustainability: an overview. Chem Biol Tech Agric 4(1):2. https://doi.org/10.1186/s40538-016-0085-1
Steggerda FR, Dimmick JF (1966) Effects of bean diets on concentration of carbon dioxide in flatus. Am J Clin Nutr 19(2):120–124. https://doi.org/10.1093/ajcn/19.2.120
Sutjaritjai N, Wangpakapattanawong P, Balslev H, Inta A (2019) Traditional uses of Leguminosae among the Karen in Thailand. Plan Theory 8(12):600. https://doi.org/10.3390/plants8120600
Temba MC, Njobeh PB, Adebo OA, Olugbile AO, Kayitesi E (2016) The role of compositing cereals with legumes to alleviate protein energy malnutrition in Africa. Int J Food Sci Technol 51(3):543–554. https://doi.org/10.1111/ijfs.13035
Tharanathan RN, Mahadevamma S (2003) Grain legumes—a boon to human nutrition. Trends Food Sci Technol 14:507–518. https://doi.org/10.1016/j.tifs.2003.07.002
Thompson P (n.d.) Ayurveda: science to self healing through recipes, yoga, aromatherapy and meditation. Paul Thompson
Townsend N, Wilson L, Bhatnagar P, Wickramasinghe K, Rayner M, Nichols M (2016) Cardiovascular disease in Europe: epidemiological update 2016. Eur Heart J 37(42):3232–3245. https://doi.org/10.1093/eurheartj/ehw334
Uauy R, Gattas V, Yáñez E (1995) Sweet lupins in human nutrition. World Rev Nutr Diet 77:75–88
United Nations (2019) World population prospects 2019: data booklet. UN https://doi.org/10.18356/3e9d869f-en

2.1 Introduction

The term "nutraceutical", consists of the words *nutrient* and *pharmaceutical*. It was introduced in 1989 by Dr. Stephen De Felice, the founder and chairman of the Foundation for Innovation in Medicine (FIM), Cranford, NJ an American organization which encourages medical health. He defined "nutraceuticals" as "food or part of a food that provides medical or health benefits, including the prevention and/or treatment of a disease" (Chaturvedi et al. 2011). In common terms we can say nutraceuticals are the products produced from foods but sold in medicinal forms of either, a capsule, tablet, powder, solution, or potion, which is not generally associated with the food and have demonstrated physiological benefits and/or provide protection against chronic diseases (Sohaimy 2012).

Nutraceuticals, ranging from isolated nutrients, dietary supplements, and herbal formulations and processed products like cereals, soups and beverages, have the potential to provide bioactive compounds from a food, and can be used with the purpose of enhancing health in the form of dosages. The concept of nutraceuticals has been proposed as a modern approach to food science, which is basically "beyond the diet, but before the drugs" (Sohaimy 2012). Nutraceuticals being a diverse product category have various synonyms such as "Functional foods," "pharmaconutrients," and "dietary integrators" that have used internationally either incorrectly or indiscriminately for nutrients or nutrient-enriched foods that can prevent or treat diseases (Palthur et al. 2010).

Foods contain various dietary components with an array of health benefits that offer an excellent opportunity to improve public health and well-being (Gul et al. 2016). The demand for foods with a positive impact on human health and wellness has exploded globally over the past two decades. Modern food preferences and progress made in the food industry have led to a completely new definition of nutrition and health through eating habits. Eating habits have a great impact on environment, industry, economy and human health. Food can even help to reduce the risk of endemic to modern society such as, obesity, osteoporosis, cancer, diabetes, allergies, and dental problems, which can occur at an early age and could be related to eating habits and preferences (Cencic and Chingwaru 2010).

Now a day's nutraceuticals are in high demand as it is expected to increase life expectancy, and improves the quality of life in older adults (Bigliardi and Galati 2013). Social, economic development in lifestyle severely affects food habits whereas, obesity and malnutrition are major problems associated with daily diet. Several articles have been reported which shows the direct relation of diet to health (Das et al. 2012). The principle, ***"Let food be thy medicine, and medicine be thy food"***, advocated by Hippocrates (460–377 BC), the well-recognized father of modern medicine, and the concept of "*Medicine and food are isogonics*" emphasize the association between nutrition and human health, and conceptualized the relationship between the use of appropriate foods for health and their therapeutic benefits (Palthur et al. 2010). The Indians, Egyptians, Chinese and Sumerians are few civilizations that provide evidences on effective use of foods as medicine and on how food can be helpful to treat and prevent disease (Sohaimy 2012).

This concept of using food as medicine is receiving a lot of interest today as food scientist and consumers realize many health benefits of certain foods. These foods contain ingredients that aid specific body function, improving our health and well-being.

2.1.1 Classification of Nutraceuticals

According to Das et al. (2012), nutraceuticals can be classified depending upon their application, natural sources (products obtained from plants, animals, minerals, or microbial sources), mode of action (antioxidation, antibacterial, hypotensive, anti-inflammatory, etc.) or as per chemical nature of the products (amino acid-based substances, carbohydrates and derivatives, fatty acids and structural lipids, isoprenoid derivatives, phenolic substances, microorganisms: probiotics, prebiotics, minerals). Their sources may range from isolated food *nutrients* (vitamins, minerals, amino acids, and fatty acids), *herbals* (herbs or botanical products as concentrates or extracts), *dietary supplements* (reagents derived from other sources), and diets to genetically engineered "designer" foods and processed products such as cereals, soups, and beverages.

Nutraceuticals can be broadly classified into dietary supplements and functional foods as shown in Fig. 2.1 (Swaroopa and Srinath 2017).

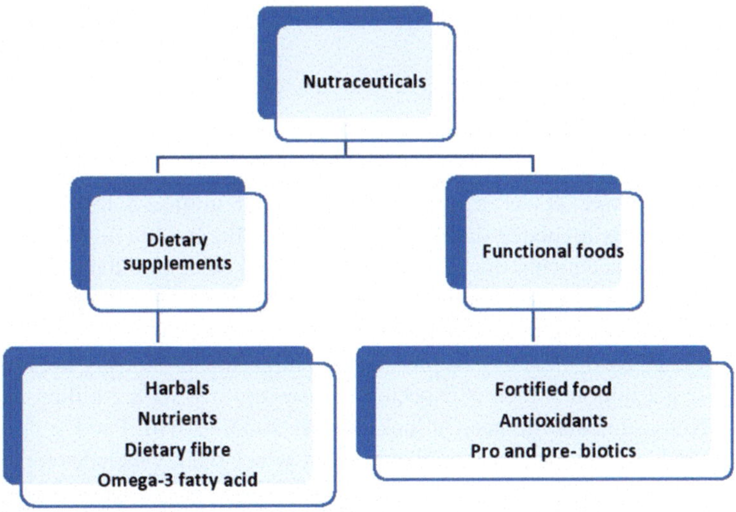

Fig. 2.1 Broad classification of Nutraceuticals

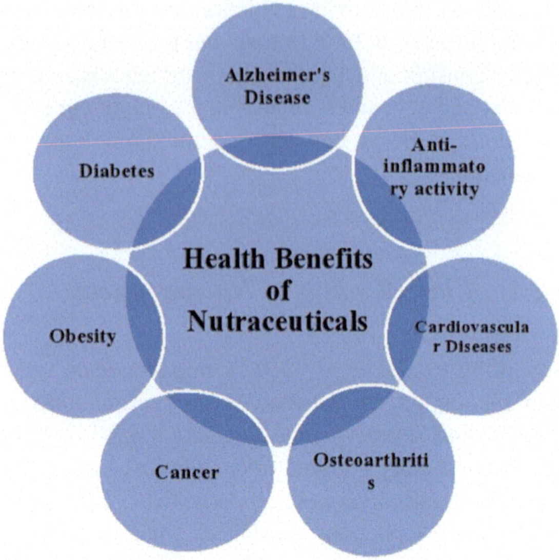

Fig. 2.2 Heath benefits of nutraceuticals

2.1.2 Applications of Nutraceuticals in Disease Management

Nutraceuticals are basically derived from food sources that are supposed to provide extra health benefits, in addition to the basic nutritional values found in food. Based on this power of providing health benefits, nutraceutical food or food components claim to prevent human from chronic diseases, improve health, delay the aging process, increases life expectancy, and provide basic support for building the structure or maintain proper functioning of the body as shown in Fig. 2.2 (Swaroopa and Srinath 2017).

2.1.3 Nutraceutical Market

The nutraceutical market has developed and changed to million-dollar industry at a global level, that is regulated under laws throughout the world (Box 2.1). Global marketing of nutraceuticals was accounted for \$128.4 billion in 2008, \$171.8 billion in 2014 and is expected to reach \$295 billion by 2022. Globally the market is estimated to be growing at a compound annual growth rate (CAGR) of 7% from 2014 to 2022 (Basu et al. 2007). At present, the nutraceuticals industry in India is of about USD 5.2 billion, which is expected to grow up to USD 8.2 billion in 2022 ("Indian Nutraceuticals Market Outlook: Vision 2022," 2016). Food supplement covers around 60% nutraceutical market in India, which mainly includes food products and supplements, malted beverages, fruit-based products, clinical products, pediatric nutrition, protein powder, sport products. Out of all these 50%

Box 2.1: Regulatory Aspects of Nutraceuticals

International Regulatory Laws (Singh and Sinha 2012)

- *Watershed Legislation Law or Dietary Supplement Health and education Act (DSHEA) (1994)*
- *Food and Drug Administration Modernization Act (FDAMA) (1997)*
- *Foods for Specified Health Uses (FOSHU) (1993)*
- *Foods with Health Claims (FHC) along with Foods with Nutrient Function Claims (FNFC) (2001)*

National Regulatory Laws (Gupta et al. 2010)

- *Prevention of Food Adulteration Act (1954)*
- *Fruit Products Order (FPO) (1955)*
- *Milk and Milk Products Order (MFPO) (1973)*
- *Standards of Weights and Measures Act (1976)*
- *Meat Food Products Order (MMPO) (1992)*
- *Infants Milk Substitutes, Feeding Bottles and Infants Foods Act (1992) further amended by central government in 1993*
- *Vegetable Oils Products (Regulation) Order (VOP) (1998)*
- *The Edible Oils Packaging (Regulations) Order (1998)*
- *Standards of Weights and Measures (Packaged commodities) Amendment Rules (SWMA) (2006)*

consumption is of food products and supplements (*Brief Report on Nutraceutical Products in India* 2015). Observing the growth of food-based nutraceutical industry, more types of food supplements can be introduced in the market based on their nutraceutical potential. Horse gram is one of those legumes which has high potential for marketing as nutraceutical due to the presence of different types of bioactive molecules.

2.2 Nutraceutical Attribute of Legumes: Horse Gram (*Macrotyloma uniflorum*)

Legumes are the seeds of dicotyledonous plants belonging to the family Fabaceae (Leguminosae). Being called "poor man's meat", legumes compensate the nutritional gap by supplying protein, carbohydrates, vitamins, minerals, dietary fiber and fatty acids. Food legumes constitute an important foodstuff in tropical and subtropical countries, and they are second in importance only to cereals as a source of protein and often they represent a necessary supplement to other protein sources (Duranti and Gius 1997; Roy et al. 2010). The protein content in legumes ranges from 20–36% by weight. Other contents are ranging from 0–65 g carbohydrates,

18–32 g total dietary fibre, 2–21 g fat and micronutrients, such as calcium, phosphorus, potassium, sodium, iron, zinc etc. per 100 g weight of food item (Brijesh Tiwari et al. 2011; Sreerama et al. 2010a, b). Besides their nutritional properties, legumes have been beneficial to human health which includes prevention of diseases including cardiovascular diseases, diabetes, obesity etc. (Guillon and Champ 2002; Venter and Van Eyssen 2001).

Horse gram (*Macrotyloma uniflorum*) is one of underutilized legumes, which is mostly cultivated in Asian and African continents, enclosing the countries particularly, India, China, Philippines, Bhutan, Pakistan and Sri Lanka under low soil fertility conditions. Its remarkable ability to thrive in parched environments makes the horse gram shrub a highly capable candidate to meet food and nutritional requirements in malnourished populations. Therefore above all the other legumes we have considered to draw the focus of scientific community towards the nutritional and nutraceutical aspects of horsegram (Durga 2012).

2.2.1 Scientific Classification and Nomenclature

Kingdom	:	Plantae
Order	:	Fabales
Family	:	Fabaceae
Genus	:	*Macrotyloma*
Species	:	*uniflorum*

Macrotyloma uniflorum or previously known *Dolichos biflorus* is commonly known as horse gram (English), *Dolic biflore* (French), Kerderkorn (German), Dolico cavallino (Italian), Frojol verde (Spanish), *Faveira* (Portuguese), *Kulattha* (Sanskrit), *Kurti-kalai* (Bengali), *Kollu* (Tamil), *Ullavallu* (Telgu), *Muthira* (Malyalam), *Kolatha* (Oriya), *Gahot* (Uttarakhand) (Bhartiya et al. 2015; Rana Smita and Agnihotri Vasudha 2018).

2.2.2 Physical Features

It is a short day and day neutral plant (Fig. 2.3a) maturing 120–180 days after planting and requires an annual temperature ranging from 18–27 °C and low to moderate annual rainfall of 200–1000 mm. It can be grown in wide range of soil such as black cotton soil, deep red loam and clay loam paddy soils are considered best for its cultivation. Although it is drought tolerant crop, but does not tolerate flooding or waterlogging (Rana Smita and Agnihotri Vasudha 2018). It has nitrogen fixation ability and helps in soil conservation (Prasad and Singh 2015).

Fig. 2.3 (**a**) Horse gram (*Macrotyloma uniflorum*) plant with pods; (**b**) Raw seeds of Horse gram *Macrotyloma uniflorum*

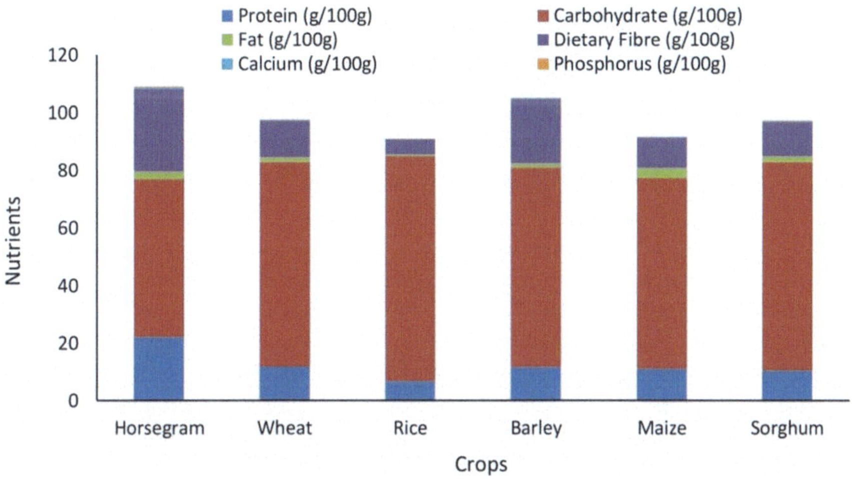

Fig. 2.4 Nutritional composition of Horse gram seeds (Nath and Vijayalakshmi 2014)

2.2.3 Nutritional Properties of Horse Gram

Horse gram being highly nutritive is not only used as food but can also be used as fodder for cattle's and horses (Marimuthu and Krishnamoorthi 2013; Morris et al. 2013). Its seeds are consumed as whole (Fig. 2.3b) as well as in dehulled state in the form of various recipes. Nutritionally it is very rich and equivalent to other commonly grown cereal crops in all aspects (Fig. 2.4).

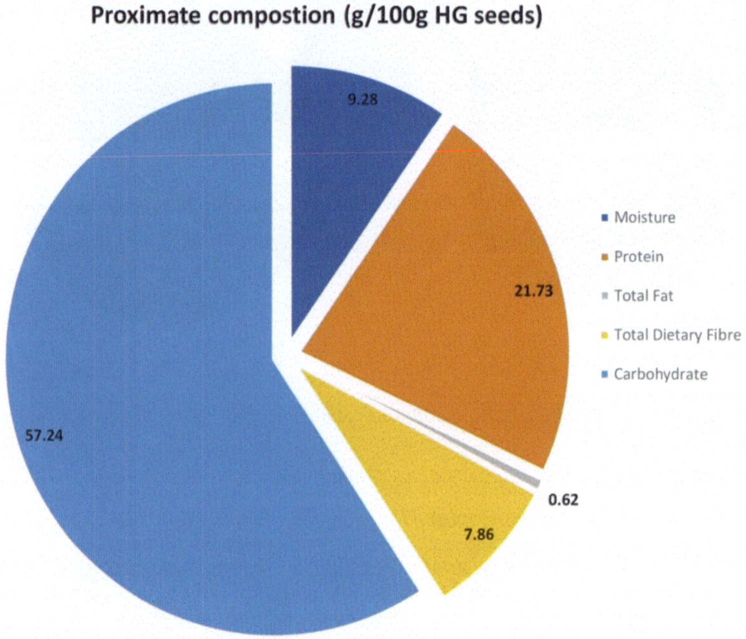

Fig. 2.5 Proximate composition of Horse gram (Longvah et al. 2017)

2.2.3.1 Protein and Amino Acid

Proteins can be classified on basis of their solubility as water soluble albumins, salt-soluble globulins, alcohol-soluble prolamins, and acid- or alkaline-soluble glutelins (Utsumi 1992). Horse gram seeds contain approximately 21.73 g protein per 100 g dry wt (Fig. 2.5). The protein content of powdered seeds of horse gram varies widely from 22.6% in cotyledon to 9.1% in seed coat fraction, and 18.6% in embryonic axe (Sreerama et al. 2010a, b). The dehulled seeds exhibit higher protein content (18.4–25.5%) than the whole (17.9–25.3%) (Laura Bravo et al. 1998). The major proteins in legume seeds are glutelins (10–20%) albumins (10–20%) (Duranti 2006) and globulins (50–90%). Out of the total protein content, in horse gram albumin-globulin protein fraction contributes from 75.27% to 78.76%, while glutelin and residual protein varies from 9.93–17.52% to 6.96–11.30%, respectively (Yadav et al. 2004). Digestibility of raw and cooked legume protein varies from 15–80% to 50–90%, respectively, as compared to cereal seed protein (75–90%) (Eggum and Beames 1983).

Horse gram seeds contain all the essential amino acids along with other conditionally essential and non-essential amino acids (Fig. 2.6). Among the essential amino acids, phenyl alanine content is highest i.e. 1997.6 mg/100 g dry weight of horse gram seeds (Longvah et al. 2017). Recommended daily consumption of amino acids by the humans are decided on the basis of their weight in kilogram (Fig. 2.7)

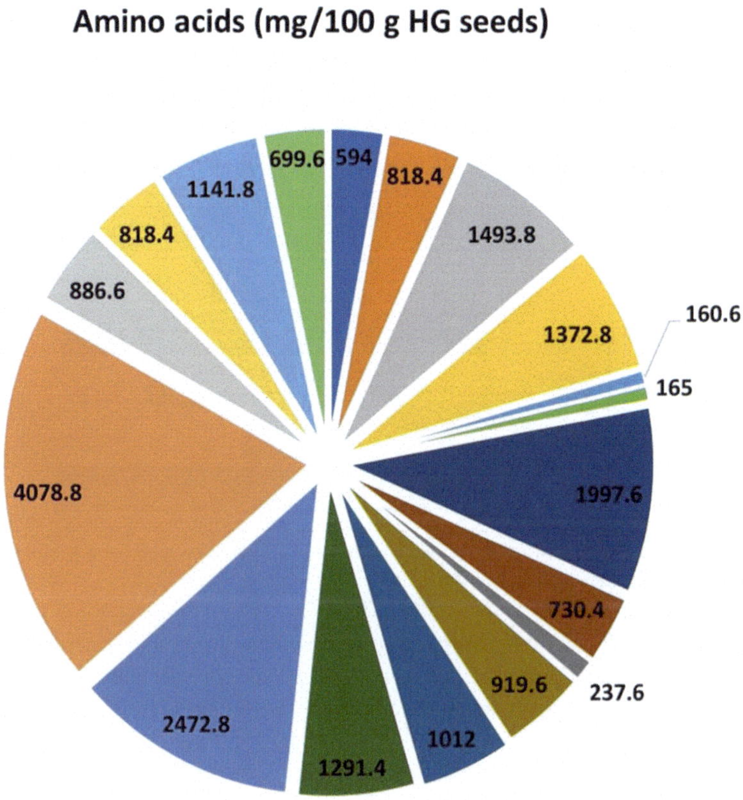

Fig. 2.6 Amino acid composition of Horse gram seeds (Longvah et al. 2017)

(FAO/WHO/UNU Expert Consultation on 2007). Through the amino acid content present in Horse gram seeds, it is observed that, the seeds can contribute efficiently for daily amino acid consumption required by an adult.

2.2.3.2 Carbohydrates

Like protein, carbohydrates are also the major component of legumes, constitute from 50% to 70% of the dry matter (Laura Bravo et al. 1998). Carbohydrates can be classified as digestible and non-digestible carbohydrates. Starch and non-starch polysaccharides are the most abundant form of carbohydrates in legume seeds with smaller amounts of oligosaccharides. Starch is categorized into partly digestible (Bravo et al. 1999) and resistance starch i.e. non- digestible carbohydrate (Asp et al. 1996). Horse gram seeds contain 57.24 g carbohydrate per 100 g of horse gram seeds (Fig. 2.5) (Longvah et al. 2017). Bravo et al. reported 43.4% resistant starch out of total carbohydrate present in horse gram seeds (Bravo et al. 1999). Starch, if

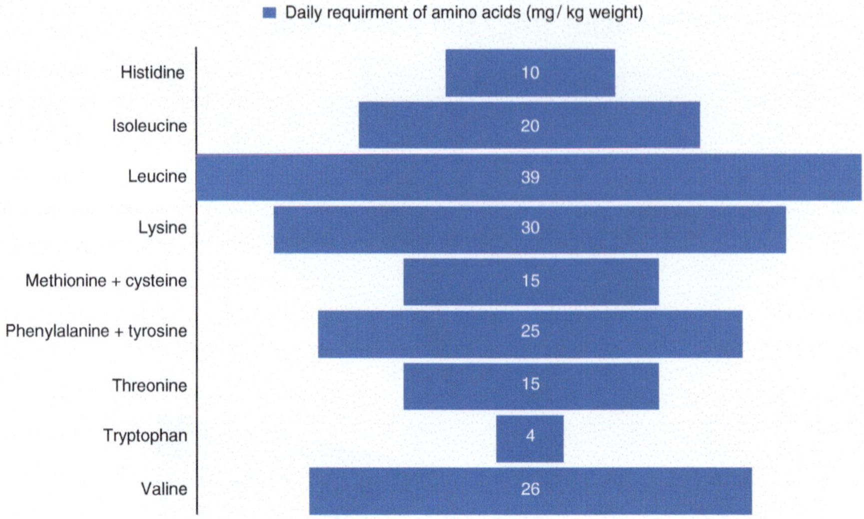

Fig. 2.7 Daily requirement of amino acids by human beings based on their per Kg weight (FAO/WHO/UNU Expert Consultation on 2007)

undigested in small intestine, reaches to large intestine where its fermentation takes place, which causes the production of butyrate. This butyrate has the capacity to prevent the large interstine from colorectal cancer by reducing the risk of malignant changes in cells (Scheppach et al. 1995), by increasing fecal bulking and lowering the pH of fecal content (Prasad and Singh 2015).

Monosaccharides (such as glucose, galactose, arabinose, fructose and inositol), disaccharides such as (sucrose and maltose) and oligosaccharides are the form of soluble sugar basically present in horse gram. Based on horse gram seeds collected from throughout India, Longvah et al. has suggested edible sugar content in its seeds. Per 100 g of Horse gram seeds contain edible sugars which consists of 48.31 g total available sugar (total starch 47.96 g, fructose 0.1 g, glucose 0.15 g and sucrose 0.1 g) and 0.35 g total free sugar (Longvah et al. 2017). Carbohydrate content varied in whole and dehulled seeds. Whole seed contains 51.9–60.9% carbohydrate while dehulled seed contain 56.8–66.4%. Carbohydrate of raw horse gram seeds comprises 36 ± 1.17 g starch per 100 g dry matter in which approximately 85% digestible, 14.47% resistant and 3.38% resistant starch associated to insoluble dietary fibres (Bravo et al. 1999).

2.2.3.3 Fat and Fatty Acids

The fat content of horse gram ranges from 0.6–2.6%. Longvah et al. described 0.62% fat content (Fig. 2.4) in horse gram seeds, collected from various regions of India (Longvah et al. 2017). Dehulled horse gram seeds exhibit higher crude fat content (0.81–2.11%) than the whole (0.70–2.06%) seeds. These seeds are also good source of fatty acids and contains 27.5% saturated fatty acids (21.97% palmitic, 2.85% arachidic, 2.32% stearic acid and 0.36% myristic), 72.49% unsaturated fatty acids (42.78% linoleic, 16.15% oleic and 13.56% linolenic acid). Among unsaturated fatty acids linoleic acid is useful for the treatment of diabetes and cardiovascular diseases (Bhartiya et al. 2015). Longvah et al. documented the high content of mono unsaturated fatty acid (MUFA) and poly unsaturated fatty acid (PUFA) i.e. 68.89 and 258 mg/100 g horse gram seeds (Fig. 2.7) (Longvah et al. 2017). These fatty acids are becoming very popular now days in cooking oils due to their different types of health benefits especially for cardiovascular diseases (Ander et al. 2003; Trautwein et al. 1999). Horse gram lipids have anti-ulcer activity due to presence of phytosterol ester which imparts protective and healing effect on acute gastric ulceration produced by alcohol (Mishra and Pathan 2011; Sudha et al. 1995) (Fig. 2.8).

2.2.3.4 Dietary Fiber

Plant foods are the only sources of dietary fibers. The fiber content of human foods, from plant sources ranges from trace amounts to almost 50% of dry weight (Anderson and Bridge 1998). Dietary fiber refers to all the polysaccharides which are not completely digested and absorbed in small intestine but undergo complete fermentation in large intestine such as inulin (Jones et al. 2006; Prosky and DeVries 1991). The Major dietary fibers which are predominantly present in cell wall of plant includes cellulose, hemicelluloses, pectin, arabinoxylans, betaglucan,

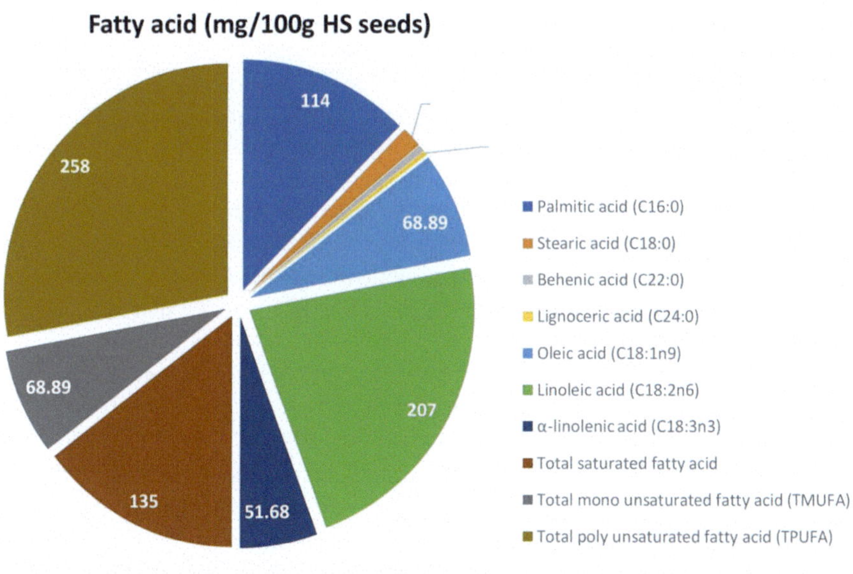

Fig. 2.8 Fatty acid composition of horse gram seeds (Longvah et al. 2017)

glucomannans, plant gums and mucilages and hydrocolloids (Cummings and Stephen 2007). Bravo et al. showed 28.8% total dietary fibers in horse gram seeds which mainly consists of insoluble dietary fibre (IDF) (27.82%) and soluble dietary fibre (SDF) (1.13%) and the ratio of IDF and SDF was 24.6 whereas horse gram flour contains 16.3% total dietary fibre (14.9% insoluble and 1.4% soluble and 2.2% resistant starch). Longvah et al. reported 7.86 g per 100 g, estimated in Indian horse gram seeds.

Total Dietary fiber composition has highest percentage of total neutral sugar (39%) followed by glucose (21%), klason lignin (16%) and arabinose (12%). Higher content of insoluble dietary fiber is important as for lower intestine functions in humans. Seeds of horse gram (*Macrotyloma uniflorum*) contain more insoluble dietary fiber than kidney bean (*Phaseolus aconitifolius*) (Bhartiya et al. 2015). Crude fiber content was higher in seed coat fractions than in embryonic axe and cotyledon fractions. Thus the presence of higher amount of fibre in seed coat can be helpful in the value addition of food products thus helpful in improving the gastro-intestinal health (Sreerama et al. 2010a, b). The major functions associated with the intake of dietary fibers are reduce constipation, modulation of blood glucose level cholesterol reduction, prebiotic effects, prevention of certain cancers (Redgwell and Fischer 2005) cardiovascular diseases (CVD), diverticulosis, obesity, lowers blood pressure reduces the risk of heart attack and colon cancer (Sharma and Kawatra 1995).

2.2.3.5 Polyphenols

The potential health benefits of legume are due to the presence of secondary metabolites such as phenolic and flavonoid compounds that possess antioxidant properties. Horse gram can also supply many bioactive substances such as phenolics, flavanoid in small quantities which have significant metabolic and/or physiological effects. Sreerama et al. reported the major flavonoid compounds present in horse gram seed are quercetin, kaempferol, myricetin, daidzein, and genistein (Fig. 2.9) (Sreerama et al. 2010a, b). Flavonoids are known to have antioxidant activities (Segev et al. 2010) and are used as antibiotics, antidiarrheal, anti-ulcer, anticancer, anti-allergic and anti-inflammatory agents, and also for treatment of diseases such as hypertension, vascular fragility, allergies, and hypercholesterolemia (Bravo et al. 1999; Sreerama et al. 2012). The dietary antioxidants possess many therapeutic properties of neurodegenerative diseases, including Alzheimer's and Parkinson's diseases (Daniel 2003).

Phenolic compounds having good antioxidant property are beneficial as it protects our body from oxidative damage from free radicals (Ramesh et al. 2011; Sundaram et al. 2013). Marathe et al. reported the variation of phenolic content of legumes n the range of 0.325–6.378 mg GAE (gallic acid equivalent)/g (Marathe et al. 2011). Sreerama et al. reported benzoic acid and cinnamic acid derivatives of phenolic compounds (Fig. 2.9) (Sreerama et al. 2010a, b). Phenolic acids are believed to work synergistically to promote human health through a variety of different mechanisms such as impacting cellular processes associated with apoptosis,

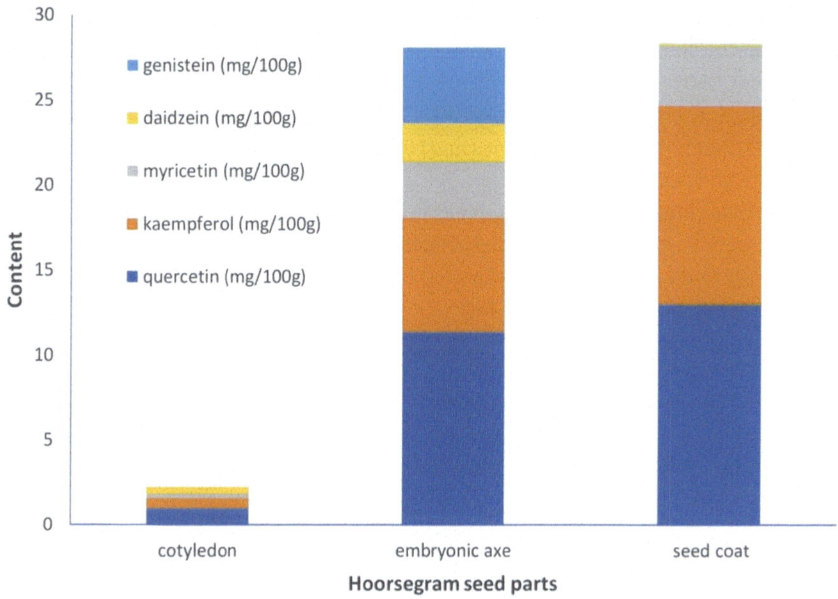

Fig. 2.9 Flavonoids of horse gram seed parts (Sreerama et al. 2010a, b)

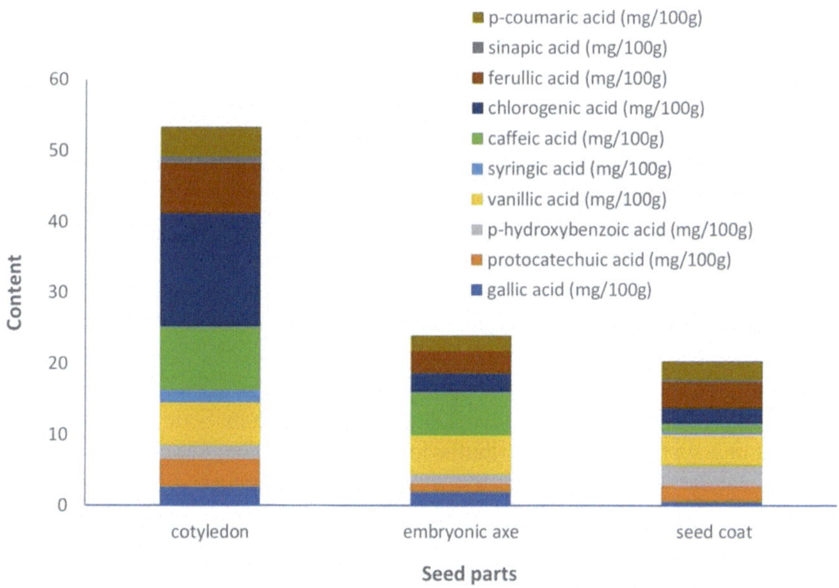

Fig. 2.10 Phenolic acids of horse gram seed parts (Sreerama et al. 2010a, b)

platelet aggregation, blood vessel dilation, enzyme activities associated with starch, protein, and/or lipid digestion, carcinogen activation, and detoxification (Mcdougall and Stewart 2005).

The phenolic acid contents (mg/100 g dw) were found higher in cotyledons (Fig. 2.10) while flavonoid content (mg/100 g dw) was found least in cotyledons (Fig. 2.8). So based on the target antioxidant nutraceuticals, the seed part may be selected. In addition, to their antioxidant properties, isoflavones and lignans exert a weak estrogenic activity and are associated in several mechanisms protecting the human body and also have been suggested to reduce the risks of cancer and to lower serum cholesterol (Barnes et al. 2000).

2.2.3.6 Minerals and Trace Elements

Horse gram seed contains minerals (mainly calcium, potassium, magnesium, Phosphorus, sodium and iron) and trace elements (Aluminium, Chromium, cobalt, copper, leas, lithium, manganese, molybdenum, nickel, zinc and mercury) as shown in Fig. 2.11a, b as documented by Longvah et al. (2017). Horse gram contains good amount of calcium, potassium, phosphorus and iron content but it has significantly lower amount of sodium content which is nutritionally advantageous for the low sodium intake recommendation in the diet (Borhade et al. 1983; Kadam and Salunkhe 1985; Kadwe et al. 1974).

2.2.3.7 Vitamins

Vitamins and co-enzyme play important role in metabolism of carbohydrates and branched chain amino acids, numerous oxidation and reduction reactions, hydrogen transfer with numerous dehydrogenases, glycogen, and sphingoid bases (Hedges et al. 1997; Martone et al. 1994; Ordauiez-garcia et al. 1996) so it is essential for us to take them from external sources to fulfill the demand of our body. Legumes are superior source of vitamins particularly vitamin B-complex (i.e. pyridoxine, niacin, riboflavin, folic acid and thiamine), but are a poor source of vitamin C and fat soluble vitamins (Dias 2012). The low B-complex content of diets causes eruption of peripheral neuropathy and visual loss in the adult population (Hedges et al. 1997; Martone et al. 1994; Ordauiez-garcia et al. 1996). Horse gram is rich source of several vitamins such as thiamine, riboflavin, niacin, pantothenic acid, total folate, γ-tocopherol etc. (Fig. 2.12) (Bhartiya et al. 2015; Gopalan et al. 1989; Longvah et al. 2017).

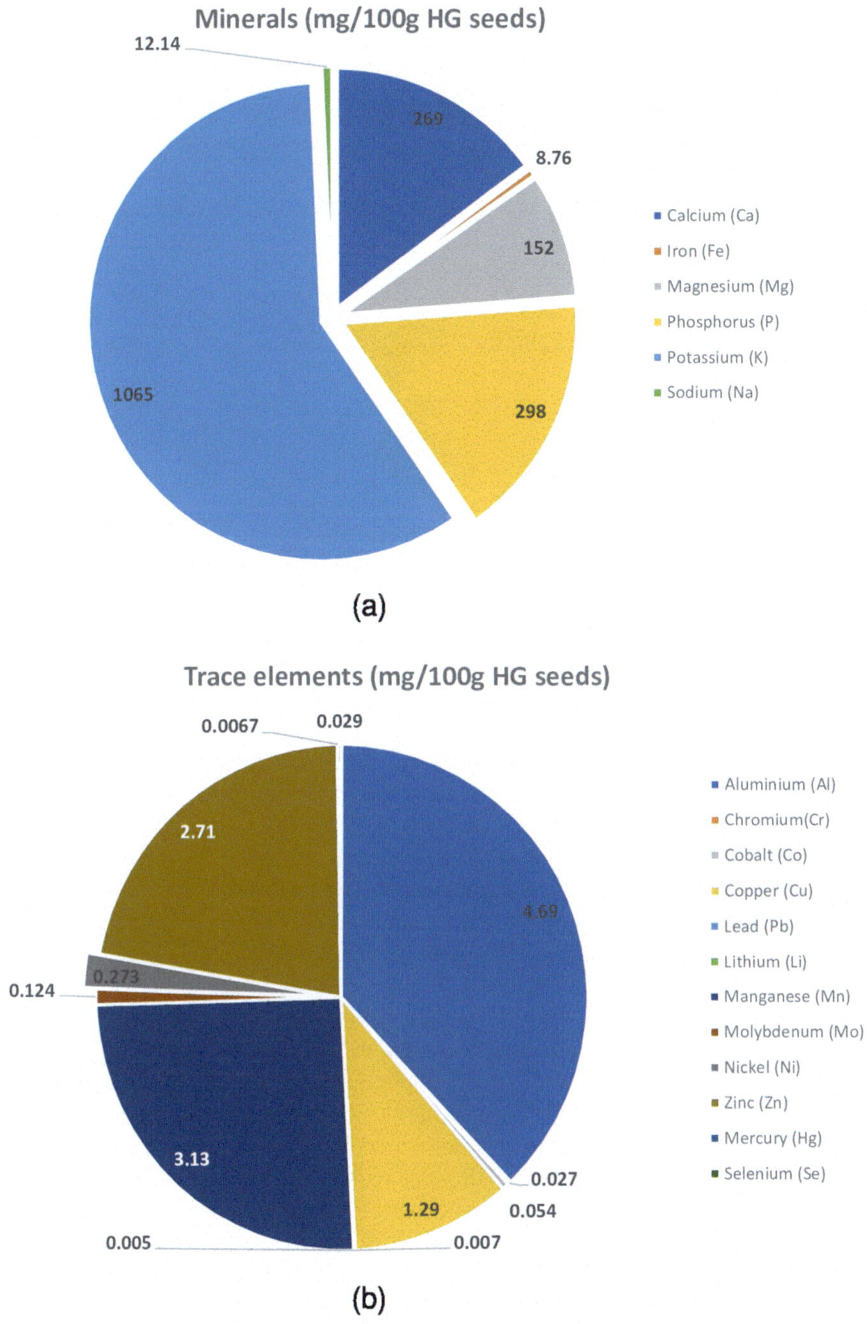

Fig. 2.11 (**a**) Minerals and (**b**) trace element found in Horse gram seeds (Longvah et al. 2017)

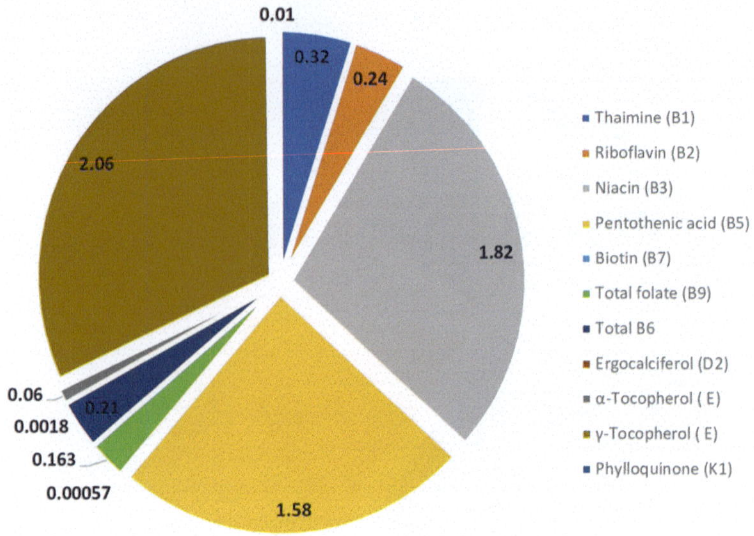

Fig. 2.12 Vitamin contents in mg/100 g Horse gram seeds (Longvah et al. 2017)

2.2.3.8 Other Nutritional and Antinutritional Factors

Horse gram contains some amounts of antinutrients also which can reduce the bioavailability of nutrients such as phytosterol (including campesterol, stigmasterol, β-sitosterol), phytate, total saponin, oligosaccharides etc. Longvah et al. reported 3.41 mg campesterol, 11.57 mg stigmasterol and 47.56 mg β-sitosterol in 100 g of horse gram seeds, whereas phytate content was 3.39 mg/100 g seed weight (Longvah et al. 2017). Total saponin was reported at a leven of 2.30 g/100 g seed weight. The Oligosaccharide content in horse gram was also reported at the level of 1.63 g per 100 g of edible portion i.e. seeds which includes 0.04 g raffinose and 1.59 g stachyose, which is the average value of the samples collected throughout India. Sreerama et al. reported much higher value of flatulence causing oligosaccharide in horse gram seeds collected from Mysore market where verbascose was also reported along with raffinose and stachyose (Sreerama et al. 2010a, b). Its flour contains trypsin inhibitor activity (9246 ± 18 TIU/g), phytic acid (10.2 ± 0.4 mg/g), polyphenols (14.3 ± 0.4mgGA/g). and oligosaccharides (26.8 mg/g) (Morris et al. 2013; Sreerama et al. 2012). Antinutrients have their certain roles in human physiological activities, but in higher concentration they can reduce the bioavailability of different mineral present in legumes. Food processing generally reduces their negative effects. Longvah et al. reported different types of organic acids in horse gram seed including total oxalate (181 mg/100 g dw), fumaric acid (37.04 mg/100 g dw), mallic acid (0.37 mg/100 dw) and succinic acid (0.68 mg/100 g dw) (Longvah et al. 2017), many of which are essential for cellular cycles.

2.3 Therapeutic Attributes of Horse Gram

The health benefits of legumes have been known for millennia due to their inherent capacity to produce several secondary metabolites which serve to cure several diseases. Use of legume in day to day diet not only enhance the nutritional status but also reduces the risk of chronic diseases such cardiovascular diseases, cancer and diabetes (Venter and Van Eyssen 2001). Legumes plays an important role in prevention and treatment of type 2 diabetes mellitus, cardiovascular disease, cancer diseases (e.g. breast and prostate cancers), overweight and obesity, osteoporosis, hypertension, gastrointestinal disorders; and results in lowering of low density lipoprotein cholesterol and elevated high density lipoprotein cholesterol levels (Boye et al. 2010; Kalogeropoulos et al. 2010; Kushi et al. 1999; Xu and Chang 2008).

Sprouted legumes are very popular in diet industry. Similar to other legumes, sprouted seed of horse gram become increasingly popular due to its high nutritive value as well as for being the excellent source to reduce the risk of various diseases and exerting health promoting effects (Pasko et al. 2009). As per *Charak Samhita*, the seeds of horse gram are useful for the cure of piles, hiccup, abdominal lump, bronchial asthma, in causing and regulating perspiration. In the *Sushruta Samhita*, it is mentioned that the seed powder is useful in stopping excessive perspiration. Horse gram seeds as well as extracts has excellent hypolipidaemic, hypoglycaemic (Senthil 2009), therapeutic properties and traditionally used to cure kidney stones, piles, urinary troubles, acid peptic disorder (gastritis), constipation, sun-burn, female diseases (leucorrhoea, menstrual troubles, bleeding during pregnancy, post-partum excessive discharges), colic caused by wind, rheumatism, hemorrhagic disease, intestinal worms, bronchitis, leucoderma, asthma, inflamed joints, sudation therapy, fever, musculoskeletal disorder, breast milk purifier, sinus wounds, tumours, ascites and localized abdominal tumor (Thirumaran and Kanchana 2000).

Besides, it also possesses high tolerance against salinity, drought, and heavy metals, horse gram species possess different medicinal properties such as anti-diabetic, antiulcer activity antimicrobial, antioxidant activity, and helps in dietary management of obesity due to the presence of beneficial bioactive compounds. It is prescribed for persons suffering from jaundice, water retention, as part of a weight loss diet, iron deficiencies and also helpful for maintaining body temperature in the winter season (Ramesh et al. 2011). Linoleic acid, present in the seeds, is useful for the treatment of diabetes and cardiovascular diseases. Horse gram lipids have anti-ulcer activity due to presence of phytosterol ester which imparts protective and healing effect on acute gastric ulceration produced by alcohol. Furthermore, extract of the horse gram seeds with spices is considered to be a potential remedy for the common cold, throat infection, fever and the soup said to generate heat (Thirumaran and Kanchana 2000) and horse gram powder or liquid act as expectorant to extract phlegm from the body (Siddhuraju and Manian 2007).

The therapeutic properties of medicinal plants are possibly due to existence of various phytochemical components. Several phytochemical constituents identified/isolated from horse gram seed, responsible for its therapeutic properties are listed in

Table 2.1 List of phytochemicals identified from horse gram (Ranasinghe and Ediriweera 2017)

Class	Phytochemical constituents
Anthocyanins	Cyanidin, Delphinidin, Malvidin, Petunidin
Flavanoids	Daidzein, Genistein, Kaempferol, Myricetin, Quercitine
Phenolic acids	Gallic acid, Protocatechuic acid, 4 Hydroxybenzoic acid, 3,4 dihyroxybenzoic acid, Syringic acid, Vanillic acid, Caffeic acid, p-Coumaric acid, Ferulic acid, Sinapic acid
Enzyme	Amylase and Glucosidase, b- N-Acetylglucosaminidase, Urease, Trypsin
Haemagglutinins	Agglutinin and Lectins

Table 2.1 (Ranasinghe and Ediriweera 2017). The salient features of multifunctional physiological effects of horse gram are summarized below:

Anti-diabetic potential: Horse gram has been found to be effective in improving the glycemic status in diabetic patients. Due to high fibre and low-fat content, its consumption has the capacity to delay gastric emptying by direct interference with glucose absorption.

It is digested slowly and produce low blood glucose response resulting in low glycemic index (Kalogeropoulos et al. 2010; Tiwari et al. 2011; Venter and Van Eyssen 2001).

Anti-urolithiatic effect: Urolithiasis is a multifaceted process that occurs from series of quite a few physicochemical event including super saturation, nucleation, growth aggregation and retention within the kidneys (Sharma et al. 2019). Horse gram has the capacity to perform against kidney stone (Bhartiya et al. 2015; Yadava and Vyas 1994).

Cardiovascular Diseases: Low dietary intake of sodium and high dietary intake of potassium, calcium, and magnesium is associated with a reduced risk of cardiovascular disease, by helping to lower blood pressure, in epidemiologic studies (Anderson 2004; Gebrelibanos et al. 2013). The low glycemic index values of legumes means that they are less likely to raise blood glucose and insulin levels, which may also decrease cardiovascular disease risk (Anderson 2004). Along with this, the seed also contains good amount of MUFA and PUFA (Fig. 2.7), which might be responsible for its activity in decreasing cardiovascular disease risk, which needs to be further explored.

2.4 Conclusion

The study of nutritional and medicinal properties of horse gram reveals that it is a rich source of nutrients. The nutritional value of horse gram is comparable with other daily consumed legumes. It is also known to have good therapeutic potential whereas the information related to specific health beneficial components is lacking. Horse gram seeds have lot of potential to be used as nutraceutical due to the presence of different types of phytochemicals. This legume is being used for making various recipes, but it needs to be explored as nutraceutical/as dietary or food supplements. Research on its medical benefits is also required as very less amount of research is available on its health benefits. So more scientific research needs to be carried out on Horse gram seeds.

Acknowledgement DST-NRDMS is duly acknowledged for project funding and Director, GBPNIHESD for providing guidance and other facilities.

References

Ander BP, Dupasquier CMC, Prociuk MA, Pierce GN (2003) Polyunsaturated fatty acids and their effects on cardiovascular disease. Exp Clin Cardiol 8(4):164–172
Anderson J (2004) Legumes. Linus Pauling Institute
Anderson J, Bridge S (1998) Dietary fiber content of selected foods. Am J Clin Nutr 47:440–447
Asp N, Van Amelsvoort J, Hautvast J (1996) Nutritional implication of resistant starch. Nutr Res Rev 9:1–31
Barnes S, Boersma B, Patel R, Kirk M, Darley-usmar VM, Kim H, Xu J (2000) Isoflavonoids and chronic disease: mechanisms of action. BioFactors 12:209–215
Basu SK, Thomas JE, Acharya SN (2007) Prospects for growth in global nutraceutical and functional food markets: a Canadian perspective. Aust J Basic Appl Sci 1(4):637–649
Bhartiya A, Aditya JP, Kant L (2015) Nutritional and remedial potential of an underutilized food legume horsegram (Macrotyloma uniflorum): a review. J Anim Plant Sci 25(4):908–920
Bigliardi B, Galati F (2013) Innovation trends in the food industry: the case of functional foods. Trends Food Sci Technol 31:118–129. https://doi.org/10.1016/j.tifs.2013.03.006
Borhade V, Kadam S, Salunkhe D (1983) Solubilization and functional properties of moth. J Food Biochem 8:229–235
Boye J, Zare F, Pletch A (2010) Pulse proteins: processing, characterization, functional properties and applications in food and feed. Food Res Int 43(2):414–431. https://doi.org/10.1016/j.foodres.2009.09.003
Bravo L, Siddhuraju P, Saura-calixto F (1998) Effect of various processing methods on the in vitro starch digestibility and resistant starch content of Indian pulses. J Agric Food Chem 46:4667–4674
Bravo L, Siddhuraju P, Calixto F (1999) Composition of underexploited Indian pulses: comparison with common legumes. Food Chem 64:185–192
Brief report on nutraceutical products in India (2015)
Cencic A, Chingwaru W (2010) The role of functional foods, nutraceuticals, and food supplements in intestinal health. Nutrients 2:611–625. https://doi.org/10.3390/nu2060611
Chaturvedi N, Sharma P, Shukla K, Singh R, Yadav S (2011) Cereals nutraceuticals, health ennoblement and diseases obviation: a comprehensive review. J Appl Pharm Sci 01(07):6–12
Cummings JH, Stephen AM (2007) Carbohydrate terminology and classification. Eur J Clin Nutr 61:5–18. https://doi.org/10.1038/sj.ejcn.1602936
Daniel S (2003) An investigation into the neuroprotective prospective of curcumin. Rhodes University
Das L, Bhaumik E, Raychaudhuri U, Chakraborty R (2012) Role of nutraceuticals in human health. J Food Sci Technol 49(2):173–183. https://doi.org/10.1007/s13197-011-0269-4
Dias JS (2012) Nutritional quality and health benefits of vegetables: a review. Food Nutr Sci 3:1354–1374
Duranti M (2006) Grain legume proteins and nutraceutical properties. Fitoterapia 77:67–82. https://doi.org/10.1016/j.fitote.2005.11.008

Duranti M, Gius C (1997) Legume seeds: protein content and nutritional value. Field Crop Res 53:31–45

Durga KK (2012) Variability and divergence in horsegram (Dolichos uniflorus). J Arid Land 4(1):71–76. https://doi.org/10.3724/SP.J.1227.2012.00071

Eggum BO, Beames RM (1983) The nutritive value of seed proteins. In: Gottschalk W, Muller HP (eds) Seed proteins biochemistry, genetics, nutritive value. Martmus Nijhoff/Dr W Junk Publishers, The Hague, p 499. https://doi.org/10.1007/978-94-009-6801-1

El Sohaimy SA (2012) Functional foods and nutraceuticals-modern approach to food science. World Appl Sci J 20(5):691–708. https://doi.org/10.5829/idosi.wasj.2012.20.05.66119

FAO/WHO/UNU Expert Consultation on (2007) WHO technical report series, protein and amino acid requirements in human nutrition. In World Health Organization technical report series, vol 935

Gebrelibanos M, Tesfaye D, Raghavendra Y, Sintayeyu B (2013) Nutritional and health implications of legumes. Int J Pharm Sci Res 4(4):1269–1279

Gopalan C, Sastri B, Balasubramanian S (1989) Nutritive value of Indian foods. ICMR. National Institute of Nutrition

Guillon F, Champ MM (2002) Carbohydrate fractions of legumes: uses in human nutrition and potential for health. Br J Nutr 88:293–306. https://doi.org/10.1079/BJN2002720

Gul K, Singh AK, Jabeen R (2016) Nutraceuticals and functional foods: the foods for the future world. Crit Rev Food Sci Nutr 56:2617–2627. https://doi.org/10.1080/10408398.2014.903384

Gupta S, Chauhan D, Mehla K, Sood P, Nair A (2010) An overview of nutraceuticals: current scenario. J Basic Clin Pharm 1(2):55–62

Hedges T, Hirano M, Tucker K, Caballero B (1997) Epidemic optic and peripheral neuropathy in Cuba: a unique geopolitical public health problem. Surv Ophthalmol 41(4):341–353

Indian Nutraceuticals Market Outlook: Vision 2022 (2016) 2nd national symposium-cum-awards on "Nutraceuticals, herbals, & functional foods" With, 1–29. https://doi.org/10.1017/CBO9781107415324.004

Jones JR, Lineback DM, Levine MJ (2006) Dietary reference intakes: implications for fiber labeling and consumption: a summary of the International Life Sciences Institute North America Fiber Workshop, June 1–2. Nutr Sci Policy 64(1):31–38. https://doi.org/10.1301/nr.2006.jan.31

Kadam S, Salunkhe D (1985) Nutritional composition, processing, and utilization of horse gram and moth bean. Crit Rev Food Sci Nutr 22(1):1–26

Kadwe R, Thakare K, Badhe N (1974) A note on the protein content and mineral composition of twenty five varieties of pulses. Indian J Nutr Diet 11:83–85

Kalogeropoulos N, Chiou A, Ioannou M, Karathanos VT, Hassapidou M, Andrikopoulos NK (2010) Nutritional evaluation and bioactive microconstituents (phytosterols, tocopherols, polyphenols, triterpenic acids) in cooked dry legumes usually consumed in the Mediterranean countries. Food Chem 121(3):682–690. https://doi.org/10.1016/j.foodchem.2010.01.005

Kushi LH, Meyer KA, Jacobs DR Jr (1999) Cereals, legumes, and chronic disease risk reduction: evidence from epidemiologic studies 1–3. Am J Clin Nutr 70:451–458

Longvah T, Ananthan R, Bhaskarachary K, Venkaiah K (2017) Indian food compostion tables. National Institute of Nutrition

Marathe SA, Rajalakshmi V, Jamdar SN, Sharma A (2011) Comparative study on antioxidant activity of different varieties of commonly consumed legumes in India. Food Chem Toxicol 49(9):2005–2012. https://doi.org/10.1016/j.fct.2011.04.039

Marimuthu M, Krishnamoorthi K (2013) Research article nutrients and functional properties of horse gram (Macrotyloma uniflorum), an underutilized south Indian food legume. J Chem Pharm Res 5(5):390–394

Martone S, Muci-Mendoza R, Reyes L, Silva J, Caballero B (1994) Epidemic optic neuropathy in Cuba: eye findings. Arch Ophthalmol 112:691–699

Mcdougall GJ, Stewart D (2005) The inhibitory effects of berry polyphenols on digestive enzymes. Biofactors 23:189–195

Mishra H, Pathan S (2011) Fatty acid composition of raw and roasted kulthi seeds. Adv J Food Sci Technol 3(6):410–412
Morris JB, Wang ML, Grusak MA, Tonnis B (2013) Fatty acid, flavonol, and mineral composition variability among seven Macrotyloma uniflorum (Lam.) Verdc. Accessions. Agriculture 3(1):157–169. https://doi.org/10.3390/agriculture3010157
Nath KG, Vijayalakshmi D (2014) Neutraceutical features of finger millet. In: Neutraceuticals: challenges and opportunities in 21st century. Agrotech Publishing Academy Udaipur, pp 116–125
Ordauiez-garcia P, Javiernieto F, Espinosa-brito AD, Caballero B (1996) Public health then and now Cuban epidemic neuropathy, 1991 to 1994: history repeats itself a century after the "Amblyopia of the blockade" 86(5)
Palthur MP, Palthur SSS, Chitta SK (2010) Nutraceuticals: a conceptual definition. Int J Pharm Pharm Sci 2(3):19–27
Pasko P, Barton H, Zagrodzki P, Gorinstein S, Folta M, Zachwieja Z (2009) Anthocyanins, total polyphenols and antioxidant activity in amaranth and quinoa seeds and sprouts during their growth. Food Chem 115:994–998. https://doi.org/10.1016/j.foodchem.2009.01.037
Prasad SK, Singh MK (2015) Horse gram – an underutilized nutraceutical pulse crop: a review. J Food Sci Technol 52(5):2489–2499. https://doi.org/10.1007/s13197-014-1312-z
Prosky L, DeVries J (1991) Controlling dietary fiber in food products. In: Gottschalk W, Muller HP (eds) Proteins. Van Nostrand Reinhold, New York
Ramesh C, Abdul R, Prabhakar B, Vijay A, Rao A (2011) Antioxidant potentials in sprouts vs. seeds of Vigna radiata and Macrotyloma uniflorum. J Appl Pharm Sci 01(07):99–103
Rana Smita, Agnihotri Vasudha (2018) Horsegram: nutritional and remedial properties. Everyman's Sci LII(6):391–393
Ranasinghe R, Ediriweera E (2017) Medicinal and nutritional values of Macrotyloma uniflorum (Lam.) Verdc (Kulattha): a conceptual study. Glob J Pharm Pharm Sci 1(2). https://doi.org/10.19080/gjpps.2017.01.555559
Redgwell RJ, Fischer M (2005) Review dietary fiber as a versatile food component: an industrial perspective. Mol Nutr Food Res 49:521–535. https://doi.org/10.1002/mnfr.200500028
Roy F, Boye JI, Simpson BK (2010) Bioactive proteins and peptides in pulse crops: pea, chickpea and lentil. Food Res Int 43(2):432–442. https://doi.org/10.1016/j.foodres.2009.09.002
Scheppach W, Bartram HP, Richter F (1995) Role of short-chain fatty acids in the prevention of colorectal cancer. Eur J Cancer 31:1077–1080
Segev A, Badani H, Kapulnik Y, Shomer I, Oren-Shamir M, Shmuel G (2010) Determination of polyphenols, flavonoids, and antioxidant capacity in colored chickpea (Cicer arietinum L.). Soc Food Sci Technol 75(2):2–6. https://doi.org/10.1111/j.1750-3841.2009.01477.x
Senthil E (2009) Evaluation of Dolichous bifiorus L. on high fructose diet induced alternations in Albino rats. J Cell Tissue Res 9(1):1727–1730
Sharma M, Kawatra A (1995) Effect of dietary fibre from cereal brans and legume seedcoats on serum lipids in rats experimental design. Plant Foods Hum Nutr 47:287–292
Sharma N, Bisht S, Gupta S, Kumar A (2019) Investigation the antilithiatic potential of horse gram (Macrotyloma uniflorum) seeds extracts on experimental kidney stones from high altitude of Uttarakhand (India). Eur J Pharam Med Res 6(1):521–526
Siddhuraju P, Manian S (2007) The antioxidant activity and free radical-scavenging capacity of dietary phenolic extracts from horse gram (Macrotyloma uniflorum (Lam.) Verdc.) seeds. Food Chem 105:950–958. https://doi.org/10.1016/j.foodchem.2007.04.040
Singh J, Sinha S (2012) Classification, regulatory acts and applications of nutraceuticals for health. Int J Pharm Biol Sci 2(1):177–187
Sreerama YN, Neelam DA, Sashikala VB, Pratape VM (2010a) Distribution of nutrients and antinutrients in milled fractions of chickpea and horse gram: seed coat phenolics and their distinct modes of enzyme inhibition. J Agric Food Chem 58(7):4322–4330. https://doi.org/10.1021/jf903101k

Sreerama YN, Sashikala VB, Pratape VM (2010b) Variability in the distribution of phenolic compounds in milled fractions of chickpea and horse gram: evaluation of their antioxidant properties. J Agric Food Chem 58(14):8322–8330. https://doi.org/10.1021/jf101335r
Sreerama Y, Sashikala V, Pratape V, Singh V (2012) Nutrients and antinutrients in cowpea and horse gram flours in comparison to chickpea flour: evaluation of their flour functionality. Food Chem 131(2):462–468. https://doi.org/10.1016/j.foodchem.2011.09.008
Sudha N, Begum J, Shambulingappa K, Babu C (1995) Nutrients and some anti-nutrients in horsegram (Macrotyloma uniflorum (Lam.) Verdc). Food Nutr Bull 16(1):100
Sundaram U, Marimuthu M, Anupama V, Gurumoorthi P (2013) Comparative antioxidant quality evaluation of underutilized/less common south Indian legumes. Int J Pharma Biosci 4(2):117–126
Swaroopa G, Srinath D (2017) Nutraceuticals and their health benefits. Int J Pure Appl Biosci 5(4):1151–1155
Thirumaran A, Kanchana S (2000) Role of pulses in human diets. In: Pulses production strategies in Tamil Nadu. Centre for Plant Breeding and Genetics, Tamil Nadu Agricultural University, Coimbatore
Tiwari BK, Gowen A, McKenna B (2011) Pulse foods: processing, quality and nutraceutical applications, 1st edn. Academic, Amsterdam
Trautwein EA, Rieckhoff D, Kunath-Rau A, Erbersdobler HF (1999) Replacing saturated fat with PUFA-rich (sunflower oil) or MUFA-rich (rapeseed, olive and high-oleic sunflower oil) fats resulted in comparable hypocholesterolemic effects in cholesterol-fed hamsters. Ann Nutr Metab 43(3):159–172. https://doi.org/10.1159/000012782
Utsumi S (1992) Plant food protein engineering. Adv Food Nutr Res 36:189–208
Venter C, Van Eyssen E (2001) More legumes for better overall health. South Afr J Clin Nutr 14(3):32–38
Xu B, Chang SKC (2008) Effect of soaking, boiling, and steaming on total phenolic content and antioxidant activities of cool season food legumes. Food Chem 110:1–13. https://doi.org/10.1016/j.foodchem.2008.01.045
Yadav S, Negi K, Mandal S (2004) Protein and oil rich wild horsegram. Genet Resour Crop Evol 51:629–633
Yadava ND, Vyas NL (1994) Horsegram. In: Arid legumes. Agro Botanical Publishers, Bikaner, pp 64–75

Chapter 3
Grain Legumes and Their By-Products: As a Nutrient Rich Feed Supplement in the Sustainable Intensification of Commercial Poultry Industry

Nirmala Nalluri and Vasavi Rama Karri

Abstract Poultry farming is probably the rapid growing and most adaptable practice which have a major role in improving the global food security. In tropics, a large number of small scale farmers adopt poultry rearing for domestic consumption and contribute significantly to food security. In poultry farming, feeding is an important aspect, which needs major input cost. The banning of diets containing bone meal and meat for animals besides increase in the feed cost of regular feed components like fish meal and maize led to search for an alternative low cost protein and energy rich sources that economically benefits the farmers. Due to high levels of protein in grain legumes, they are considered as potent meal to meet the demand of plant proteins for poultry diets.

In the present review, we view the present level of awareness of using grain legumes in poultry nutrition. It presents and discusses the potential nutritional importance of broad spectrum of grain legumes as poultry feed. The effect and impact of dietary fiber and anti-nutritional factors present in different legumes for consumption, digestion and growth performance of poultry species were also reviewed in this study. Subsequently, different methods like heat, chemical or biological treatments to minimize the impact of different anti-nutritional factors of plant origin were discussed to improve their efficacy for low input farming systems. In this chapter, utilization of by-products produced during the processing of pulses as potential source of poultry feed was also described. Finally, pulses along with their by-products serve as nutrient rich feed to support organic poultry farming system that indirectly confer global food security and perform a crucial role in the establishment of endurable growth because of their economic and ecological values.

N. Nalluri · V. R. Karri (✉)
Department of Biotechnology, GITAM Institute of Technology, GITAM (Deemed to be University), Visakhapatnam, Andhra Pradesh, India

P. Guleria et al. (eds.), *Sustainable Agriculture Reviews 51*, Sustainable Agriculture Reviews 51, https://doi.org/10.1007/978-3-030-68828-8_3

Keywords Anti-nutritional factors · By-products · Broilers · Chemical composition · Digestibility · Egg quality · Grain legumes · Layers · Poultry feeds · Production efficiency

3.1 Introduction

Universal need of animal products, especially eggs, meat and milk has significantly grown worldwide due to increased population, urbanization and rise in income (FAO 2004, 2006). In developing nations, switch to monogastric production was intensified with expanded pigs and poultry farming which is contributing about 77% of growth in contrary to the stagnation and consolidation of poultry activity in the EU (Geers and Madec 2006). In order to support this large scale increase of poultry farming, higher proportions of poultry meal (FAO 2006) i.e. either as natural resources or formulated compounds are required. Increased industrial and vertically integrated poultry operations known as landless systems demands increased purchase of compound feed products. It is estimated that by 2050 more animal products will be needed globally (between 60–70%) (Makkar et al. 2016) which increases the global demand for legume protein as poultry feed. The expansion of the animal feed resource base by developing new additives or identifying new raw materials that enhance the effectiveness of their use can act as a major part in the satisfactory growth of the livestock farming (Makkar et al. 2016).

Globally, poultry is considered as the main protein source in the form of eggs and meat. It contributes to improved human nutrition by offering high nutrient sources like eggs and meat and indirectly generates revenue for small-scale farmers, particularly women and decreases financial vulnerability. As demand for poultry continues as an important addition to the world food supply, animal feed have become a crucial part of the integrated food chain. By 2050, the global population is assumed to be 9.1 billion, which is 34% higher than the present population and the global demand for poultry meat will increase up to 85%, where there may be 30% increase in egg production by 2020. To produce a dozen eggs in laying hens and to gain 1 kg body mass in broilers, approximately a quality feed of 2 kg protein is needed. So, quality feeding is essential in poultry farming, where protein and energy rich feed should be provided to attain good amount of eggs and meat. To attain high quality eggs and meat they mainly depends on certain key ingredients like cereal grains, grain legumes and oilseed cakes as they are considered as the principal basis of protein and energy for effective feed management. In poultry and livestock farming, the cost of the feed will be approximately 60–80% of the overall price. So, due to feed deficiency and high feed cost, it is necessary to give attention on alternative protein additives. As per the reports by Shi et al. (2012), soybean meal, which is an essential and mostly used protein source of poultry meal is getting more costly and so it is essential to choose other economically cheaper and easily available sources of protein as a feed.

In accordance to FAO (Food and Agriculture Organization), from 2011 to 2020, the cost of grains would rise by 20%, giving rise to 30% rise in the cost of poultry meal (AVEC Annual Report 2010–2014). Since, feed is one of the important factors to decide the yield potential, efforts must be made to enhance its nutrient value. The main cereal grains used for poultry diets are barley, wheat, sorghum and triticale. Whereas, the important legumes like soybean meal, peas, lupines, meal of canola and beans were utilized as crucial protein-rich sources. The industries always choose for the low-cost ingredients to increase the gain (Batal 2009). The most important problem facing by the poultry industry is expensive and insufficient supply of animal feed besides the common plant protein (soybean) used mainly for poultry nutrition (Farrell 1996). Poultry production depends mainly on the diets excessive in protein supplements and edible grains, where soybean meal is the most popular among the others. But, it is usually obtained from genetically transformed varieties which led to choose other vegetable proteins with the increasing demand of organic food. The restriction of meat and bone meal in many regions is also a major reason to switch to other vegetable sources, which do not cause any health hazards upon their consumption. Further, the amino acid requirement of different poultry species varies, where a single plant protein source cannot provide the exact amino acid profile they require. So, they must be supplemented with immense range of plant protein sources which complements each other. Due to these aspects utilization of other sources of feed is rising day by day in poultry farming. In this concept, various types of grain legumes contribute a crucial part in imparting both protein and energy rich diet to the poultry.

Legumes of *Leguminosae or Fabaceae* family are the third largest species of soil plants consisting of nearly 751 genera and 19,000 species. The *Leguminosae* family is again sub divided into three subfamilies, which are *Cesalpinioideae*, *Papilionoideae* and *Mimosoideae.* Legumes are economically and socially important plants due to their exceptional diversity, edible nature as vegetables and their multiple applications in the fields of agriculture and horticulture as food and medicine. These are versatile crops utilized directly as food or in different types of processed materials or consumed in many agricultural systems as feed. Use of legumes as a staple food and as a potent protein source was started earlier in 6000 BC in the countries of Asia, America and Europe. India shares approximately 35% of the global area in production and is the leading producer as well as consumer of various pulses. The major pulse crops cultivated in India are pigeon pea, chick pea, black gram, green gram, lentil and field pea. Many farmers cultivate legume crops in the process of crop rotation to replenish the soil depleted of nitrogen as these crops are involved in biological nitrogen fixation and are generally used as natural fertilizers. The legumes used by the humans are called as food legumes or grain legumes and are further divided into pulses and oilseeds. Pulses are dried and edible seeds that are been consumed from many years and an emphasis to enhance quality of legumes

Fig. 3.1 Different types of protein and energy rich grain legumes acting as an important source of feed in poultry farming. (**a**) Pigeonpea, (**b**) Cowpea, (**c**) Faba beans, (**d**) White Lupins, (**e**) Mung beans, (**f**) Field peas, (**g**) Lentils, (**h**) Soybean, (**i**) Common Vetch, (**j**) Yellow Lupins, (**k**) Velvet beans

in nutritional aspects were carried out (Jain et al. 1980) (Fig. 3.1). Singh et al. (1984) and Singh and Eggum (1984) reviewed the factors affecting their nutritional profile and found that legumes and cereal grains are the potential protein source for both humans and animals.

Grain legumes were been utilized as both animal meal and human food and are been cultivated for mature or immature grains. More than 40 types of grain legumes (also termed as pulses) were present which are been historically planted for dry seeds rich in protein and energy. Crops that are harvested green to use as vegetables, forage, green manure and grazing were excluded from the grain legumes group. In addition to them, seeds like clover and alfalfa used particularly for sowing are also excluded (FAO 2010). Pulses are easily digestible and act as a wonderful protein source containing balanced amino acid profile and are also rich in vitamins and iron. The crude protein content of grain legumes varies, ranging from

27% in peas and faba peas to nearly 50% in soybeans. Among the grain legumes, soybean is globally used as the predominant source of protein in the poultry diet. Additionally, pulses protein and protein isolates also have outstanding functional characteristics that help in the development of innovative food products. Such characteristics involve solubility, water and oil binding ability, properties of imparting foaming, gelation and viscosity. Proteins of pulses are predominate in albumins soluble in water and globulins soluble in salt (Marcone et al. 1998), with lower levels of prolamins and glutelins. Since pulses are the source of high levels of globulins, they are used as a reliable source of commercial protein concentrate & isolates. The table given below (Table 3.1) illustrates an overview of the protein profile of different pulses.

Grain legumes also play a significant role in the production of useful by-products for animal feed and hence indirectly provide food security. Further, there is an immense scope to utilize crop by-products (straw and other plant parts) as ruminant feed. Additional by-products like chunies (blend of endosperm parts and seed coats) and husks collected during pulse processing also act as animal feed. These legumes by-products are potent energy and protein sources that do not compete with the human food, however contribute to reduction in the levels of soybean and cereals in the livestock diets in intensive livestock farming. These legume by-products are utilized by smallholder framers mainly in Asia to mitigate the feed scarcity in mixed crop or livestock farming. So, now-a-days, inclusion of grain legumes as potential protein source in poultry meal has been increased rapidly.

Considerable work has been carried out on the utilization of grain legumes and their by-products as poultry feed resulting in a massive source of published and unpublished reports and a scholarly review was lacking in this field. So, to address this gap and to create awareness on the usage of grain legumes, we have assembled and incorporated the data available in this book chapter review. This chapter mainly discusses about different kinds of grain legumes, their production, consumption as well as their nutritional values, anti-nutritional factors and methods of eliminating their negative effects to use them finally as significant animal feed source. Furthermore, we also discussed the advance aspects of the use of by-products produced during the processing of pulses to prepare nutrient rich healthy products at economic level.

Table 3.1 Typical protein profile of different pulses

Protein	Peas	Beans	Chickpeas	Lentils
Crude protein (Dry basis whole seed)	15–32%	18–25%	~22%	27.9–32.1%
Globulins	65–85%	55–80%	42%	51%
Albumins	20–35%	10–20%	16%	11–16%
Glutelins	12%	10%	9.9	11%
Prolamins	~1%	~1%	0.48%	3.5%

Source: Pulse Canada (2011)

3.2 Area, Yield and Production of Grain Legumes in Developing Countries

Among the other countries, India is the biggest producer (25%), consumer (27%) and importer (14%) of pulses in the world. During the period of 2017–2018, over 29 million ha of area was cultivated with pulses and produced the record of 25.23 million tons (Mt) of the highest yield ever at 841 kg/ha production rate. Among the different states, 12 states were identified as the major producers of pulses contributing a share of more than 90%. They were in the following order as Madhya Pradesh as the first (> 8 Mt) followed by Rajasthan (>3 Mt), Maharashtra (>3 Mt), Uttar Pradesh (>2 Mt) Karnataka (2 Mt) and Andhra Pradesh (>1 Mt). Whereas in, Gujarat, Jharkhand, Tamil Nadu, and Chhattisgarh the production is <1.0 Mt each. In the developing countries, the demand for pulses will continue to increase with the growing population, rise in income and increased demand for livestock feed. Further, requirement of soybean in the bio-diesel production has also led to an increase in the demand for legumes as an alternative feed due to current policies in the US and Europe. The global share of pulses in protein consumption is only 5%, but in many developing countries, the pulses input ranges between 10–40% and in low income countries they play a key role in food and nutrition as they are consumed frequently along with rice or wheat. Further, pulses help in reducing the rate of poverty and hunger by improving the quality of nutrition. The total global acreage under pulses recorded during the year 2014 was around 851.91 lakhs/ha with a productivity of 774.73 lakh tones and yield of 909 kg/ha (Table 3.2). According to the data recorded in 2014, pulses are globally cultivated in 198 countries and the data of their global area, production and yield was given in Table 3.2. The total production share of different pulses in the world is: dry beans 31.64%, Chickpea 17.72%, dry peas 14.44%, pigeon pea 6.531%, Lentil 6.23% and others 23.66%. Even though, they are produced seasonally, varieties that produce throughout the year would produce high yields and thus can be readily available for the feedstock (FAO 1989).

The production of different grain legumes viz. pigeon pea, chick pea, faba bean, dry bean, soybean, cowpea, groundnut and lentil has increased since 1980 at the ratio of more than 1% per year. Among the other crops, cowpea has shown the maximum production i.e. 4.7% per year between the years 1994–1995 and 2008–2010.

Table 3.2 Global area, production and yield of various grain legumes

Crop	Area	Percentage to total	Production	Percentage to total	Productivity
Chickpea	139.81	16.41	137.31	17.72	982
Lentil	45.24	5.31	48.27	6.23	1067
Pigeonpea	70.33	8.26	48.90	6.31	695
Pea	69.32	8.14	111.86	14.44	1614
Beans	306.13	35.93	245.16	31.64	801
Total pulses	851.91		774.73		909

Source: FAO Statistics (2014)

Table 3.3 Global area, production and yield of grain legumes (Nedumaran et al. 2015)

	1994–1996			2008–2010		
Crop name	Area harvested (Mh)	Production (Mt)	Yield (Kg/ha)	Area harvested (Mh)	Production (Mt)	Yield (Kg/ha)
Dry bean	25.9	17.1	659	27.4	21.7	793
Chickpea	10.9	8.1	739	11.5	9.9	864
Cowpea	8.2	2.7	339	10.4	5.3	508
Pigeonpea	4.1	2.8	684	4.7	3.7	800
Lentil	3.4	2.8	822	3.7	3.7	1003
Faba bean	2.3	3.6	1565	2.5	4.2	1701
Soybean	62.0	131.2	2115	99.4	238.6	2399
Groundnut	22.1	29.4	1331	24.0	37.4	1558

Pigeon pea and lentil production has also risen within the 15 years period, which is showing the growth rate of 2.1–2.0% per year. Pigeon pea, chick pea, dry bean and lentil have shown high yields around 850 kg/ha. Cowpea, which is a food and feed crop, has relatively low yield of 0.5 tons per hectare, but its yields are rising at an average rate of 2.9% per year (Table 3.3). In comparison to the developed countries, the production of pigeonpea, chickpea, lentil, dry bean and soybean exceeded the yield levels in developing countries.

3.3 Use of Grain Legumes in Poultry Feed as an Alternative Food Components

Use of alternative foods in poultry feed is an economically important stand point. Changes in the feed cost and unavailability of imported feed components, which are not easily available at all locations has increased the production cost of egg and meat and led to choose for locally available sources. Application of feed ingredients in biofuel industry and adverse climatic conditions are also some of the reasons for choosing other feed formulations. Thus, the poultry nutritionists are opting for other feed ingredients, which do not have any economic and environmental issues. Pulses also called as grain legumes are one of the best alternative feed ingredients to include in poultry diets. However, their usage is mainly confined by the anti-nutritional factors present in them, which can be decreased by various processing techniques and enzymatic treatments. The anti-nutritional factors present in them are: lectins, phytate, trypsin inhibitors, protease inhibitors, oligosaccharides, tannins, L-canavanine, isoflavones and tannins. The grain legumes are widely classified as protein-rich feedstuffs for the formulation of traditional and "organic" poultry diets and they represent a viable alternative to solvent-extracted soybean meal (SBM). In this context, it is necessary to outline the data available on the nutritional value of different grain legumes to assess their potential use as an effective feedstuff for poultry. Usage of different grain legumes as effective poultry feed was explained in brief.

3.3.1 Chickpea

Chickpea (*Cicer arietinum L.*) is one among the globally essential grain legumes (FAO 1993) as it acts as a potential source of vitamins, protein & minerals in human nutrition. Chickpea is mostly grown as legume in Mediterranean countries, as it can be acclimatized in semi-arid climatic environment and few irrigated varieties yield approximately 3.5 tons/hectare (Iliadis 2001). Chickpea is cultivated as an important crop in various countries like Asia, Europe, Africa and the Americas and India is one of the largest producer and consumer of it. Even though most of them are cultivated for human consumption, they are also useful to poultry as an alternative feed source of energy and protein. In chickpea, the crude protein content varies from 124 to 306 g/kg of dry matter (DM), which relies upon the farming system (Chavan et al. 1989) (Table 3.4). It is also a strong source for the supply of digestible form of dietary minerals like Ca, Mg, P, K and Fe. Friedman (1996) identified that protein characteristics of chickpea protein were identical to soybean meal.

Table 3.4 Nutrient and mineral composition of raw chickpeas (Chavan et al. 1989)

Nutrient content	Gms/Kg as fed basis
Crude protein	124–306
Total carbohydrates	506–709
Starch	372–508
Total sugars	48–93
Crude fiber	12–135
Cellulose	71–97
Hemicellulose	35–87
Pectic substances	15–38
Lignin	22–59
Crude fat	31–74
Ash	20.4–46.7
Calcium	930–2590
Phosphorous	2440–4580
Magnesium	917–1680
Iron	30–106
Copper	6–21
Zinc	15–42
Sodium	98–1501
Potassium	6923–10,284

In chickpeas the carbohydrate content is 506–709 g/kg, where crude fiber, complete sugars and starch are the key elements. Some amino acids are limited in chickpea, where first is sulfur containing amino acids and then valine, threonine and tryptophan in the next. In chickpea, the total lipid content varies from 31 to 74 g/kg (Chavan et al. 1989). The main elements of the neutral lipids are triglycerides, where as the main element of polar lipids is lecithin. Amongst the fatty acids, unsaturated fatty acids comprises of 671.3 g/kg (linoleic acid 432.9 g/kg, oleic acid 218.4 g/kg and linolenic acid 20.0 g/kg), whereas saturated fatty acids constitutes 104.2 g/kg (stearic acid 12.0 g/kg and palmitic acid 92.2 g/kg).

Similar to various other legumes, chickpeas too consists of few anti-nutritional factors, like amylase inhibitors, protease, polyphenols, lectins and oligosaccharides, which affects the nutrient absorption (Singh 1988; Chavan et al. 1989; Saini 1993; Perez-Maldonaldo et al. 1999). To enhance their nutrient content and to simultaneously increase the level of effective utilization of chickpeas in poultry diets, the anti-nutrient factors should be removed to enhance their utilization potential (Vander Poel 1989). Anti-nutritional elements are generally removed by heat treatment, where different combinations of process temperatures and heating time will be applied for various anti-nutrient factors (Vander Poel 1989; Gatel 1994). Heat treatment involves, infrared radiation, extrusion, micronizing, steam processing which is also called as flaking and autoclaving. Among all the processing methods for heat treatment, extrusion gives better outcome to eliminate the anti-nutritional factors present in legumes. The anti-nutritional factors of chickpea are inactivated when the grounded chickpeas (2 mm) were ejected wet at 120°C for 20 s (Christodoulou et al. 2006b). Marquez et al. (1998) stated that, trypsin inhibitor activity (66%) in chickpea was inactivated by dry heating at 140°C for 6 h. Moreover, Monsoor and Yusuf (2002) also stated that, heating of chickpeas at 100°C for 5 min in boiling water will enhance the *in vitro* digestibility by 9% along with the elimination of protease inhibitors. Christodoulou et al. (2006) reported that, no variations have been observed in broiler performance among raw and heat-treated chickpeas and stated that the union of the heating time and applied process temperature did not inactivated most of the anti-nutritional factors.

3.3.1.1 Use of Chickpea in Poultry Nutrition

Viveros et al. (2001) stated that when high amount of chickpeas in raw condition were included in the feed then there will be an adverse effect on the gain of body weight (BW), feed conversion ratio (FCR), DFC and on the other hand substitution of raw with autoclaved chickpeas enhanced bodyweight gain & Daily Feed Consumption (DFC). While no alteration was observed in Feed Conversion Ratio (FCR) related to the control diet. They also recommended that the addition of raw chickpeas at the rate of 300 kg/t to the broiler chicken diets has decreased the digestibility of protein and starch by 18% & 3% and probable metabolisable energy by 9% related to those provided with control diet. Supplying raw chickpeas to broiler chickens up to a level of 450 kg/t of diet has caused increase in the weight of liver, gizzard and pancreas but, when they are fed with autoclaved chickpeas up to150 kg/t of diet, decreased the liver weight and increased weight of the gizzard (Viveros et al. 2001). Christodoulou et al. (2006) reported that, chickpeas may be utilized as an alterntaive source to soybean meal in broiler chicken diets upto 120 kg/t, without affecting final BW, DFC and FCR and thus providing enhanced non-GM chicken meat production, but negative effect was seen with the addition of 240 kg/t to the diet. They further presented that, partial substitution of soybean meal with chickpeas upto 120 kg/t of concentrate mixture, as the major protein source in the broiler chicken diets developed equal effective performance. Irene Katogianni et al. (2008) reported that, slow replacement up to 75% of chickpeas for soybean expeller in the production of organic broilers upto 81 days of age is appropriate since, the production of lighter birds, carcass and carcass parts, feed conversion ratio and proportion of carcass parts were similar to control birds.

3.3.2 Pigeonpea

Pigeonpea (*Cajanus Cajan* L.) is a major grain legume, cultivated in both tropical and subtropical regions of the world (Jambunathan et al. 1991). In 2014, the total global area of pigeonpea crop was 46,000 km^2, where 82% was cultivated in India (Emefiene et al. 2014). It is grown for both human food and animal meal. It is wide spread as a green manure in the tropics due to its immense dry matter content. Based on the dry matter, pigeonpea seed consists of 22–27% crude protein, 7.3–10% crude fiber, 1.7–2.1% ether extract, 61.2% nitrogen free extract, 7.59% lysine and 3.1–4.2% ash (Akande et al. 2010; Arif et al. 2017). The important vitamins present in it are riboflavin, thiamine, niacin and choline. In the study performed by Sugui et al. (2007) in Philippines, pigeonpea was found to be an economically low cost feed for poultry. So, the use of pigeonpea in poultry diets as a protein and energy base would be an alternative when compared to the costly cereal grains and oil seed meals. But, the application of pigeonpea as poultry feed is restricted by the anti-nutritional factors present in them like other grain legumes, which reduce their efficiency, intake and nutritional value. The major anti-nutritional factors present in the legumes are trypsin and chymotrypsin inhibitors (Ganzon-Naret 2014a, b). Wisaniyasa et al. (2015) stated that compared to soybean, pigeonpea has low trypsin and chymotrypsin inhibitor activities. Their seeds consists of four amylase iso-inhibitors and nine trypsin inhibitors (Gemede and Ratta 2014), which were considered as anti-nutritional factors and as promising agents towards pests. They further reported that, their effect can be decreased by boiling, processing, grinding, soaking, fermentation and cooking. Further, treatments like soaking of pigeonpea seeds for 24 h, continued by 20 min cooking has been stated to eliminate the negative effect of the trypsin and chymotrypsin inhibitors. Swathi et al. (2014) reported that, cooking at 37°C for 12 h increased the protein digestibility of pigeonpea seeds. Study done by Tekale et al. (2016) stated that, cooking of dehulled and soaked seeds reduced the total phenols content by 49%, while sprouted pigeonpea seeds enhanced the protein content and decreased the composition of carbohydrates. Soaking the seeds for 12 h followed by boiling reduced the trypsin inhibitor activity and improved the starch and protein (pepsin) digestibility. Based on the studies conducted on the impact of various kinds of processing techniques of pigeonpea, Wisaniyasa et al. (2015) stated that, the amount of phytic acid was decreased to 50% by germination, while roasting and autoclaving were more efficient to decrease the phytic acid content. They also reported that by heat treatment lectin, trypsin inhibitor, tannin and HCN content of guar seeds was reduced. Arif et al. (2017) stated that decortications of legume seeds increased the dry matter (DM) levels, sulphur and nitrogen free extract (NFE) but decreased the seed crude fiber, crude protein and decreased the fat as ether extract (EE). The amount of tannin was also reduced to 0.02%, compared to 0.32% in corticated seeds. Swathi et al. (2014) noticed that, chymotrypsin and trypsin inhibitor activities can be reduced by soaking in salt solution related to those soaked in sterile water. The two techniques changed the legumes content without change in the carbohydrate fraction. On the other hand, Abdulla et al. (2017) stated that digestibility of lipids, amino acids and starch in the broilers provided with diet having tannin-free faba bean were less compared to those provided with control diet and there is no reduction in any anti-nutritional factors by dehulling, as lecithin and trypsin inhibitors were aggregated in the cotyledons.

3.3.2.1 Use of Pigeonpea in Poultry Nutrition

Many studies have been done to observe the impact of pigeonpea as an alternative protein source in poultry diet. Amaefule et al. (2006) reported that addition of 10% of pigeonpea seed meal in poultry diets can act as a better protein source. The experiment was carried out by taking three treatments, in the primary treatment, the seeds were toasted for 30 m, next they were boiled for 30 m and at last they were soaked for 24 h in water. At the final they observed that, no major changes were noticed among the treatments with raw or processed pigeonpea seeds diet. Finally the experiment stated that, daily weight, live weight, FCR and average mortality did not statistically vary among the groups provided with the feed of processed or raw pigeonpea seed meal. But, protein utilization was reduced in case of chickens provided with raw pigeonpea seed meal; even though it reduced the daily feed expenses. Tusar et al. (2015) stated that, the augmentation of 100 g/kg of pigeonpea seed in broilers diet enhanced the feed intake, protein, lysine, methionine, FCR and total serum lipids in the birds. Castillo et al. (2016) studied the impact of pressure cooked pigeonpea seed meal in place of soybean meal and corn in the starter diet for boilers which are supplied at the proportion of 0%, 20%, 30%, 40% and 50%. By the experiment it was observed that, rising the dietary inclusion of pigeonpea seed meal, the level of crude protein present in it showed an apparent reduction in weight gain (Castillo et al. 2016). Poultry birds provided with 15% of pigeonpea seeds produced broiler mass of 85% and resulted in higher and heavier daily weight gain, effective feed conversion and better quality carcasses. George and Elliott (1986) recommended that raw pigeonpea seeds may be provided to commercial layer diet at the level of 400 g/kg, which do not affect the production of eggs, layers health and intake of feed by the birds. Arif et al. (2017) reported that, decortications may decrease anti-nutritional factor levels in the feed. Based on his study, it was observed that feeding Lohman broiler chicks with the diet of 7.5% decorticated pigeonpea seeds, showed improvement in FCR and body weight. Amaefule and Nwagehara (2004) reported that crude protein and dry matter digestion was increased with the processed pigeonpea seeds. As per the study of Ogbu et al. (2015), the crude fat, crude protein and ash were lower in cooked pigeonpea seed meal and the nitrogen-free extract (NFE) and ether extract (EE) were higher in the crude pigeonpea seed meal. Amaefule et al. (2011) stated that, 40% addition of pigeonpea meal in the starter diet of broilers enhanced the productive performance compared to 30% addition, but this ratio has enhanced the price of the diet. Based on these reports, the optimum level of feeding pigeonpea protein source to broilers diet may be 15%. Improvement in nutritional quality and broiler growth efficiency could be attained by the processing of pigeonpea seeds, especially by boiling and fermentation. Percentage of addition of pigeonpea for broiler diets in substitution to soybean may vary from 7.5% in total or up to 50%. Pigeonpea can enhance the meat yield and growth, which can be utilized to decrease the feeding rate, without showing any adverse impact on its growth.

3.3.3 Cow Pea

Cowpea (*Vigna unguiculata*) is an essential annual legume among the grain legumes cultivated in tropical and subtropical regions (Ameen et al. 2005). It is erect, climbing and also creeping and prostrate based on the type of cultivar. In Africa, it is one of the most popular grain legumes and is also grown in few parts of Asia & America. It could be cultivated in different types of soils, like marginal soils of low resource farmers. It is called as the “hungry-season crop” as it is harvested prior to the other cereal crops. Different parts of this crop like pods, seeds and leaves are usually utilized as human food. It is a flexible crop and relying on the climatic and economic constraints the farmers can harvest this crop for the need of grains or for providing forage to their livestock (Gomez 2004). For providing both grains and forage, dual purpose types have been cultivated according to different crop systems adopted in Africa (Tarawali et al. 1997). During 2008–2010, the yield of cowpea was over 3.4 t/ha in developed countries that are greater than the normal production of cowpea in developing areas. Among the developed areas, around 1/3rd of the cowpea yields are from Asia & Latin America and 1/10th of the yields are from Africa. The reason behind this may be because of its plantation in marginal soils of Africa.

Cowpea is a great source of proteins, fatty acids, essential amino acids, minerals and vitamins (Fageria et al. 1990). Cowpeas nutrient composition is identical to lupines and field peas and they can act as a best source of dietary protein in animal diet (Igbasan and Guenter 1997), particularly at the time of expensive and short supply of animal proteins (Wiryawan and Dingle 1995). The amino acid profile is identical to soybeans, but these are low in sulfur containing amino acids (methionine & lysine) and rich in lysine (Coertze and Venter 1996). Even though there is a change in the protein content of cowpeas (Canon and Carre 1989), they could act as substitute to soya oilcake meal without lipids. Sharma and Singhania (1992) observed that, in cowpea the dry matter content was 82%, on the other hand Fawzi and A/Rahmim (1967) and Silanino et al. (1981) stated that, in cowpea dry protein content was 25%, fiber 28.9%, ether extract 2.4%, nitrogen free extract 43.1% and the digestibility of proteins was greater compared to other legumes. Thus the inclusion of cowpea in poultry diets has decreased the cost of feed and further it also showed improvement in production and growth (Teguia et al. 2007; Defang et al. 2008; Chakam et al. 2008). But, like other grain legumes, consumption of raw cowpeas was restricted due to its anti-nutritional factors like pectins, protease inhibitors, phenolic compounds and non-starch polysaccharides, which may decrease their nutrient digestibility and protein quality (Arora 1995). Like chymotrypsin and trypsin, the protease inhibitors reduce the pancreatic enzymes activity. Some processing methods like soaking, roasting, boiling, dehulling and autoclaving decreases the anti-nutritional factors of cowpea (Amaefule et al. 2004; Teguia et al. 2007; Chakam et al. 2008; Defang et al. 2008; Medugu et al. 2012). Further to these, cooking also reduces the anti-nutritional factors of different grain legumes and improves the protein quality.

3.3.3.1 Use of Cowpea in Poultry Nutrition

According to the studies conducted by Wondifraw (2018), cowpea can be effectively used in place of soybean grains in SASSO x RIR crossbred chicks diet of approximately 75% with no effect on body weight gain, feed intake and feed conversion ratio (FCR). But, when soybean grains were completely replaced with grains of cowpea no decrease in chicks body weight gain was noticed. So, replacement of soybean with 22.5% of treated cowpea grains of Totally Mixed Ration (TMR) could be suggested as the optimum level of feed for rearing chicks. So, it was stated that, addition of treated cowpea grains (22.5%) of Totally Mixed Ration for chicks is economically feasible and decrease the production cost without affecting the weight gain, feed intake and chicks feed conversion efficiency related to the chicks fed with control diet. Akanji et al. (2016) studied about the use of cowpea diets on the characteristics of performance and hematology of broiler chickens, where good response was noticed in the broilers fed with dehulled and cooked cowpea. Eljack et al. (2010) examined the effect of feeding graded levels (0%, 10%, 20% and 30%) of cowpea on broiler chickens performance, carcass relative weight, dressing percentage and observed that in 6- weeks-old broiler chicks, improvement ($P < 0.05$) in case of total & weekly weight gain and feed conversion ratio were observed with all ratios of cowpea correlated to the control diet. At the same time, carcass relative weights showed an apparent gain in the apparent weight of thighs and breast of birds carcasses, fed with 30% cowpea related to various other treatments. Tsigab Nirayo Embaye et al. (2018) also studied on the impact of cowpea grain on the performance of broiler chickens and stated that, inclusion of cowpea grains up to a level of 20% to broiler chickens diet didn't showed any negative impact on the development of chickens. Thus, they finally reported that inclusion of about 5–20% of cowpea grains may be effectively utilized as a source of protein in poultry diets.

3.3.4 Mung Bean

Mung bean (*Phaseolus aereus*: *Vigna radiate*) is one of the plant protein that can be used as poultry feed. It is commonly known as green gram, moong dal, mung dal, munggo, mash bean, green soy and golden gram (Daghir 2008). It is native to India and is a drought tolerant annual tropical legume plant cultivated in various tropical and Asian countries for human and animal consumption (Park and Yang 1978; Hussain and Burhanddin 2011). Raw mung bean consists of 20–30% crude protein, 0.63–16.3% crude fiber, 0.07–1.9% crude fat, 6.26–12.07% moisture, > 45% starch, 6.6% sugars, 53.38–67.68% nitrogen free extract and 3.3–5% ash (Deshpande 1992; El-Adawy 1996; USDA 2000; Robinson and Singh 2001; Mubarak 2005; Habibullah and Shah 2007; Agugo and Onimawo 2009; Paul et al. 2011; Oburuoga and Anyika 2012; Ganzon-Naret 2014a, b; Padmashree et al. 2016).

Studies carried out by various researchers reported that, raw mung bean consists of about 3035–3470 kcal metabolisable energy/kg feed (Creswell 1981; USDA 2000; Padmashree et al. 2016). El-Adawy (1996) and Robinson and Singh (2001) stated that the amino acid profile of raw mung bean is similar to soybean and is high in tryptophan and lysine but lacks methionine and cysteine. The amino acid content of raw mung bean are 24.6 g histidine, 41.6 g lysine, 134.5 g aspartic acid, 63.5 g arginine, 215 g glutamic acid, 31.2 g threonine, 41.5 proline, 49.4 g serine, 42.2 g glycine, 53.0 g valine, 43.2 g alanine, 32.2 g tyrosine, 83.2 g leucine, 47.4 g isoleucine, 19.5 g methionine, 9.5 g tryptophan and 56.4 g phenylalanine/kg. On the other side, the raw mung bean consists of about 67.4–97.0 mg iron, 840–1320 mg calcium, 10.35–17.0 mg manganese, 556–1890 mg magnesium, 36.2 mg potassium, 3670–3910 mg phosphorous, 26.8 mg zinc and 120 mg sodium/kg (USDA 2000; Mubarak 2005). According to Robinson and Singh (2001), raw mung bean is high in vitamins like B_1, B_2, A, C and niacin. Similar to the other legumes, mung bean also contains of some anti-nutritional factors like amylase inhibitor, chymotrypsin and trypsin inhibitors, protease inhibitor, tannins, phytic acid, lectins, flatulence, saponin, anti-thiamine factor and gossypol (Wiryawan et al. 1997). Without proper processing of raw mung beans, the presence of anti-nutrient factors reduces its biological value and nutrient utilization and thus limits its usage as poultry feed (Kay 1979).

3.3.4.1 Use of Mung Bean in Poultry Nutrition

Mung bean has more energy rate compared to other grain legume seeds (Wiryawan and Dingle 1995). Supplementing 100 g mung bean to boilers of 15–35 days old and 28–35 days old showed no impact on the weight gain, feed intake, feed efficiency, price of feed per kg of broiler produced and broiler production efficiency (Binalay 2012). Research was carried out by Sherif M. Hassan (2016) to study the impact of serving various levels of raw mung bean on the laying hens egg quality and stated that, raw mung bean can be supplemented up to 15% in laying hen feed. It shares approximately 22.21% of the total dietary protein and almost 48.05% of the soybean protein used without having negative effects on the egg quality of 55–63 week-old laying hens. Singh et al. (2013) noticed that providing sprouted mung bean of about 10 g per bird per day could give protection towards coccidiosis in broiler chickens and lead to changes in growth, parasitological and haematological parameters. Rungcharoen et al. (2010) have done a chain of experiments to detect the distinct metabolisable energy of mung bean hull in broilers and impact of its supplementation in broiler diets on nutrient digestibility and efficiency of growth. This study suggested that, supplementation of mung bean hull in the broiler diets must be lower than 5% to attain optimal nutrient digestibility and growth efficiency in broilers. In case of layers, addition of raw mung beans in the diet at the rate of 15% or 30% did not decrease feed efficacy or egg production. But, production of egg was consequently low at the inclusion rate of 45%. Pelleting diets has no impact with 15% or 30% addition rate; however has a promising impact on the production at 45% level (Robinson and Singh 2001). In both instances, body weight was marginally reduced by the introduction of mung beans to the diet. They further reported that, inclusion of raw mung bean up to 45% did not show any affect in the aspects of feed consumption, egg specific gravity, colour of the egg yolk and egg weight. The general advice is to use mung beans at rates of up to 30% in the diet of the layers, provided that the diet is balanced properly, particularly with the amino acids. Vinh et al. (2013) done experiments to analyze the impact of mung bean hulls in maize based feed for laying (20–38 week old) and pre laying (9–10 weeks) efficiency of Ri x Luong Phuong hens. By their study it was observed that, growth rate was not affected in the pre laying stage through dietary energy rate (2650–2450 kcal Me/Kg range) in the time period of 10–16 weeks but were decreased by 12% and 26% in 16–20 weeks period with mung bean hulls diet of 14% and 18% (Vinh et al. 2013). It was further stated that supplementation of 14% or 18% mung bean hulls in the laying time didn't affect the egg quality or egg production.

3.3.5 *Faba Beans*

Faba beans (*Vicia faba* L.) called as horse beans, broad beans or field beans are considered as a potent protein source in poultry meal, as their amino acid profile is identical to soybean, besides its low methionine content (Bond et al. 1985; Wiryawan and Dingle 1999; Gatta et al. 2013). As the cost of soybean and corn (mostly utilized for poultry feed) always increases, there is a need to choose for an alternative nutritious feed. So, faba bean acts as an alternative protein source for poultry feed as a substitute to soybean. It is consumed as a predominant food source since many years in India, Mediterranean area, China & Pakistan, due to their tendency to cultivate in different climatic areas. Two kinds of faba beans are cultivated, one is '*V. faba* major', or broad beans with broad flattened seeds mostly grown in southern regions of Europe, utilized for human nutrition and the second one is round shape '*V. faba* minor', field beans or horse beans, with medium to small seeds, utilized for human or animal nutrition. Medium size seeded varieties have been generated around the Middle East and North Africa, mostly in Egypt, while small seeded types are found in Ethiopia.

Faba beans are abundant in proteins, fiber, carbohydrates, vitamins and have a hypocholesterolemic effect (Macarulla et al. 2001). So, due to its high nutrient profile it could be effectively supplied as an alternative source of protein instead of soybean feed (Castanon and Perez-Lanza 1990; Nalle et al. 2010). In developing countries, it is mostly consumed by the humans, as it is comparatively cheaper than meat products. But, the biological value of faba beans is adversely affected by anti-nutritional factors present in them like condensed tannins, phytic acid, trypsin inhibitors, lectins, saponins & favism-inducing elements. So, the elimination of anti-nutritional factors is essential for the productive usage of food legumes in both human and animal nutrition. Some factors like hemagglutinins, trypsin inhibitors & condensed tannins could be eliminated through heat processing (cooking, boiling, autoclaving) or can be removed by pretreatments like germination, soaking, dehulling, fermentation and augmentation with different enzymes and chemicals. Various previous reports suggests that, different methods like genetic selection, heat treatment, exogenous fiber degrading enzyme addition will help to mitigate the negative impact of anti-nutritional factors of this legume (Van der Poel et al. 1991; Cowieson et al. 2003; Woyengo and Nyachoti 2012).

3.3.5.1 Use of Faba Beans in Poultry Nutrition

Przywitowski et al. (2016) conducted studies on the impact of faba beans replaced with soybean meal at various quantities of tannins in condensed condition on the growth, breast meat features and carcass traits in the finisher turkeys of the age 13–18 weeks. They were fed with 100, 200 and 300 g/kg of low-tannin and high-tannin faba beans and control diet containing soybean meal. Irrespective to the faba bean variety, dietary supplementation of faba bean up to 300 g/kg instead of soybean meal had no negative effects on carcass traits of finisher turkeys or their productive parameters and has no impact on the physicochemical characteristics and their content. Thus they concluded that, both low tannin and high tannin faba beans (having 0.76 & 7.84 g/kg of tannins) equal to 300 g/kg can be supplemented in turkey's diets as an alternative to soybean meal. Mahmoud Alagawany et al. (2019) conducted an experiment on Hi-sex Brown laying hens to study the impact of various quantities of dietary faba bean with or without enzyme cocktail additives on egg quality, fertile performance and blood metabolites. In that experiment, the hens were supplemented with five dietary treatments of faba bean, which are 0% (control), 25%, 50%, 75% or 100% and two treatments of enzyme cocktail (Gallazyme) 0 or 250 mg/kg diet. No variations were noticed in the hen performance with or without enzyme cocktail and finally they concluded that, 50% faba bean supplemented with Gallazyme in place of soybean meal could be added in the laying hens diets to improve their performance. Liga Proskina and Sallija Cerina (2017) stated that, inclusion of 200 g/kg field peas and faba beans in the poultry diets can reduce the feed conversion ratio and increase the live weight gain and finally enhance the production ability.

3.3.6 Field Peas

Similar to other grain legumes, field peas (*Pisum sativum* L.) could supply a rich share of dietary energy and protein for poultry. These are primarily cultivated in regions of temperate climate, besides globally using as a food source (Castell et al. 1996). Normally, only the screenings left from milling field peas were used in the poultry industry, however by the ban of bone meal and meat in poultry industries along with the entry of GM-soybean meal in the poultry feed chain, usage of field peas as poultry meal has turn into famous in Canada and Europe (Cowieson et al. 2003). Now-a-days, different varieties of field peas are being cultivated especially for the use of poultry meal and the nutritional content of the harvested peas is affected by the various factors like time of seeding, genetic and agricultural conditions, where the field peas were cultivated (Nikolopoulou et al. 2007). Compared to the cereal grains, field peas are rich in starch, which is an important nutritional element of energy. According to the studies carried out by the FAO and the European Commission's authoritative institutions, usage of local European protein crops (beans, peas) for domestic feed provides higher independence of the farming industry from fluctuations of price of the feed in the world market and facilitates viable use of the agricultural area. Subjected to cost, field peas could be opted as one of the most preferable pulses in the poultry farming owing to the following reasons: high amino acid availability and better protein quality, high energy content and good digestibility, reduced anti-nutritional factors levels, lack of difficulty in storage and milling and less endosperm/hull ratio (needs less processing). The fat composition in peas is very low, like the fiber and lignin concentrations, while the soluble carbohydrates (mostly starch) content is high. Compared to the other crops of grain legumes, these are considered as one of the best sources to replace soybean oil meal usage in poultry diets as its protein is well absorbed (Jeroch 1998; Morkunas 2002). Even though the protein level in field peas is less compared to soybeans, the protein consumption is highly divergent (Bennett 2002). Jeroch (1998) mentioned that, peas can provide required protein to the adult birds. The lysine content in the peas is relatively high, but they are lack of sulphur containing amino acids and tryptophan

(Gatel 1994). So, they are not appropriate to feed for young birds as a sole source of protein (Bennett 2002).

In field peas, the normal crude protein level is 22.6% (90% dry matter basis) and depending upon the species, the digestibility of crude protein in field peas is comparatively more with digestibility coefficients of 83–88%. The importance of peas is because of their lysine composition, mostly essential by mono-gastric animals. On the other hand, canola meal and cereal grains are less in lysine and are abundant in cystine and methionine. So, field pea protein, cereals and canola meal proteins are nutritionally equivalent to each other and thus improve one another's importance while supplemented as poultry diets.

3.3.6.1 Feeding Field Peas to Poultry

Now-a-days field peas can be utilized as an effective protein and energy rich source to various kinds of poultry on the grounds of careful measurements to assure that the diet meets the required nutrients. Usually, conventional usage of peas in poultry diets was limited by the waste material obtained at the time of grinding field peas used for human intake. For better results, this type of feed of above 15% is not considered for poultry diets (Igbasan and Guenter 1996). They could be supplemented in the diets of various poultry classes, where proposed percentages may vary from each other. Initial studies suggested the inclusion of field peas up to 30% in case of broilers (Farrell et al. 1999), 25% for layers (Perez-Maldonado et al. 1999), 55% for turkey finisher diets & 25% to turkey starter diets (Savage et al. 1986). To enhance the digestibility of proteins in diet having maximum levels of field peas, commercial feed enzymes can be added to it (Cowieson et al. 2003). Robertas Juodka et al. (2016) stated that inclusion of higher levels of peas in females i.e. from 10% to 40% in the turkey's diets produced 0.09–0.31% of highest fat in abdomen and in males 0.28–0.61% lower shoulder muscles content. Further, use of more levels of field peas in diet didn't affect the chemical proportion of the meat & their physical index. Liga Proskinaa and Sallija Cerinab (2017) carried out studies to identify the effect of inclusion of various amounts of beans and peas (cultivated in Lativa) in poultry diet and finally suggested that 200 g kg^{-1} results in rise of the live weight gain and reduction in the feed conversion ratio. It also leads to lesser production cost and lesser feed utilization per unit of products generated. Laudadio et al. (2012) reported that, dehulled micronized peas can be a cost effective and suitable diet for laying hens.

3.3.7 Lupines

Lupines are the distinct legume species propagated globally from the past thousand years in various soil types. For many years, lupines are being a sustainable feedstuff, but not been preferred commercially on a large-scale by virtue of many anti-nutritional factors that have reduced animal growth and productivity. According to Schuster (1992), lupines are rich in alkaloids (up to 4%), which made this plant species ineffective to use in poultry feed diets. Later different varieties of sweet lupines are produced with low or negligible levels of alkaloids over genetic selection, which increased their use in the diets of different monogastric and ruminant species as safe animal feed. In the case of limited or costly soybean meal, they can be used as poultry feed and one more advantage of using lupines is processing is not required before feeding. So, the domestic feeders can purchase lupines that are available locally and may include them in their feeds after grinding. Different species of lupines are cultivated in temperate regions and among them white lupines are the most common type and contribute to more profits (Gladstones 1998). Von Sengbusch (1942) developed new genotypes with zero alkaloid content and are unlike the native bitter lupine species with high alkaloid content (1–4%) and the 14 new genotypes developed were mentioned as sweet lupines and their alkaloid content was reduced to less than 0.02–0.05% (Jansen et al. 2015).

Globally, lupines are cultivated in almost 1 million hectares and most of it is planted in Australia (FAOSTAT 2015) and in this country farmers mostly cultivate blue lupine (*Lupinus angustifolius L.*). Lupine seeds are locally processed and are exported to other countries, mostly European Union (ex. the Netherlands). In Australia, blue lupine is a common ingredient of poultry diets (Choct 2004). Among the different types of lupines, yellow lupine seeds contain high crude protein content than white and blue lupine seeds. Lupines are devoid of starch, which is very uncommon for edible bean species. They have a thick seed coat, which comprises of mainly cellulose and it is removed during the first step of processing. The kernel of lupine is rich in protein (40%), fiber (40%) and reduced fat (8%) and is mostly made up of unsaturated fatty acids. They also consists of subside quantities of carotenoids, β-carotene, zeaxanthin, tocopherols and lutein. Australian sweet lupine contains high calcium and phosphate content than cereals with varied trace element content with respect to the soil where the lupine is cultivated. Lupines are less in sulfur containing amino acids and more in manganese and the levels of manganese might be high in few cultivars and are likely to balance the content of trace elements to enhance mineral nutrition.

3.3.7.1 Feeding Lupines to Poultry

In poultry nutrition lupines are mostly employed for their constant quality and reduced anti-nutritional factors. To use them as poultry feed, they should be supplemented with artificial sources of methionine and lysine. Application of lupines in poultry and pig diets can be increased by the addition of enzymes, which enhance the utilization of non-starch polysaccharides to promote their breakdown into sugars in small intestine. According to the previous study, lupine protein is employed in the same proportion as soybean meal protein (Steenfeldt et al. 2003; Mieczkowska et al. 2005) in poultry industry. But, the usage of narrow-leafed lupine (*Lupinus angustifolius*) as a potential protein source for poultry feed was restricted because of the presence of non-starch polysaccharides and high alkaloids that negatively influence the feed absorption, growth and utilization of nutrients (Oliverll and Jonker 1997; Hughes et al. 2000). To overcome this, new varieties have been developed like sweet lupines, which are confirming to be an important protein and amino acid source to pig and poultry diets. In this aspect, Kaczmarek et al. (2014) studied on the nutrient profile of four varieties of *Lupinus angustifolius L.* seeds to include them as poultry feed. During the recent years, plant breeders have successfully developed lupine varieties with very less alkaloid levels (Jezierny et al. 2010; Coboru 2011) that enhanced nutrient profile for monogastric animals. Further, Kaczmarek et al. (2014) also studied on the alterations in digestibility of nutrients and metabolisable energy (AMEN) quantity of narrow-leafed lupines polish cultivars in broiler chickens. They have concluded that, narrow leafed lupine varieties are potential protein sources; however these are the low source of apparent metabolisable energy for chickens (Kaczmarek et al. 2014). They stated that the nutrient profile of narrow leafed lupine seeds to feed for broilers rely upon the raffinose content & viscosity. The *L. angustifolius* seeds containing raffinose content less than 1.2% and less water extract viscosity are more nutritious (Kaczmarek et al. 2014). The present polish *L. angustifolius* variety is low in alkaloid levels and has no effect on broiler performance. Lee et al. (2016) studied on the potentiality of blue lupins as a source of protein for laying hens diets. By their study, they stated that, inclusion of blue lupins in the laying hen diet in the proportion of 150 g/kg DM has exhibited no negative effects on hen health or egg production and can be utilized in a balanced diet with the addition of non-starch polysaccharide degrading enzymes to decrease the soya protein dependence. Nalle et al. (2011) also noted promising outputs with an addition of 200 g/kg intake of white lupine as feed, which generated weight gain in broilers. Daniel mierlita (2015) studied the impact of lupine seeds in diets of broilers on various perspectives like fatty acid profile and animal performance and expressed that, lupine seeds free of alkaloids can be supplemented as substitute to provide vegetable proteins in poultry diets. Their results have shown that, the seeds of lupines enhanced the intramuscular fat nutritional quality in the chest, reduced the saturated fatty acids with simultaneous increase in the percentage of polyunsaturated fatty acids n-3 and n-6, making meat favorable for human nutrition. Steenfeldt et al. (2003) stated that the inclusion of blue lupine at 200 g/kg in diets of broilers has didn't reduced the feed intake related to the control but the inclusion has discouraged the feed conversion efficiency and weight gain. They further stated that, the utilization of an external source of enzyme extract, consisting of galactanase and lactase activity has automatically enhanced weight gain & digestibility (Steenfeldt et al. 2003).

3.3.8 Lentil

Lentil (*Lens culinaris or Lens esculenta*) is a grain legume grown mainly for their edible seeds (Ford et al. 2007; Bejiga 2006). It is an annual, herbaceous, bushy plant which can grow up to 60–75 cm height. They are cultivated in temperate countries as a summer crop and in tropical and subtropical areas as winter crop; where as in the cool season they could grow at higher altitudes (above 1800 m) (Bejiga 2006). It is tolerant to drought environment and is cultivated universally. Among the important grain legumes around the world, lentils rank 5th and are highly essential in the diets of majority of people living in India & the Middle East (FAO 2012; Gohl 1982). They are categorized into two groups: the macrosperma (big seeds), primarily grown in North Africa, Europe and America and the microsperma (small seeds) cultivated in Egypt, Asia and Ethiopia. Both the groups are grown in south-eastern Europe and western Asia. Their seeds differ in color relying on the type of cultivar and they are available in green, grey, pale red, brownish or black and red lentils are usually cultivated in Canada and United States. Lentils are mostly consumed as food and in addition to this; they can also be effectively used as meal to domestic livestock because of their maximum percentage of protein (22–29%) and negligible digestive inhibitors. Further, the crop residues and by-products obtained after its processing can be efficiently used by livestock, especially in poultry industry. Compared to other legumes, lentils have low sulphur containing amino acids, so, they must be combined with other protein sources to provide balanced diet with sufficient quantity of amino acids. The lentils oil content is low and it is about 20 g/kg with linolenic acid (C18:3) and linoleic acid (C18:2) constituting about 12% and 44% respectively out of the total fatty acids (Castell and Cliplef 1990) content. Castell and Cliplef (1990) finally concluded that the nutrient content of the lentils is similar to the nutrient profile of peas and can act as a nutrient feed. The major anti-nutritional factors of lentils are protease inhibitors, but their levels are not considered for feeding poultry.

3.3.8.1 Feeding Lentil to Poultry

Farhoomand (2006) added processed (heated succeeded by boiling) or raw lentils in broilers feed at the concentrations of 10%, 20% and 30%. Best growth was observed with 10% and growth was low with 30% and whereas, processing did not show any improvement. They further recommended that lentil seeds up to 20% could be applied in broiler diets however could not be used as a single protein source. In case of broilers, lentils must be used in carefully composed feed, by giving attention to the amino acids concentrations. In an experiment conducted by Parviz (2006) in Iran, broiler diets were supplemented with raw or processed lentils at the concentrations of 100, 200 and 300 g/kg. Better development was observed with 100 g/kg and growth was reduced at 300 g/kg level. Further, the author suggested that, lentil seeds can be included at a level of 200 g/kg but not as the only protein source. Kilicalp and

Benli (1994) stated that the usage of lentils in the diets of layers resulted in reduced egg production often at low addition rates i.e. at 5%.

3.3.9 Soybean

Soybean (*Glycine max*. L) originated from East Asia is an annual legume crop cultivated worldwide in vast area of geographical conditions. The main leading producers of this crop were the United States of America, Brazil & Argentina and were primarily used for production of oilseed meal & vegetable oil for feeding of animals. Even though, the United States rank as the maximum soybean producer, Argentina occupies the rank of maximum exporter of soybean meal succeeded by Brazil and the United States. The production of soybean in EU is low and is one of the main markets for the import of soybean meal (Soystats 2012). The increase in soybean meal usage as an alternative source of protein for animal feeding is the increased production of soybean. In addition to its role as a source of better quality edible oil for humans, it can also act as a source of fine quality vegetable based protein feed to different animals globally. It has a tremendous importance in food and animal feed industries in preparation of various food formulations. Besides being used as edible oil for cooking, soybean meal (by-product) obtained after oil separation (extracted either mechanically or by solvent means) has significant importance as feed ingredient for various animals in different concentrations, even in fish farming. It is widely used due to its significant crude protein & energy levels (Waldroup 1982). Soybean proteins have high biological value, high fat and unsaturated fatty acids (with 50% linoleic acid) (Leeson and Summers 2005). It is universally accepted as an animal feed due to its significant aspects like: more protein levels & appropriate amino acid profile (except methionine), small variations in nutrient quantity, accessibility though out the year and inadequacy of anti-nutritional factors if processed perfectly. Soybean consists of approximately 37–42% of protein (Yaklich 2001; Medic et al. 2014) and is one among the legumes to consider as a balanced protein diet. There are two major types of proteins in soybean, which are 11S glycinin & 7S β-conglycinin and are structurally globular. Presently, most of the soybeans supplied to the poultry are in the condition of extruded soybeans and soybean meal (SBM) (Karr-Llienthal et al. 2005). These forms are subjected to pressure & heat to decrease the impact of different anti-nutritional factors like antigenic proteins, trypsin inhibitors, lectins, oestrogens, non-starch polysaccharides & saponins (Liener and Kakade 1980). Sometimes, thermal treatment can damage various other important nutrients (Chunmei et al. 2010). So, by maintaining appropriate processing conditions like heating time, moisture content and temperature can reduce other side effects and inactivate anti-nutritional factors like lectins and trypsin inhibitors that give better results while supplied as feed to monogastric animals (Araba and Dale 1990).

During soybeans treatment different types of soybean products were produced like soybean meal, full-fat soybeans, soybean protein concentrate (SPC) and soybean oil. To eliminate the anti-nutritional factors, the full fat soybeans also called as whole soybeans are produced by various techniques like autoclaving/cooking, extruding, toasting/roasting, jet-sploding and micronizing techniques. Generally to feed the pigs and poultry, soybeans are converted to defatted meals. Now-a-days, the quantity of full fat soybeans usage is more in the livestock industry because of the improvement of new varieties containing less anti-nutritional factors (Gu et al. 2010). In addition, due to their high energy content, well treated full fat soybeans are used as a potential feed element for poultry feeding. Among the soybean products, soybean meal occupies 78.7% of the total soybean yield and the remaining is soybean oil (18.6%) and wastes (FEFAC 2007). Two different types of soybean meal are there, which are soybean meal and dehulled soybean meal, relying on either the seed coat is excised or not. The soybean amino acid profile is near to fishmeal profile, except the amino acid methionine (INRA 2004), which could be balanced in poultry feed by supplementing with artificial methionine source. In terms of crude protein, soybean meal is better than other vegetable protein sources and is superior in both digestible and total amino acid content. In poultry, soybean meal protein digestibility is about 85% (Woodworth et al. 2001), where the individual amino acid digestibility ranges between 82–94%. Kerley and Allee (2003) stated that, over the other vegetable protein sources, soybean meal meets the animal's requirement by complementing the limited cereal based amino acid (maize) diets and is generally preferred as the most potential amino acid source. In monogastrics, carbohydrates present in the soybean meal are not completely absorbed by the colonic microbiota (Kerley and Allee 2003); so by the elimination of stachyose & raffinose sugars with suitable methods the metabolisable energy content could be enhanced by 12% (Graham et al. 2002).

Soybean protein concentrate is synthesized from the defatted flakes by removing the carbohydrates of soluble nature either by enzymatic degradation or ethanol extraction. Coming to the oil extracted from soybean, it is mainly utilized for human consumption. But, its usage is increased as a fat source for animals by the reason of developing energy rich diets for modern breed varieties. In poultry, soybean oil of feed grade is commonly utilized as energy rich feed, due to its metabolisable energy content and high digestibility compared to alternative vegetable oils/fats. This is mostly supplied as feed grade fat in rations for growing turkeys and broiler chickens to enhance the effectiveness of feed usage and to increase energy density of feeds (Sell et al. 1979). Huyghebaert et al. (1988) reported that, the high energy rate of soybean oil is associated to its high ratio of poly unsaturated fatty acids, which are better utilized & absorbed as energy source by the animals.

3.3.9.1 Using Soybean as Poultry Feed

Soybean meal is a promising protein rich source to feed layers and broilers while they require high quantity of protein in their diets. It is normally supposed that around 46% of soybean meal prepared for feeding animals is utilized for poultry meal like broilers, layers & turkeys and remaining 25% is utilized for feeding pigs.

In the United States, soybean meal around 50% is utilized for poultry, 25% for swine and 12% for beef cattle. Presently, most of the soybean forms used in poultry industry is extruded soybeans & soybean meal (Karr-Llienthal et al. 2005). Vojtech Rada et al. (2017) replaced meal of soybean with raw full-fat soybean (RFFS) and studied its effect on the growth conditions, apparent ileal amino acids digestibility (AIAAD), intestinal histometric characteristics, weight of pancreas and activity of trypsin in broiler diets. For the examination they have taken 208 × ROSS 308 male chickens and fed with four different types of dietary proportions like control (without RFFS) and batches fed with 4%, 8% and 12% of RFFS and finally concluded that, inclusion of RFFS in broiler diets in a range of 8% is feasible with no negative effects on feed conversion ratio (FCR) & growth rate. On the other hand, both these features were adversely affected with RFFS of 12% added to the diet. Pitala et al. (2016) conducted a study to assess the impact of using soy in roasted form to the Isa brown hens breed and demonstrated that, the addition of roasted soy has shown improvement in egg weight, laying rate, fat content and enhanced the energy levels. Mu and Kitts (2018) demonstrated that, use of soy lecithin in place of tallow in the broiler chickens diets effectively improved the complete polyunsaturated quality of both leg and breast muscles. With the replacement of lecithin with tallow in the diet, the thiobarbituric acid reactive substances content (TBARS) in the cooked leg and breast muscles were decreased correlated to the increased polyunsaturated fatty acid components. But, the leg muscle's TBARS quantity was usually identified to be higher than that of muscle of the breast. Based on these studies, soybean acts as a potential protein and energy source to the poultry.

3.3.10 Velvet Beans

Velvet bean (*Mucuna pruriens*) is commonly referred to as Bengal bean native to southern China and Eastern India, where mostly propagated once as green vegetable crop (Duke 1981). In the late nineteenth century, it was introduced to the southern states of the United States of America and from there in the early twentieth century; it was again introduced to the tropics (Eilitta and Carsky 2003). It has historically been used as a feed for livestock and from the past few years, *Mucuna* species is potentially been used in tropical and subtropical areas as cover crop, fodder and green manure. It was stated to have been utilized as a cattle feed in Mauritius before the start of the twentieth century (Piper and Tracy 1910). In the year 1895, scientists at the Florida Agricultural Experimental Station, considered it as a promising feed to include in various animal diets (Clute 1896).

Velvet bean seeds are high in protein (24–30%, DM basis), starch (28%, DM basis) and gross energy (10–11 MJ/kg, DM basis) (Siddhuraju et al. 2000; Pugalenthi et al. 2005). Bressani (2002) compared velvet bean dietary quality with common beans (*Phaseolus vulgaris* L.) and found that structure, protein quality and digestibility, amino acid and micronutrient content of velvet beans are generally similar to them. Velvet bean yields vary from 10 to 35 tons of green material/ha/year and

0.25–3.3 tons of seed/ha/year relying on the circumstances of farming (Ecocrop 2011). Similar to other legumes, velvet beans also contain many anti-nutritional factors (ANFs) like L-3,4-dihydroxyphenylalanine (L-dopa), tannins, haemagglutinin, no phenolics, chymotrypsin and trypsin inhibitors, anti-vitamins, dimethyltryptamine, phytic acid, saponins, protease inhibitors, flatulence factors and hydrogen cyanide (Vadivel and Janardhanan 2000). L-dopa levels in seed ranges from 1.6% to 7% (Cook et al. 2005). These ANFs content can be effectively reduced by various treatments like boiling in water for 1 h, 48 h of soaking and 30 min boiling, autoclaving for 20 min, or 24 h cracked seeds soaking in 4% $Ca(OH)_2$ solution (Cook et al. 2005; Pugalenthi et al. 2005). Sandoval-Castro et al. (2003) and Ayala-Burgos et al. (2003) studied the impact on the degeneration mechanism of rumen micro-organisms by the anti-nutritional factors found in velvet beans. Different studies conducted on ruminants exhibited no adverse effects on health and animal performance by the consumption of velvet bean grain or foliage.

Sandoval-Castro et al. (2003) discovered that, the *in vitro* dry matter digestibility of husks and beans were respectively maximum at 97.9% and 79.0% and indicated that there was no detrimental impact on *in vitro* fermentation by these anti-nutritional factors. Other *in vitro* analysis conducted by Adesogan et al. (2004) also stated that velvet bean dry matter digestibility is higher compared to soya. Those findings indicate that, in ruminant diets velvet bean can substitute standard sources of energy (corn and sorghum) and their husks are highly digestible and can be included in their diets.

3.3.10.1 Using Velvet Bean Seeds as Poultry Feed

Velvet bean seeds are efficient protein additives when soybean usage in poultry rations is prevented by their price or accessibility. There seeds could be made into a wealthy protein feed and could be given to all livestock categories (Chikagwa-Malunga et al. 2009). The elevated concentrations of energy and protein levels in seeds of velvet beans act as a valuable feed to apply in poultry diets but is restricted by the anti-nutritional factors present and they cannot be used unless suitable treatments are applied to decrease them (Carew and Gernat 2006). Regarding this, different treatments were tested to reduce the concentrations of anti-nutritional factors like soaking, autoclaving, boiling, dry roasting and combination of the above treatments (Vadivel et al. 2011). Treated seeds (soaking plus boiling or dry roasting) may include up to 10% in the diet, but they should be carefully used and must be apparently avoided in starters. It has been found that, dry roasting of velvet beans is an effective approach to restrict the adverse reactions in Japanese quails & broilers (Emiola et al. 2007; Ukachukwu and Obioha 2007; Tuleun et al. 2009). But, previous researchers combined different treatments and stated that roasting was less effective in broilers and Guinean fowls than boiling (Emenalom et al. 2005; Dahouda et al. 2009). It was stated that, it is not effective to soak with or without additives in water separately (Tuleun and Dashe 2010; Vadivel et al. 2011) and thus they must be coupled with heat processing methods like autoclaving or boiling. Carew and

Gernat (2006) stated that heat treatment may aid to decrease the negative effects of velvet beans and the duration of heat treatment may affect their usage efficiency. For example, velvet bean seeds boiling for 20 min produced lower rates of growth than 40 or 60 min boiling (Tuleun and Igba 2008).The optimal treatment suggested by various researchers is soaking in sodium bicarbonate or water, succeeded by boiling (60–90 min) and drying. The previous process effectively eliminated the anti-nutritional factors (Vadivel et al. 2011) and the quality of broilers was retained by including 10–20% of velvet beans in the broilers diet. (Akinmutimi and Okwu 2006; Emenalom et al. 2006; Farougou et al. 2006; Ukachukwu and Obioha 2007; Ani 2008).

Different studies have been stated that, seeds of velvet beans in raw condition shall not be combined in the feed of broiler and layers as their performance has been considerably reduced. Iyayi et al. (2006) observed that a 25% reduction in broiler quality is caused by the addition of 5% raw velvet bean. More levels (20% of diet) of these seeds have lead to apparent fatalities in guinea fowl (Dahouda et al. 2009). Velvet beans tend to be more harmful to growth than the feed intake, even though the results vary among classes (Trejo et al. 2004; Tuleun and Igba 2008). In case of layers, including 12.5% raw velvet bean seeds in their diets dropped the day-to-day production of eggs from 78.5% to 65.5% (Tuleun and Igba 2008). Applying different types of seed treatments decreased the adverse impact of velvet beans; however no similar efficiency was shown compared to the control diets despites the low concentrations of velvet beans. In case of egg production, feeding laying hens with 20% toasted velvet bean seeds showed 74% per day and boiled seeds exhibited 59% per day, whereas the control diet exhibited 84% (Tuleun and Igba 2008). In Japanese quails (laying), supplementing 15% heated seeds resulted in a substantial decrease in efficiency, but still reduced the cost of feed/egg produced as well as cost of feed/bird and thus promoted the use of velvet bean seeds (Tuleun and Dashe 2010). Although they are economically beneficial and cost-effective, even processed velvet bean seeds are not suggested for production of eggs commercially. Finally, it was suggested to use velvet beans for broiler diets of appropriately 20%, beside it is not suggested for layer diets.

3.3.11 Common Vetch

Common Vetch (*Vicia sativa*) is one among the most economically significant globally grown annual seed legume (Mikic et al. 2009). Even though it is introduced in the arid regions of the Middle East (Ford et al. 2008), it is currently grown commonly in different parts of the globe (Francis et al. 1999) as it is a wealthy mineral & protein source and (Akpinar et al. 2001; Mao et al. 2015). *V. sativa* is mostly consumed by many people (Enneking 1994; Akpinar et al. 2001) particularly by the poor vegetarians (Uzun et al. 2011) and it also used as a cattle feed (Kaya et al. 2013). Totally the common vetch and other vetches are cultivated roughly around 573,7 69 ha, corresponding to 926,982 tons yield per year and a normal annual yield of 1616 kg /ha (FAO 2013) and is mostly cultivated in Mexico, Ethiopia, the Russian Federation, Turkey and Spain (FAO 2013). As the regular vetch plant could be cut for silage management, for grazing (Rihawi et al. 2010; DePeters et al. 1989; Garduno-Castro et al. 2009) or for hay production (Cakmakci and Acikgoz 1994; Abbeddou et al. 2011; Larbi et al. 2011), their seeds could be harvested at low price and used as an alternative food to incorporate in the diets of animals since it is potential in minerals and highly digestible protein (Collins et al. 2002; Seymour et al. 2002). But, common vetch seeds generally consists of specific neurotoxic compounds like b-cyano L-alanine (BCA), c-glutamyl-b-cyano-alanine (GCA), which might be harmful to feed monogastric animals (Ford et al. 2008; Megıas et al. 2014). The common vetch seeds crude protein (CP) content of dry matter is 283.63 g/kg similar to 52.43% of soybean meal (Hess et al. 2006). The difference in crude protein level in various varieties ranges from 266 to 316 g/kg DM. (Larbi et al. 2011), while the CP content slightly varies from 289 to 298.9 g/kg DM under different substantial circumstances (Larbi et al. 2011; Mao et al. 2015). Common vetch seeds consist of 8.95–38.00 g/kg dry matter of crude fat. The unsaturated to saturated fatty acids ratio ranges from 3.0:1.0 to 3.7:1.0 (Pastor-Cavada et al. 2009; Mao et al. 2015). The main element of saturated fatty acid is palmitic acid (including total 15.79–17.42%), whereas the main elements of unsaturated fatty acids are linoleic acid (including 52.56–54.39% of the total), α-linolenic acid (including 9.03–12.12%) and oleic acid (including 8.33–11.54%) (Mao et al. 2015). Based on their fatty acid structure, Pastor-Cavada et al. (2009) suggested that seeds of common vetch may be utilized as a significant base of nutritional fats. The complete amino acids present in the vetch seeds varies from 173.00 to 286.05 g / kg DM, where glutamic acid (22.10–55.20 g/kg DM) & leucine (16.38–21.31 g/kg DM) are the vital and non-essential amino acids in common vetch plants, respectively.

Common vetch seeds, like other seed legumes, comprise a range of anti-nutritional variables, like GCA, BCA, vine, trypsin inhibitor, phenolics, tannins and convicin. To avoid the negative impacts of anti-nutritional factors in the vetch plants, it was suggested to incorporate it at peak concentrations (< 0.1% of complete protein or 0 toxin concentrations) in monogastric animal diets (Ford et al. 2008). But, the study conducted on it reports that, high levels of common vetch seeds included in the diets of ruminants don't cause any adverse effects (Koumas and Economides 1987; Matic et al. 2007; Fırıncıoglu et al. 2007) and can have the advantage of improved absorption of nutrients (Koumas and Economides 1987).

3.3.11.1 Using Common Vetch Seeds as Poultry Feed

Common vetch seeds have been utilized as a protein source in poultry diets (Darre et al. 1999). Based on the reports of Saki et al. (2008), dietary vetch seeds does not have any adverse effects on broilers but 30% dietary vetch seeds cause turbid and untidy bones, may be by the impact of anti-nutritional factors (ANFs) on metabolic sulfur containing amino acids, particularly methionine (Tate 1996). He further stated that, comparative weights of breasts, broiler thighs, pancreas and livers are not damaged by adding increased levels of heated or raw common vetch plants compared to the control diet containing corn-SBM. Saki et al. (2008) finally concluded that, inclusion of 10% vetch seeds in the broilers diets will not affect the broilers performance. In addition, processed and raw vetch seeds don't cause any undesirable reactions in broiler development and feed consumption (Darre et al. 1998). Farran et al. (1995) noticed that egg production, feed intake and feed conversion ratio could be reduced by using 25% vetch seed levels in layers diets. Finally, the results of this study have shown that 10% of vetch seeds have no adverse effects on broiler performance and organ reflex. Therefore, the vetch seeds can be an appropriate plant protein substitute to broilers.

Addition of untreated common vetch seeds of more than 20% in layer diets and 15% in broiler diets reduced their performance (Gul et al. 2005). In the study conducted by Yalcin et al. (1998) it was noticed that, raw common vetch seeds could partly substitute soybean meal diet by the addition of 50 g/kg in the feed with no effect on BW gain, FCR, DFC or carcass yield. But, addition of higher concentrations like 100 & 150 g/kg negatively affected body weight and FCR, implying that when offered at these higher concentrations, birds could be vulnerable to antinutritional factors present in the common raw vetch plants. In addition, Darre et al. (1998) stated that the limited substitution of corn -SBM (rate of 100 g/kg feed) in place of heated or raw vetch seeds (processed by decantation-cooking technique) didn't influence the broiler chickens total body weight, body weight gain or feed conversion ratio compared to the corn-SBM control diet. In contrary Saki et al. (2008) analyzed raw versus heated common vetch seeds (processed by soaking followed by cooking in water at 100 °C for 60 min and drying for 48 h) nutrient value on the growth of broiler chickens. They included vetch seeds in 1–21 and 21–42 days age old male broiler chicken meals at separate proportions with maximum of 300 g/kg and recorded an adverse impact in both the treatments on the aspects of DFC consumption and body weight gain. But, except in the 300 g/kg raw common vetch treatment, no impact was noted on feed conversion ratio in the two treatments. Likewise, in the research done by Sadeghi et al. (2011) on the dietary quality of raw and heat-treated common vetch plants supplemented in the diet of 22–42 day broiler chickens in different levels like 0, 100, 200 or 300 g/kg feed, it was noted that rising the percentage of common vetch seeds in the meal adversely affected body weight profit however considerably enhanced feed conversion ratio. They further observed that, processing the seeds (soaking seeds at room temperature in water for 10 h, followed by washing in tap water & cooking for 90 min at 95 °C, then again washing & cooking for 30 min at 95 °C) enhanced both body weight gain and feed conversion ratio, may be as the result of effect of heat in improving the availability and use of nutrients (Sadeghi et al. 2011).

In addition, Sadeghi et al. (2011) also concluded that no negative impacts were exhibited by the feed treatments on carcass features, serum parameters or small intestinal morphology relative to the control diet. Earlier studies also stated that, nutritional augmentation of raw common vetch plants to the broilers at a supplementation rate of 300 g/kg resulted in reduced fat in abdominal fat concentrations & greater relative weight of liver related to those fed with control diet (Sadeghi et al.

2011). Castanon and Perez-Lanzac (1990) examined the impact of the addition of various levels of raw common vetch plants up to 469 g/kg on laying hens productivity. Based on their study, they indicated that DFC consumption, egg manufacturing and feed conversion ratio were adversely impacted by enhancing the raw common vetch seed ratio in the diet, which could be because of the aggregation of toxins in the bird's body tissues. Farran et al. (1995) studied the effect of the inclusion of raw common vetch seeds (75, 150 or 225 g/kg) as a partial replacement of corn-SBM on the growth performance and egg quality of Comb White Leghorn laying hens. Based on those studies, it was concluded that, raw common vetch seeds can partly substitute corn-SBM at a level of 150 g/kg of diet by not influencing DFC consumption, egg protein, egg weight, egg quality or feed conversion ratio, still a greater incorporation of 225 g/kg diet affected the reproductive performance. Those results were inconsistent with Gul et al. (2005), who noted that raw seeds may partly substitute corn-SBM in layer diets; however the Lohman laying hen egg laying efficiency is affected at a rate of 220 g/kg.

Kaya et al. (2013) conducted an experiment for determining the impact of physical & raw treated common vetch seed as a supplement, on the laying hen performance, quality of egg, liver histopathology & blood parameters. From the study it was found that, there is a risk of high damage to the liver by the usage of raw vetch seeds and is also noticeable that the increased processing of vetch seeds could induce reduced liver damages. Similarly, Farran et al. (2001) stated that, there is an effect on the egg production by the augmentation of 600 g/kg common vetch seed to the diet. They also mentioned that, all the autoclaved, soaked and soaked-autoclaved common vetch seeds enhanced the Comb White Leghorn laying hens DFCs, egg production and feed conversion ratios compared to raw common vetch plants. Kaya et al. (2011) studied the impact of the augmentation of 250 g/kg soaked, soaked & boiled or raw common vetch seeds in the hen diet during the maximum egg production time. Diets comprising of cooked and soaked common vetch seeds showed no effect on final body weight, weight change, egg protein or feed conversion ratio, where diets comprising of raw, frozen common vetch seeds adversely affected final body weight, change in weight, egg protein and FE in comparison to the control. Based on their results, researchers suggested addition of boiled and soaked common vetch seeds to the hen diet at a maximum rate of 250 g/kg feed. In another study, they additionally reported that characteristics like DFC consumption and quality of egg of laying hens supplied with 250 g/kg of raw or autoclaved common vetch seeds were identical to the hens fed with common vetch diet in roasted condition; however the hens showed decreased egg production. They further noticed that in the common vetch treatment groups, DFC consumption, egg protein and egg weight were all decreased compared to the hens provided with a normal control diet (Kaya et al. 2013). In contrast, specifications of egg quality like shell strength, shell thickness, shape index, shell weight, yolk index, yolk colour and albumen index did not differ among the treatments (Kaya et al. 2013). There can be an improvement in the performance of layers and broilers by the augmentation of heat treated common vetch seeds. Overall, the literature cited indicates that, supplementing common vetch plants in poultry diets do not influence the efficiency of broiler chickens development or laying hens egg output. Further, the processed or raw (soaked and cooked) seeds could be combined up to 10% and 20% for broilers diets and the treated seeds can be securely utilized up to 25% in diets of laying hens.

The recommended levels and the effect of feeding extreme levels of various grain legumes to both broilers and layers is given in a tabular form (Table 3.5).

3.4 Pulses and Their By-Products Management

Monogastric animals, rich in proteins and starch/sugars, are the most effective transformers of raw ingredients into meat and eggs and the ruminants produce only milk. But, monogastric animal's normal feeding strategies include the use of cereals and expensive protein meals as the basis for their diet. This often leads to direct competition with human consumption in many developing countries and the cost of the feed increases with the import of raw materials from other countries. This is why an effort should be made to use locally produced legume by-products as a feed to monogastric animals, even at the cost of lower animal productivity levels. To decrease the cost of feed for poultry, wider levels of by-products will assist in achieving better nutrients utilization. Only if we increase the digestibility of conventional feed materials (wheat, maize, soy, etc.) and by-products, improved usage and feed conversion ratio could be expected. In addition, pulses and legume plants are the main elements of sustainable farming and climate-resilient crop systems and play a significant role in offering valuable animal feed not only as complete grains but also as by-products and thus contribute indirectly to food safety. By-products of legumes are the cheap potent protein & energy rich sources, which can be utilized efficiently without simply dumping them.

Before consumption, pulses are subjected to various primary & secondary processes like dehulling, puffing, grinding and splitting and the processing methods differs from cultivar to cultivar and from country to country. For hulling grain legumes, the traditional home-scale technique is to pound them in a mortar with a pestle, either after a few hours of spreading the grains in the sun, or after mixing them with a little water. Then the hull scatters off to generate smooth cotyledons. They are usually reformed to dal by breaking the entire seed and over 75% of overall legumes cultivated in the country are made to dal. They consists of 2–5% germ and 11–14% husk and remaining is endosperm of seed and the processing recovery rates vary from 70% to 88% of the raw material. The major by-products obtained after milling of pulses were in different forms like mixture of germ and powder (7–12%), brokens (6–13%) & husk (4–14%). The by-products like husk & small brokens were utilized as cattle feed and can be supplied to others like elephants & swans. Likewise, lentil husks are utilized in poultry feed and bengal gram brokens are used as horse feed and also utilized in besan preparations. The additional by-products like chunies (blend of endosperm & seed coats divisions) and husks produced at the time of pulses processing (for human utilization) also act as an effective animal feed. Pulses are the potent amino acids sources, minerals, fiber, energy,

Table 3.5 The recommended levels of various grain legumes to poultry and the effect of feeding them in excessive amounts

Type of grain legume	Recommended level in the diet	Effect of feeding extreme amounts
Chickpeas	Heat processed chickpeas could be added up to 20% in both broiler and layer diets. Inclusion level of chickpeas at the rates of 25–40% in layer diets were shown to sustain egg production.	Substitution of soybean meal with chickpeas at high addition rates negatively affects the productive performance.
Pigeonpea	Raw pigeon pea seeds could be added up to 10%, while processed (toasted) seeds could be added up to 20% in both broiler and layer diets.	Enhancing the inclusion level of pigeonpea to 30% or above decreases the body weight, which might be due to rise in antinutritional factors present in pigeonpea.
Cow pea	For broiler diets, heat treated seeds could be added up to 20%. In case of layers feeding cowpea feed showed low egg production in the initial 8 weeks and afterwards improved.	Inclusion at higher levels is harmful.
Mung bean	Addition of mung bean hull in broiler diet must be lower than 5% to attain optimal nutrient digestibility and growth performance. In case of layers the general suggestion to utilize mung beans is up to 30%, ensuring the diet is well formulated with specific amino acids.	Further research has to be done to determine the effect of feeding excessive amounts of mung bean.
Faba beans	For broiler chicken diets, heat treated (steam pelleting) white flowered faba beans can be included up to 20%. In order to enhance egg production in laying hen feed, faba beans and enzyme augmentation can be included at the level of less than 50% as a substitute to soybean meal.	Addition of faba bean at high levels could adversely affect egg quality and egg production indices in laying hens.
Field peas	In both broiler and layers diets, peas can be added up to 30% if feed was balanced with artificial amino acids.	Artificial feed enzymes could be included to enhance protein digestibility in feed consisting of maximum rates of field peas.
Lupins	Addition of blue lupins at the level of 15% is suggested for poultry diet. In laying hens diet, addition of blue lupins at the level of 15% caused no negative effects in health of the hen or egg production and could be utilized as a part of a balanced portion with the addition of enzymes such as galactanase and lactase to lower the dependence on soybean protein.	High dietary addition of white lupins in poultry diets at 300 g/kg developed only a slight effect on the growth parameters, which does not restrict the use of lupin seed as poultry feed.

(continued)

Table 3.5 (continued)

Type of grain legume	Recommended level in the diet	Effect of feeding extreme amounts
Lentil	In broilers feed, lentil seeds could be added up to 20% but are not suggested for layers feed. Additionally amino acid augmentation is suggested for use in poultry diets.	For broilers, lentils must be utilized in precisely formulated feeds with attention given to amino acids rates. In case of layers, it is important to balance the amino acid proportion of the diet.
Soybean meal	In case of broilers, up to 25% is recommended for the starter feed and 20 for the finisher feed. For layers over 12–15% is recommended.	Large amounts of soybean meal may cause troubles from pasting up in which excreta sticks to the vent. This can lead to renal dysfunction due to retention of the urates which interferes with passing of the droppings. Soybean meal may also produce goiter
Velvet beans	Processed seeds could be used up to 20% in broiler diets, but these are not suggested for layer diets.	Inclusion at higher levels is not suggested.
Common vetch	Raw and processed (soaking and cooking) seeds could be added up to 10% & 20% in broilers diets. For layers diet, processed seeds could be safely used up to 25%.	Higher levels are not recommended

essential fatty acids & vitamins and the utilization of pulses by-products in livestock feeding gains social, economic and environmental benefits by decreasing the grains usage for animal feeding and not competing with the human food (Fig. 3.2). Further, they help in return of the manure to the farmland and fixes atmospheric nitrogen, thereby decreasing the spraying of chemical fertilizers. The plant by-products remained after seed harvesting could be utilized as a dry animal feed source (Solomon 2004; Tolera 2007). In Asia, they are mainly utilized by smallholder farmers in mixed crop-livestock production systems to mitigate the inadequacy of feed (Nigam and Blummel 2010). During the period of shortage of feed resources, the pulses by-products can be utilized to fill the feed gaps and shall be utilized as supplements to forages & natural pastures (Williams et al. 1997).

3.5 By-Products Utilization

By-products are the secondary by-products derived from the harvesting and processing of pulses and are currently regarded as an efficient source of functional compounds (Carle et al. 2001). These by-products also possess therapeutic properties, which can be branded as a food with neutraceutical properties (Vasavi Rama Karri and Nirmala Nalluri 2017). There is increase in demand in developing countries to analyze the significance of these by-products for proteins and

Fig. 3.2 Feeding legumes byproduct as a potential nutrient rich diet to chickens

phytochemicals in order to reduce hunger and run-over malnutrition, particularly in pregnant women and children. In India, dhal mills produce an estimate of 2.5 MT of by-products per year, consisting of approximately 40% of cotyledon and technology to retrieve and enhance their use has been established (Ramakrishnaiah et al. 2004). The by-products of various oil seeds and food grains like pods and hulls have the potency to use as an insoluble dietary fiber source. Soybean and oats are a very good source consisting of around 64–89% TDF (Cole et al. 1999; Blasi et al. 2000; Kraska et al. 2010). The lentils, pea and chickpea hulls consists of 88.9%, 86.7%, and 74.8% fiber respectively (Dalgetty and Baik 2006). Mamdouh Abbas Abbel-Monein (2013) studied the impact of using processed green beans by-products and stated that it is a potential nutrient source that allows it to use in diets of broilers. Adding 16% green bean by-product to the broilers diet showed no negative effect on the broilers performance, blood parameters & carcass characteristics. Thus, the by-products of various pulses could be suggested as a low cost protein source for poultry diets, therefore decreasing the reliability on traditional protein sources and decreasing the expense of feed formulation.

3.6 Conclusion

This chapter has explored the nutritional profile of various grain legumes and their inclusion in the poultry diets at various levels. Increased use of alternative sources in poultry feed has increased the demand for grain legumes supplements that indirectly plays a major role in organic production. Because of their nutritional profile, grain legumes production and consumption continues to increase as an alternative feed supplement to poultry industry. It is apparent from various studies that the anti-nutritional factors present in legumes could be decreased by various processing techniques like heat and mechanical treatments and can be substantially included in the poultry diets. Demands for animal production under environmental safety regulations, food safety and production economics call for the development of different types of highly nutritious and low anti-nutritional factors containing legumes, that not only enable animal production systems to combat the demand under sustainable conditions and also to increase benefits of production. These improvements are essential for balancing the increasing demand of grain legumes as animal feed. Those efforts require both breeding and gene modification technologies to enhance the food production for animals. This review also gives awareness on the nutritional values of different grain legumes and their appropriate percentage of inclusion in poultry diets. Besides this, use of pulses by-products in poultry feed was also discussed, which is a cheap protein rich source and indirectly contributes to food security. Information available on this topic was collected and incorporated to emphasize the potential role of grain legumes and their by-products as nutrient rich poultry feed. The present study mainly desires to improve the usage of grain legumes & there by-products in poultry farming and the literature cited here can be useful for various researchers and poultry farmers.

Acknowledgements The authors are thankful for the consent and endorse the encouragement of Department of Biotechnology, G. I. T., GITAM (Deemed to be University), Visakhapatnam in successful completion of this study.

References

Abbeddou S, Rihawi S, Hess HD, Iniguez L, Mayer AC, Kreuzer M (2011) Nutritional composition of lentil straw, vetch hay, olive leaves, and saltbush leaves and their digestibility as measured in fat-tailed sheep. Small Rumin Res 96:126–135. https://doi.org/10.1016/j.smallrumres.2010.11.017

Abbel-Monein MA (2013) Effect of using green beans processing by-products with and without enzyme supplementation on broilers performance and blood parameters. J Agrobiol 30(1):43–54. https://doi.org/10.2478/agro-2013-0005

Abdulla JM, Rose SP, Mackenzie AM, Pirgozliev VR (2017) Feeding value of field beans (*Vicia faba* L. var. minor) with and without enzyme containing tannase, pectinase and xylanase activities for broilers. Arch Anim Nutr 71:150–164. https://doi.org/10.1080/1745039X.2017.1283823

Adesogan AT, Chikagwa-Malunga SK, Salawu MB, Kim SC (2004) The *in vitro* digestibility, gas production and fermentation characteristics of Mucuna pruriens, and soybean mealated with or without L-dopa. J Anim Sci 82(1):212–216

Agugo UA, Onimawo IA (2009) Heat treatment on the nutritional value of green gram. Electron J Environ Aaric Food Chem 8:924–930

Akande KE, Abubakar MM, Adegbola TA, Bogoro SE, Doma UD (2010) Chemical evaluation of the nutritive quality of pigeon pea [*Cajanus cajan* (L.) Millsp]. Int J Poult Sci 9:63–65. https://doi.org/10.3923/ijps.2010.63.65

Akanji AM, Fasina OE, Ogungbesan AM (2016) Effect of raw and processed cowpea on growth and heamatological profile of broiler chicken. Bangaldesh J Anim Sci 45(1):62–68. https://doi.org/10.3329/bjas.v45i1.27490

Akinmutimi AH, Okwu ND (2006) Effect of quantitative substitution of cooked Mucuna utilis seed meal for soybean meal in broiler finisher diet. Int J Poult Sci 5(5):477–481. https://doi.org/10.3923/ijps.2006.477.481

Akpinar N, Akpinar MA, Turkoglu S (2001) Total lipid content and fatty acid composition of the seeds of some *Vicia* L. species. Food Chem 74:449–453. https://doi.org/10.1016/s0308-8146(01)00162-5

Alagawany M, Elnesr SS, Farag MR, El-Hack M EA, Khafaga AF, Taha AE, Dhama K (2019) Omega-3 and Omega-6 fatty acids in poultry nutrition: effect on production performance and health. Animals 9(8)573. https://doi.org/10.3390/ani9080573

Amaefule KU, Nwagehara NN (2004) The effect of processing on the utilisation of pigeon pea (*Cajanus cajan*) seed meal and pigeon pea seed meal based diets by pullets. Int J Poult Sci 3:543–546. https://doi.org/10.3923/ijps.2004.543.546

Amaefule KU, Ojewola GS, Ironkwe MC (2006) Inclusion level and prolonged feeding of raw or processed pigeon pea (*Cajanus cajan*) seed meal as protein source for pullets. Int J Poult Sci 4:289–295. https://doi.org/10.3923/ijps.2006.289.295

Amaefule KU, Ukpanah UA, Ibok AE (2011) Performance of starter broilers fed raw pigeon pea [*Cajanus cajan* (L.) Millsp.] seed meal diets supplemented with lysine and or methionine. Int J Poult Sci 10:205–211. https://doi.org/10.3923/ijps.2011.205.211

Ameen OM, Fatope OM, Usman LA, Adebayo SA (2005) Bioactive metabolites in improved cowpea seeds. Afr J Biotechnol 4:513–516

Ani AO (2008) The feeding value of processed velvet bean (*Mucuna pruriens*) for pullet chicks. J Trop Agric Food Environ Ext 7(2):149–155. https://doi.org/10.4314/as.v7i2.1597

Araba M, Dale NM (1990) Evaluation of protein solubility as an indicator of under processing soybean meal. Poult Sci 69:1749–1752. https://doi.org/10.3382/ps.0691749

Arif M, Rehman A, Saeed M, Abd EL-Hack ME, Alagawany M, Abbas H, Arian MA, Fazlani SA, Abbasi IHR, Ayasan T (2017) Effect of different processing methods of pigeon pea (*Cajanus cajan*) on growth performance, carcass traits, and blood biochemical and hematological parameters of broiler chickens. Turk J Vet Anim Sci 41:38–45

Arora SK (1995) Composition of legume grains. In: D'Mello JPF, Devendra C (eds) Tropical legumes in animal nutrition. D'Mello Cab International, Wallingford, pp 67–93

Ayala-Burgos AJ, Herrera-Diaz PE, Castillo-Caamal JB, RosadoRivas CM, Osornio-Munoz L, Castillo-Caamal AM (2003) Rumen degradabiility and chemical composition of the velvet bean (Mucuna spp.) grain and husk. Trop Subtrop Agroecosyst 1:71–75

Batal AB (2009) How much DDGS for poultry? Feed Int 30:18–19

Bejiga G (2006) *Lens culinaris* Medik. Record from Protabase. In: Brink M, Belay G (eds) PROTA (Plant Resources of Tropical Africa/Ressources vegetales de l'Afrique tropicale). PROTA Foundation, Wageningen

Bennett C (2002) Organic diets for small poultry flocks [Internet]. C 2002–2014. [cited 2014 Jul 14]. Available from: http://en.engormix.com/MA-poultry-industry/nutrition/articles/organic-diets-small-poultry-t105/p0.html

Binalay JP (2012) Effect of ground mung bean on the performance of broiler. Benguet State University, La Trinidad

Blasi DA, Drouillard J, Titgemeyer EC, Paisley SI, Brouk MJ (2000) Soybean hulls, composition and feeding value for beef and dairy cattle. Contribution No. 00-79-E. Kansas State University. https://www.bookstore.ksre.ksu.edu/pubs/MF2438.pdf

Bond DA, Lawes DA, Hawtin GC, Saxena MC, Stephens JS (1985). http://www.hort.purdue.edu/newcrop/cropfactsheets/fababean.html. Accessed June 2005

Bressani R (2002) Factors influencing nutritive value in food grain legumes: Mucuna compared to other grain legumes. In: Flores M, Eilitta M, Myhrman R, Carew LB, Carsky RJ (eds) Food and feed from Mucuna: current uses and the way forward. Proceedings of an international workshop, vol 1. CIDICCO, CIEPCA, Worlds Hunger Research Centre, Tegucigalpa, pp 164–188

Cakmakci S, Acikgoz E (1994) Components of seed and straw yield in common vetch (*Vicia sativa* L.). Plant Breed 113:71–74. https://doi.org/10.1111/j.1439-0523.1994.tb00704.x

Canon L, Carre B (1989) Effect of autoclaving on the metabolisable energy value of smooth peas (*Pisum Sativum*) in growing chicks. Anim Feed Sci Technol 26:337–345. https://doi.org/10.1016/0377-8401(89)90045-X

Carew LB, Gernat AG (2006) Use of velvet beans, *Mucuna* spp. as a feed ingredient for poultry: a review. Worlds Poult Sci J 62(1):131–144. https://doi.org/10.1079/WPS200590

Carle R, Keller P, Schieber A, Rentschler C, Katzschner T, Rauch D, et al (2001) Method for obtaining useful materials from the by-products of fruit and vegetable processing. Patent application, WO 01/78859 A1

Castanon JIR, Perez-Lanzac J (1990) Substitution of fixed amounts of soyabean meal for field beans (*Vicia faba*), sweet lupins (*Lupinus albus*), cull peas (*Pisum sativum*) and vetchs (*Vicia sativa*) in diets for high performance laying Leghorn hens. Br Poult Sci 31:173–180. https://doi.org/10.1080/00071669008417243

Castell AG, Cliplef RL (1990) Methionine supplementation of barley diets containing lentils (*Lens culinaris*) or soybean meal: live performance and carcass responses by gilts fed *ad libitum*. Can J Anim Sci 70(1):329–332. https://doi.org/10.4141/cjas90-041

Castell AG, Guenter W, Igbasan FA (1996) Nutritive value of peas for nonruminant diets. Anim Feed Sci Technol 60:209–227. https://doi.org/10.1016/0377-8401(96)00979-0

Castillo AD, Gallo SMH, Peralta RC, Solis T (2016) Economic analysis and carcasses quality of broiler chickens, fed with *Cajanus cajan*. Glob Adv Res J Agric Sci 5:008–013

Chakam VP, Teguia A, Tchoumboue J (2008) Performance of finisher broiler chicks as affected by graded levels of cooked cowpeas (*Vigna unguiculata*) in the grower-finisher diet. Bull Anim Prod Afr 56:251–258

Chavan JK, Kadam SS, Salunkhe DK (1989) Chickpea. In: Salunkhe DK, Kadam SS (eds) CRC handbook of world food legumes: nutritional chemistry, processing technology and utilization, vol I. CRC Press, Inc, Boca Raton, pp 247–288

Chikagwa-Malunga SK, Adesogan AT, Sollenberger LE, Phatak SC, Szabo NJ, Kim SC, Huisden CM, Littell RC (2009) Nutritional characterization of Mucuna pruriens. 4. Does replacing soybean meal with Mucuna pruriens in lamb diets affect ruminal, blood and tissue l-dopa concentrations. Anim Feed Sci Technol 148(2–4):124–137. https://doi.org/10.1016/j.anifeedsci.2008.03.003

Choct M (2004) The net energy value of commonly used plant ingredients for poultry in Australia. RIRDC Publication No 04/RIRDC Project No UNE-82J, pp 1–66

Christodoulou V, Bambidis VA, Hucko B, Iliadis C, Mudrik Z (2006) Nutritional value of chickpeas in rations of broiler chickens. Arch Geflugelk 70:112–118

Christodoulou V, Bampidis VA, Hucko B, Mudrik Z (2006b) The use of extruded chickpeas in diets of broiler turkeys. Czech J Anim Sci 51:416–423

Chunmei G, Hongbin P, Zewei S, Guixin Q (2010) Effect of soybean variety on anti-nutritional factors content, and growth performance and nutrients metabolism in rat. Int J Mol Sci 11:1048–1056. https://doi.org/10.3390/ijms11031048

Clute O (1896) Cassava, the velvet bean, prickly comfrey taro, Chinese yam, canaigre, alfalfa, flat pea, sachaline. Fla Agric Exp Station Bull 35:327–354

Coboru (2011) Descriptive register of varieties (in Polish). Slupia Wielka (Poland)

Coertze AF, Venter S (1996) A.3-Cowpeas. Indigenous seed crops. Agricultural Research Council, pp 1–5

Cole JT, Fahey GC Jr, Merchen NR, Patil AR, Murray SM, Hussein HS, Brent JL Jr (1999) Soybean hulls as a source of dietary fiber for dogs. J Anim Sci 77:917–924. https://doi.org/10.2527/1999.774917x

Collins CL, Henman DJ, King RH, Dunshea FR (2002) Common vetch (*Vicia sativa* cv Morava) is an alternative protein source in pig diets. Proc Nutr Soc Aust 26:S249

Cook BG, Pengelly BC, Brown SD, Donnelly JL, Eagles DA, Franco MA, Hanson J, Mullen BF, Partridge IJ, Peters M, Schultze-Kraft R (2005) Tropical forages: an interactive selection tool. CSIRO, DPI& F(Qld), CIAT and ILRI, Brisbane

Cowieson AJ, Acamovic T, Bedford MR (2003) Supplementation of diets containing pea meal with exogenous enzymes: effects on weight gain, feed conversion, nutrient digestibility and gross morphology of the gastrointestinal tract of growing broiler chicks. Br Poult Sci 44:427–437. https://doi.org/10.1080/0007166031000159829 2

Creswell DC (1981) Nutritional evaluation of mung beans (*Phaseolus aureus*) for young broiler chickens. Poult Sci 60(8):1905–1909. https://doi.org/10.3382/ps.0601905

Daghir NJ (2008) Poultry production in hot climates, 2nd edn. CABI, Oxfordshire, p 180

Dahouda M, Toleba SS, Youssao AKI, Mama Ali AA, Dangou-Sapoho RK, Ahounou SG, Hambuckers A, Hornick JL (2009) The effects of raw and processed Mucuna pruriens seed based diets on the growth parameters and meat characteristics of Benin local guinea fowl (*Numida meleagris*, L). Int J Poult Sci 8(9):882–889. https://doi.org/10.3923/ijps.2009.882.889

Dalgetty DD, Baik BK (2006) Fortification of bread with hulls and cotyledon fibers isolated from peas, lentils and chickpea. Cereal Chem 83(3):269–274. https://doi.org/10.1094/CC-83-0269

Daniel mierlita (2015) The effect of lupine seed in broiler diet on animal performance and fatty acids profile of their meat. Bull UASVM Anim Sci Biotechnol 72(2). https://doi.org/10.15835/buasvmcn-asb:11375.9

Darre MJ, Minior DN, Tatake JG, Ressler C (1998) Nutritional evaluation of detoxified and raw common vetch seed (*Vicia sativa* L.) using diets of broilers. J Agric Food Chem 46:4675–4679. https://doi.org/10.1021/jf980931i

Darre MJ, Minior DN, Tatake JG, Ressler C (1999) Nutritional evaluation of detoxified and raw common vetch seed (Vicia sativa L.) using diets of broilers. J Agric Food Chem 46:4675–4679

Defang HF, Teguia A, Awah-Ndukum J, Kenfack A, Ngoula F et al (2008) Performance and carcass characteristics of broilers fed boiled cowpea (Vigna unguiculata L Wal) and or black common bean (Phaseolus vulgaris) meal diets. Afr J Biotechnol 7:1351–1356. https://doi.org/10.5897/AJB07.816

DePeters EJ, Medrano JF, Bath DL (1989) A nutritional evaluation of mixed winter cereals with vetch utilized as silage or hay. J Dairy Sci 72:3247–3254

Deshpande SS (1992) Food legumes in human nutrition. A personal perspective. Crit Rev Food Sci Nutr 32:333–363. https://doi.org/10.1080/10408399209527603

Duke JA (1981) Handbook of legumes of world economic importance. Plenum Press, New York

Ecocrop (2011) Ecocrop database. FAO, Rome. Available at: http://ecocrop.fao.org/ecocrop/srv/en/home

Eilitta M, Carsky RJ (2003) Efforts to improve the potential of *Mucuna* as a food and feed crop: background to the workshop. Trop Subtrop Agroecosyst 1:47–55. http://www.redalyc.org/articulo.oa?id=93911288002

El-Adawy TA (1996) Chemical nutritional and functional properties of mung bean protein isolate and concentrate. Minufiya J Agric Res 21:657–672

Eljack BH, Fadlalla I, Ibrahim MT (2010) The effect of feeding cowpea (Vigna unguiculata) on broiler chickens performance and some carcass quality measurements. Coll Vet Med Anim Prod Sudan Univ Sci Technol 56(124):175–178

Embaye TN, Ameha N, Yusuf Y (2018) Effect of cowpea (*Vigna unguiculata*) grain on growth performance of Cobb 500 broiler chickens. Int J Livest Prod 9(12):326–333. https://doi.org/10.5897/IJLP2017.0424

Emefiene ME, Joshua VI, Nwadike C, Yaroson AY, Zwalnan NDE (2014) Profitability analysis of pigeon pea (Cajanus cajan) production in Riyom LGA of Plateau State. Int Lett Nat Sci 13:73–88. https://doi.org/10.18052/www.scipress.com/ILNS.18.73

Emenalom OO, Udedibie ABI, Esonu BO, Etuk EB (2005) Evaluation of processed velvet bean (Mucuna pruriens) as a feed ingredient in starter diets for broiler chickens. Jpn J Poult Sci 42(4):301–307. https://doi.org/10.2141/jpsa.42.301

Emenalom OO, Udedibie ABI, Esonu BO, Etuk EB (2006) Cooking in local alkaline solution as a method for improving the nutritive value of velvet bean (*Mucuna pruriens*) for broilers. In: Rege JEO, Nyamu AM, Sendalo D (eds) The role of biotechnology in animal agriculture to address poverty in Africa: opportunities and challenges. Proceedings of the fourth All-Africa Conference on Animal Agriculture, Arusha, Tanzania, 20–24 September 2005, pp 239–245

Emiola AI, Ologhobo AD, Gous RM (2007) Influence of processing of mucuna (*Mucuna pruriens var utilis*) and kidney bean (Phaseolus vulgaris) on the performance and nutrient utilization of broiler chickens. J Poult Sci 44(2):168–174. https://doi.org/10.2141/jpsa.44.168

Enneking D (1994) The toxicity of *Vicia* species and their utilization as grain legumes. PhD thesis

Fageria NK, Baligar VC, Jones CA (1990) Common bean and cowpea. In: Growth and minerals nutrition of field crops. Dekker, New York, pp 281–381

FAO (1989) Report of the global consultation on agricultural extension. Food and Agriculture Organization, Rome

FAO (2004) The state of agricultural commodity markets 2004. FAO, Rome

FAO (2006) Livestock's long shadow: environmental issues and options, by Steinfeld H, Gerber P, Wassenaar T, Castel V, Rosales M, de Haan C. FAO, Rome

FAO (2010) FAO statistical yearbook. Accessed on 30 August

FAO (2012) FAOSTAT. Food and Agriculture Organization of the United Nations, Rome

FAO (Food and Agriculture Organization of the United Nations) (2013) FAOSTAT online database. Available at: http://www.fao.org

FAO (Food and Agriculture Organization) (1993) FAO yearbook. Production 1992 (vol 46). FAO, Rome, pp 105–115

Farhoomand P (2006) Performance and carcass traits of lentil seed fed broilers. Indian Vet J 83(2):187–190

Farougou S, Kpodekon M, Tokannou R, Djossou VD, Akoutey A, Youssao IAK (2006) Utilisation de la farine de *Mucuna pruriens* (L.) DC dans l'aliment de croissance des pintades (Numida meleagris). Rev Med Vet 157(10):502–508

Farran MT, Uwayjan MG, Miski AMA, Sleiman FT, Adada FA, Ashkarian VM, Thomas OP (1995) Effect of feeding raw and treated common vetch seed (Vicia sativa) on the performance and egg quality parameters of laying hens. Poult Sci 74(10):1630–1635. https://doi.org/10.3382/ps.0741630

Farran MT, Dakessian PB, Darwish AH, Uwayjan MG, Dbouk HK, Sleiman FT, Ashkarian VM (2001) Performance of broilers and production and egg quality parameters of laying hens fed 60% raw or treated common vetch (*Vicia sativa*) seeds. Poult Sci 80:203–208. https://doi.org/10.1093/ps/80.2.203

Farrell DJ (1996) How effective is "effective energy" for poultry? In: Aust. Poult. Sci Symp. Feb University of Sydney, Australia, pp 169–173

Farrell DJ, Perez-Maldonado RA, Mannion PF (1999) Optimum inclusion of field peas, faba beans, chick peas and sweet lupins in poultry diets. II. Broiler experiments. Br Poult Sci 40:674–680. https://doi.org/10.1080/00071669987070

Fawzi H, Ibrahim I (1967) Economical fattening of kenana calves on Gezira agricultural by-products. Sudan J Vet Sci Anim Hus 8:12–123

FEFAC (European Feed Manufacturers Federation) (2007) Industrial compound feed production. FEFAC Secretariat General, Brussels, May 2007

Fırıncıoglu HK, Tate M, Unal S, Dogruyol S, Ozcan D (2007) A selection strategy for low toxin vetches. Turk J Agric For 31:303–311

Ford R, Rubeena RRJ, Materne M, Taylor PWJ (2007) Genome mapping and molecular breeding in lentil:lentil. In: Kole C (ed) Genome mapping and molecular breeding. Pulse, sugar and starch crops, vol 3. Springer, Heidelberg/Berlin/New York/Tokyo, pp 91–108

Ford R, Maddeppungeng AM, Taylor PWJ (2008) Vetch. In: Kole C, Hall TC (eds) Compendium of transgenic crop plants: transgenic legume seeds and forages, vol 9. Blackwell Publishing, Oxford, pp 163–170

Francis CM, Enneking D, Abd El Moneim AM (1999) When and where will vetches have an impact as seed legumes? In: Proceedings of the 3rd international food legume research conference, Adelaide, Australia, pp 375–384

Friedman M (1996) Nutritional value of proteins from different food sources. A review. J Agric Food Chem 44:6–29. https://doi.org/10.1021/jf9400167

Ganzon-Naret ES (2014a) Evaluation of graded levels of cooked pigeon pea seed meal (Cajanus cajan) on the performance and carcass composition of Asian sea bass (*Lates calcarifer*). Anim Biol Anim Hus 6:1–9

Ganzon-Naret ES (2014b) Use of raw and heat-treated mung bean seeds (*Phaseolus aureus*) as replacement for soybean meal protein in the diets for sea bass, Lates calcarifer fingerlings in tanks: effects on growth performance, nutrient utilization and survival rate. AACL Int J Bioflux Soc 7:458–467

Garduno-Castro Y, Espinoza-Ortega A, Gonzalez-Esquivel CE, Mateo-Salazar B, Arriaga-Jordan CM (2009) Intercropped oats (Avena sativa)-common vetch (*Vicia sativa*) silage in the dry season for small-scale dairy systems in the Highlands of Central Mexic. Trop Anim Health Prod 41:827–834. https://doi.org/10.1007/s11250-008-9258-7

Gatel F (1994) Protein quality of legume seeds for non ruminant animals: a literature review. Anim Feed Sci Technol 45:317–348. https://doi.org/10.1016/0377-8401(94)90036-1

Gatta D, Russo C, Giuliotti L, Mannari C, Picciarelli P, Lombardi L, Giovannini L, Ceccarelli N, Mariotti L (2013) Influence of partial replacement of soya bean meal by faba beans or peas in heavy pigs diet on meat quality, residual anti-nutritional factors and phytoestrogen content. Arch Anim Nutr 67:235–247. https://doi.org/10.1080/1745039X.2013.801137

Geers R, Madec F (2006) Livestock production and society. Wageningen Academic Publishers, Wageningen

Gemede HF, Ratta N (2014) Antinutritional factors in plant foods: potential health benefits and adverse effects. Int J Nutr Food Sci 3:284–289. https://doi.org/10.11648/j.ijnfs.20140304.18

George AS, Elliot R (1986) Nutritional value of pigeonpea (*Cajanus cajan*) in poultry diets. Potential for Pigeonpea in Thailand, Indonesia and Burma. In: Wallis ES, Woolcock RF, Byth DE (eds) Coarse Grains, Pulses, Roots and Tuber Crops (CGPRT) Centre for Research and Develoment in the Humid Tropics of Asia and the Pacific, CGPRT No.15, 75 pp

Gladstones JS (1998) Distribution, origin, taxonomy, history and importance. In: Gladstones JS, Atkins CA, Hamblin J (eds) Lupins as crop plants: biology, production and utilization. CAB International, Wallingford, p 140

Gohl B (1982) Les aliments du betail sous les tropiques. FAO, Division de Production et Sante Animale, Roma

Gomez C (2004) Cowpea: post-harvest operations. In: Mejia (ed) Post-harvest compendium. AGST, FAO, Rome

Graham KK, Kerley MS, Firman JD, Allee GL (2002) The effect of enzyme treatment of soybean meal on oligosaccharide disappearance and chick growth performance. Poult Sci 81:1014–1019. https://doi.org/10.1093/ps/81.7.1014

Gu C, Pan H, Sun Z, Qin G (2010) Effect of soybean variety on antinutritional factors content, and growth performance and nutrients metabolism in rat. Int J Mol Sci 11:1048–1056. https://doi.org/10.3390/ijms11031048

Gul M, Yoruk MA, Macit M, Esenbuga N, Karaoglu M, Aksakal V, Aksu MI (2005) The effects of diets containing different levels of common vetch (Vicia sativa) seed on fattening performance, carcass and meat quality characteristics of Awassi male lambs. J Sci Food Agric 85(9):1439–1443. https://doi.org/10.1002/jsfa.2120

Habibullah AM, Shah HU (2007) Proximate and mineral composition of mung beans. Sarhad J Agric 23:463–466

Hassan SM (2016) Effects of adding different dietary levels of raw mung bean (Phaseolus aereus) on productive performance and egg quality of laying hens. Int J Poult Sci 15:271–276. https://doi.org/10.3923/ijps.2016.271.276

Hess V, Fickler J, Fontaine J, Heimbeck W (2006) AminoDat 3.0- amino acid composition of feedstuffs. Evonik-Degussa GmbH, Health and Nutrition, Hanau

Hughes RJ, Choct M, Kocher A, Van Barneveld RJ (2000) Effect of food enzymes on AME and composition of digesta from broiler chickens fed on diets containing non-starch polysaccharides isolated from lupin kernel. Br Poult Sci 41:318–323. https://doi.org/10.1080/713654936

Hussain I, Burhanddin M (2011) Optimization of germination conditions for germinated mung bean flour by response surface methodology. Afr J Food Sci Technol 2:232–239

Huyghebaert G, Munter GD, Groote GD (1988) The metabolisable energy (AMEn) of fats for broilers in relation to their chemical composition. Anim Feed Sci Technol 20:45–58. https://doi.org/10.1016/0377-8401(88)90126-5

Igbasan FA, Guenter W (1996) The evaluation and enhancement of the nutritive value of yellow-green- and brown-seeded pea cultivars for unpelleted diets given to broiler chickens. Anim Feed Sci Technol 63(1–4):9–24. https://doi.org/10.1016/S0377-8401(96)01045-0

Igbasan FA, Guenter W (1997) The influence of micronization, dehulling and enzyme supplementation on the nutritional value of peas for laying hens. Poult Sci 76:331–337. https://doi.org/10.1093/ps/76.2.331

Iliadis C (2001) Evaluation of six chickpea varieties for seed yield under autumn and spring sowing. J Agric Sci (Camb) 137:439–444. https://doi.org/10.1017/S0021859601001502

INRA (Institut Scientifique de Recherche Agronomique) (2004) Tables of composition and nutritional value of feed materials. In: Sauvant D, Perez JM, Tran G (eds) Wageningen Academic Publishers, 2nd ed, p 186

Iyayi EA, Kluth H, Rodehutscord M (2006) Chemical composition, anti-nutritional constituents, pre-caecal crude protein and amino acid digestibility in three unconventional tropical legumes in broilers. J Sci Food Agric 86(13):2166–2171. https://doi.org/10.1002/jsfa.2592

Jain KC, Sherma P, Gupta SC, Reddy LJ, Singh L (1980) Breeding for vegetable-type Pigeon peas. In: Proceedings of the international workshop on pigeon peas (ICRISAT), vol 2, Patachera, India, pp 165

Jambunathan R., Hall SD, Sudhir P, Rajan V, Sadhana V (1991) Uses of tropical grain legumes. In: Proceedings of a consultants meeting, 27–30 Mar 1989

Jansen G, Jurgens HU, Schliephake E, Seddig S, Ordon F (2015) Effects of growing system and season on the alkaloid content and yield of different sweet L. angustifolius genotypes. J Appl Bot Food Qual 88:1–4. https://doi.org/10.5073/JABFQ.2015.088.001.Choct

Jeroch H (1998) Prufung von DL-Methionin – supplementen zu erbsenreichen futtermischungen in der broilerkükenmast [The study of DL-methionine additives in chicken broiler diets with high level of peas]. Vet Med Zoot 5:93–97

Jezierny D, Mosenthin R, Bauer E (2010) The use of grain legumes as a protein source in pig nutrition: a review. Anim Feed Sci Technol 157:11–128. https://doi.org/10.1016/j.anifeedsci.2010.03.001

Juodka R, Nainiene R, Juskiene V, Juska R, Stuoge I (2016) Effects of different amounts of field peas (*Pissum sativum* L.) in the diets for turkeys on meat qualities. J Appl Anim Res 44(1):150–157. https://doi.org/10.1080/09712119.2015.1021810

Kaczmarek SA, Kasprowicz-potocka M, Hejdysz M, Mikula R, Rutkowski A (2014) The nutritional value of narrow-leafed lupin (*Lupinus angustifolius*) for broilers. J Anim Feed Sci 23:160–166

Karri VR, Nalluri N (2017) Pigeon pea (Cajanus cajan L.) by-products as potent natural resource to produce protein rich edible food products. Int J Curr Agric Sci 7(7):229–236

Karr-Llientha LK, Kadzere CT, Grieshop CM, Fahey GC (2005) Chemical and nutritional properties of soybean carbohydrates as related to nonruminants: a review. Livest Prod Sci 97:1–12. https://doi.org/10.1016/j.livprodsci.2005.01.015

Katogianni I, Zoiopoulos PE, Adamidis C, Fegeros K (2008) Comparison of two broiler genotypes grown under the provisions of EU organic legislation. Eur Poult Sci 72:116–120. http://62.217.125.90/xmlui/handle/123456789/4125

Kay DE (1979) Food legumes. Tropical Products Institute, London, pp 273–281. https://trove.nla.gov.au/version/31204775

Kaya H, Celebi S, Macit M, Geyikoglu F (2011) The effects of raw and physical processed common vetch seed (*Vicia sativa*) on laying performance, egg quality, metabolic parameters and liver histopatology of laying hens. Asian Australas J Anim Sci 24:1425–1434. https://doi.org/10.5713/ajas.2011.11041

Kaya A, Yoruk AM, Esenbuga N, Temelli A, Ekinci O (2013) The effect of raw and processed common vetch seed (Vicia sativa) added to diets of laying hens on performance, egg quality, blood parameters and liver histopathology. J Poult Sci 50:228–236. https://doi.org/10.2141/jpsa.0120013

Kerley MS, Allee GL (2003) Modifications in soybean seed composition to enhance animal feed use and value: moving from dietary ingredient to a functional dietary component. AgBioforum 6(1&2):14–17

Kilicalp N, Benli Y (1994) Possibilities of using lentil flour in rations for layers. Hayvancılık Arastırma Dergisi 4(1):47–49

Koumas A, Economides S (1987) Replacement of soybean meal by broad bean or common vetch seed in lamb and kid fattening diets. Tech Bull Agric Res Inst Nicosia 88:1–5

Kraska RC, McQuate RS, Soni MG (2010) Oat hull fiber – oat 2 trim food usage conditions for general recognition of safety. Trim Holdings, Inc., Mundelein. http://www.accessdata.fda.gov/scripts/fcn/gras_notices/grn000366.pdf. Accessed 5th Mar 2013

Larbi A, El-Moneim AM, Nakkoul H, Jammal B, Hassan S (2011) Intra-species variations in yield and quality determinants in Vicia species: 3. Common vetch (*Vicia sativa* ssp. sativa L.). Anim Feed Sci Technol 164:241–251. https://doi.org/10.1016/j.anifeedsci.2011.01.004

Laudadio V, Nahashon SN, Tufarelli V (2012) Growth performance and carcass characteristics of guinea fowl broilers fed micronized-dehulled pea (*Pisum sativum* L.) as a substitute for soybean meal. Polut Sci 91(11):2988–2996. https://doi.org/10.3382/ps.2012-02473

Lee MRF, Parkinsona S, Fleminga HR, Theobalda VJ, Leemansa DK, Burgessb T (2016) The potential of blue lupins as a protein source, in the diets of laying hens. Vet Anim Sci 1–2:29–35. https://doi.org/10.1016/j.vas.2016.11.004

Leeson S, Summers JD (2005) Commercial poultry nutrition 3. Nottingham University Press, Nottingham, p 398

Liener IE, Kakade ML (1980) Protease inhibitors. In: Liener IE (ed) Toxic constituents of plant foodstuffs. Academic, New York, pp 7–71

Makkar HPS, Tran G, Heuze V, Giger-Reverdin S, Lessire M, Lebas F, Ankers P (2016) Seaweeds as livestock feed: a review. Anim Feed Sci Technol 212:1–17. https://doi.org/10.1016/j.anifeedsci.2015.09.018

Mao ZX, Fu H, Nan ZB, Wan CG (2015) Fatty acid, amino acid, and mineral composition of four common vetch seeds on Qinghai-Tibetan plateau. Food Chem 171:13–18. https://doi.org/10.1016/j.foodchem.2014.08.090

Marcone MF, Kakuda Y, Yada RY (1998) Salt-soluble seed globulins of various dicotyledonous and monocotyledonous plants. Isolation/purification and characterization. Food Chem 62:27–47. https://doi.org/10.1016/S0308-8146(97)00158-1

Marquez MC, Fernandez V, Alonso R (1998) Effect of dry heat on the *in vitro* digestibility and trypsin inhibitor activity of chickpea flour. Int J Food Sci Technol 33:527–532. https://www.researchgate.net/deref/http%3A%2F%2Fdx.doi.org%2F10.1046%2Fj.1365-2621.1998.00218.x

Matic R, Nagel S, Kirby G, Young I, Smith K (2007) Vetch breeding and vetch use in Australia. 11th international symposium on forage crops, Novi Sad, Serbia

Medic J, Atkinson C, Hurburgh CR (2014) Current knowledge in soybean composition. J Am Oil Chem Soc 91:363–384. https://doi.org/10.1007/s11746-013-2407-9

Medugu CIB, Saleh JU, Igwebuike RL, Ndirmbita (2012) Strategies to improve the utilization of tannin-rich feed materials by Poultry. Int J Poult Sci 11:417–423. https://doi.org/10.3923/ijps.2012.417.423

Megıas C, Cortes-Giraldo I, Giron-Calle J, Vioque J, Alaiz M (2014) Determination of b-cyano-L-alanine, c-glutamyl-b-cyano-L-alanine, and common free amino acids in *Vicia sativa* (Fabaceae) seeds by reversed-phase high-performance liquid chromatography. J Anal Methods Chem:1–5. https://doi.org/10.1155/2014/409089

Mieczkowska A, Jansman AJM, Kwakkel RP, Smulikowska S (2005) Effect of dehulling and α-galactosidase suplement on the ileal digestiblity of yellow lupin based diets in broiler chickens and adult roosters. J Anim Feed Sci 14(2):297–304. https://doi.org/10.22358/jafs/67016/2005

Mikic A, Peric V, Dordevic V, Srebric M, Mihailovic V (2009) Anti-nutritional factors in some grain legumes. Biotechnol Anim Hus 25:1181–1188

Monsoor MA, Yusuf HKM (2002) *In vitro* protein digestibility of lathyrus pea (*Lathyrus sativus*), lentil (*Lens culinaris*) and chickpea (*Cicer arietinum*). Int J Food Sci Technol 37:97–99. https://doi.org/10.1046/j.1365-2621.2002.00539.x

Morkunas M (2002) Locally procurable fodder for poultry. Vilnius (Lithuania): Lithuanian Institute of Animal Science, 157 p, Lithuanian

Mu K, Kitts DD (2018) Use of soy lecithin to improve nutritional quality of poultry meats and its effect on stability and sensory attributes. J Nutr Food Sci 8:714. https://doi.org/10.4172/2155-9600.1000714

Mubarak AE (2005) Nutritional composition and anti-nutritional factors of mung bean seeds (*Phaseolus aureus*) as affected by some home traditional processes. Food Chem 89:489–495. https://doi.org/10.1016/j.foodchem.2004.01.007

Nalle CL, Ravindran V, Ravindran G (2010) Nutritional value of faba beans (*Vicia faba* L.) for broilers: apparent metabolisable energy, ileal amino acid digestibility and production performance. Anim Feed Sci Technol 156:104–111. https://doi.org/10.1016/j.anifeedsci.2010.01.010

Nalle CL, Ravindran V, Ravindran G (2011) Nutritional value of white lupins (Lupinus albus) for broilers: apparent metabolisable energy, apparent ileal amino acid digestibility and production performance. Animal 6:579–585. https://doi.org/10.1017/S1751731111001686

Nedumaran S, Abinaya P, Jyosthnaa P, Shraavya B, Rao P, Bantilan C (2015) Grain legumes production, consumption and trade trends in developing countries. Working Paper Series No 60. ICRISAT Research Program, Markets, Institutions and Policies. Patancheru 502 324, Telangana, India: International Crops Research Institute for the Semi-Arid Tropics, 64 pp

Nigam SN, Blummel M (2010) Cultivar dependant variation in food-feed-traits in groundnut (*Arachis hypogaea* L.). Anim Nutr Feed Technol 10:39–48

Nikolopoulou D, Grigorakis K, Stasini M, Alexis MN, Iliadis K (2007) Differences in chemical composition of field pea (*Pisum sativum*) cultivars: effects of cultivation area and year. Food Chem 103:847–852. https://doi.org/10.1016/j.foodchem.2006.09.035

Oburuoga AC, Anyika JU (2012) Nutrient and anti-nutrient composition of mung bean (*Vilgna radiate*), acha (*Digitaria exilis*) and crayfish (*Astacus fluviatilis*) flours. Pak J Nutr 9:841–844. https://doi.org/10.3923/pjn.2012.841.844

Ogblu NN, Ogbu CC, Okorie AU (2015) Haematological indices of broiler chickens fed raw and processed pigeon pea (*Cajanus Cajan*) seed meal. J Anim Prod Adv 5:711–717

Oliverll MD, Jonker A (1997) Effect of sweet, bitter and soaked micronized bitter lupins on broiler performance. Br Poult Sci 38:203–208. https://doi.org/10.1080/00071669708417970

Padmashree A, Semwal AD, Khan MA, Govindaraj T, Sharma GK (2016) Effect of infrared processing on functional, nutritional, anti-nutritional and rheological properties of mung bean (*Phaseolus aereus*) seeds. Int J Adv Res 4:606–613

Park HG, Yang CY (1978) The mung bean breeding programme at the Asian vegetable research and development centre. In: Proceedings of the 1st international mung bean symposium Asian vegetable research and development centre, Shanhua, Taiwan, ROC, pp 214–216

Parviz F (2006) Performance and carcass traits of lentil seed fed broilers. Indian Vet J 83:187–190

Pastor Cavada E, Juan R, Pastor JE, Alaiz M, Vioque J (2009) Fatty acid distribution in the seed flour of wild Vicia species from Southern Spain. J Am Oil Chem Soc 86:977–983. https://doi.org/10.1007/s11746-009-1426-z

Paul T, Rubel NHM, Sayed MA, Akhtaruzzaman M (2011) Proximate compositions, mineral contents and determination of protease activity from green gram (*Vigna radiate, L. Wilczek*). Bangladesh Res Pub J 5:207–213
Perez-Maldonado RA, Mannion PF, Farrell DJ (1999) Optimum inclusion of field peas, faba beans, chick peas and sweet lupins in poultry diets. I. Chemical composition and layer experiments. Br Poult Sci 40:667–673. https://doi.org/10.1080/00071669987061
Piper CV, Tracy SM (1910) The Florida velvetbean and relatedplants. Bull 179. US Dept Agric Bureau Plant Ind, pp 1–26
Pitala W, Kulo AE, Sessi KM, Gbeassor M (2016) Effects of incorporation of roasted soybeans on the performance of laying and the financial profitability of laying hens. Int J Poult Sci 15:373–378. https://doi.org/10.3923/ijps.2016.373.378
Proskina L, Cerina S (2017) Faba beans and peas in poultry feed: economic assessment. J Sci Food Agric 97:4391–4398. https://doi.org/10.1002/jsfa.8415
Przywitowski M, Mikulski D, Zdunczyk Z, Rogiewicz A, Jankowski J (2016) The effect of dietary high-tannin and low-tannin faba bean (*Vicia faba* L.) on the growth performance, carcass traits and breast meat characteristics of finisher turkeys. Anim Feed Sci Technol 221:124–136. https://doi.org/10.1016/j.anifeedsci.2016.08.027
Pugalenthi M, Vadivel V, Siddhuraju P (2005) Alternative food/feed perspectives of an under-utilized legume *Mucuna pruriens* var. *utilis* – a review. Plant Foods Hum Nutr 60:201–218. https://doi.org/10.1007/s11130-005-8620-4
Pulse Canada (2011) Pulse proteins. http://www.pulsecanada.com/uploads/72/47/7247a8ce522273 3d6ef7da7ace146a01/Pulse-Protein.pdf
Rada V, Lichovnikova M, Safarik I (2017) The effect of soybean meal replacement with raw full-fat soybean in diets for broiler chickens. J Appl Anim Res 45(1):112–117. https://doi.org/10.1080/09712119.2015.1124337
Ramakrishnaiah N, Pratape VM, Sashikala VB, Narasimha HV (2004) Value addition to by-products from dhal milling industry in India. J Food Sci Technol 41(5):492–496
Rihawi S, Iniguez L, Kanus WF, Zakulata M, Wurzinger M, Soelkner J, Larbi A, Bomfim MAD (2010) Fattening performance of lambs of different Awassi genotype, fed under cost reducing diets and contrasting housing conditions. Small Rumin Res 94:38–44. https://doi.org/10.1016/j.smallrumres.2010.06.007
Robinson D, Singh DN (2001) Alternative protein sources for laying hens. A report Rural Industries Research and Development Corporation, Kingston, p 85
Rungcharoen P, Amornthewaphat N, Ruangpanit Y, Attamangkune S, Rattanatabtimthong S (2010) The utilization of mung bean bran in nursery pig diets. In: Proceedings of the 48th Kasetsart University Annual Conference, Kasetsart, 3–5 March
Sadeghi GH, Tabeidian SA, Toghyani M (2011) Effect of processing on the nutritional value of common vetch (*Vicia sativa*) seed as a feed ingredient for broilers. J Appl Poult Res 20:498–505. https://doi.org/10.3382/japr.2010-00306
Saini HS (1993) Distribution of tannins, vicine and convicine activity in legume seeds. In: Van der Poel AFB, Huisman J, Saini HS (eds) Recent advances of research in antinutritional factors in legume seeds. Wageningen Pers, Wageningen, pp. 95–100
Saki AA, Pourhesabi G, Yaghobfar A, Mosavi MA, Tabatabai MM, Abbasinezhad M (2008) Effect of different levels of the raw and processed vetch seed (*Vicia sativa*) on broiler performance. J Biol Sci 8:663–666. https://doi.org/10.3923/jbs.2008.663.666
Sandoval-Castro CA, Herrera P, Capetillo-Lea CM, AyalaBurgos AJ (2003) In vitro gas production and digestibility of Mucuna bean. Trop Subtrop Agroecosyst 1:77–79
Savage TF, Nakaue HS, Holmse ZA, Taylor TM (1986) Feeding value of yellow peas (*Pisum sativum* L., variety Miranda) in market turkeys and sensory evaluation of carcasses. Poult Sci 65:1383–1390. https://doi.org/10.3382/ps.0651383
Schuster WH (1992) Olpflanzen in Europa. DLG-Verlag, Frankfurt/Main
Sell JL, Tenesaca LG, Bales GL (1979) Influence of dietary fat on energy utilization by laying hens. Poult Sci 58:900–905. https://doi.org/10.3382/ps.0580900

Seymour M, Siddique KHM, Brandon N, Martin L, Jackson E (2002) Response of vetch (*Vicia* spp.) to plant density in South-Western Australia. Aust J Exp Agric 42:1043–1051. https://doi.org/10.1071/EA01198

Sharma C, Singhania D (1992) Performance of cowpea (*Vigna unguiculata* (L.) Walp.) genotypes for fodder trails. Ann Arid Zone 31:65–66

Shi SR, Lu J, Tong HB, Zou JM, Wang KH (2012) Effects of graded replacement of soybean meal by sunflower seed meal in laying hen diets on hen performance, egg quality, egg fatty acid composition, and cholesterol content. J Appl Poult Res 21(2):367–374. https://doi.org/10.3382/japr.2011-00437

Siddhuraju P, Becker K, Makkar HP (2000) Studies on the nutritional composition and antinutritional factors of three different germplasm seed materials of an under-utilized tropical legume, *Mucuna puriens var. utilis*. J Agric Food Chem 48:6048–6060. https://doi.org/10.1021/jf0006630

Silanino V, Bansul HC, Bozzini A (1981) Improvement nutritional quality of food crops, Plant production protection paper, 34. Food and Agricultural Organization, Rome

Singh U (1988) Antinutritional factors of chickpea and pigeonpea and their removal by processing. Plant Foods Hum Nutr 38:251–261. https://doi.org/10.1007/BF01092864

Singh IF, Eggum BO (1984) Factors affecting the protein quality of pigeon pea (*Cajanus cajan*). Qual plant food. Hum Nutr 34:273–283

Singh IF, Jain KC, Janbunathan R, Faris DG (1984) Nutritional quality of vegetable pigeon peas [Cajanus cajan (L.) Mill sp.]: mineral and trace elements. J Food Sci 49(2):645–646. https://doi.org/10.1111/j.1365-2621.1984.tb12489.x

Singh VS, Palod J, Vatsya S, Rajeev RK, Shukla SK (2013) Effect of sprouted mung bean (*Vigna radiata*) supplementation on performance of broilers during mixed *Eimeria* species infection. Vet Res Int 1(2):41–45

Solomon B (2004) Assessment of livestock production system and feed resource base in Sinana Dinsho District of Bale Highlands, Southeast Oromia. MSc thesis, Alemaya University, Alemaya, pp 135

Soystats (2012). http://www.soystats.com/2011/. Accessed 6 May

Steenfeldt S, Gonzalez E, Bach Knudsen KE (2003) Effects of inclusion with blue lupins (Lupinus angustifolius) in broiler diets and enzyme supplementation on production performance, digestibility and dietary AME content. Anim Feed Sci Technol 110:185–200. https://doi.org/10.1016/S0377-8401(03)00218-9

Sugui FP, Sugui CC, Pastor EC (2007) Performance of broilers fed with different levels of pigeon pea (*Cajanus Cajan* L.) seed meal. MMSU Sci Technol J 1(1)

Swathi M, Lokya V, Swaroop V, Mallikarjuna N, Kannan M, Dutta-Gupta A, Padmasree K (2014) Structural and functional characterisation of proteinase inhibitors from seeds of *Cajanus cajan*. Plant Physiol Biochem 83:77–87. https://doi.org/10.1016/j.plaphy.2014.07.009

Tarawali SA, Singh BB, Peters M, Blade SF (1997) Cowpea haulms as fodder. In: Singh BB (ed) Advances in cowpea research. IITA, Ibadan

Tate ME (1996) Vetches: feed or food. Chem Aust 63:549–550

Teguia A, Chakam VP, Tchoumboue J (2007) Effets de la cuisson ou de l'extrusion du niebe (*Vigna unguiculata*) nsur les performances de production des poulets de chair en finition. Bull Anim Prod Afr 55:118–126. https://doi.org/10.4314/bahpa.v55i2.32800

Tekale SS, Jaiwal BV, Padul MV (2016) Identification of metabolites from an active fraction of Cajanus cajan seeds by high resolution mass spectrometry. Food Chem 211:763–769. https://doi.org/10.1016/j.foodchem.2016.05.128

Teresa Macarulla M, Medina C, Aranzazu De Diego M, Chavarri M, Angeles Zulet M, Alfredo Martinez J, Noel-Suberville C, Higueret P, Mara Portillo P (2001) Effects of the whole seed and a protein isolate of faba bean (Vicia faba) on the cholesterol metabolism of hypercholesterolaemic rats. Br J Nutr 85:607–614. https://doi.org/10.1079/BJN2000330

Tolera A (2007) Feed resources for producing export quality meat and livestock in Ethiopia: examples from selected Woredas in Oromia and SNNP Regional States. Ethiopia Sanitary

and Phytosanitary Standards and Livestock and Meat Marketing (SPS-LMM) Program, Addis Ababa, Ethiopia, 77 p
Trejo LW, Santos R, Hau E, Olivera L, Anderson S, Belmar R (2004) Utilisation of mucuna beans (*Mucuna pruriens* (L.) DC ssp. deeringianum (Bart) Hanelt) to feed growing broilers. J Agric Rural Dev Trop Subtrop 105(2):155–164
Tuleun CD, Dashe NA (2010) Effect of dietary levels of toasted mucuna seed meal (TMSM) on the performance and egg quality parameters of laying Japanese quails (Coturnix coturnix japonica). Int J Poult Sci 9(12):1092–1096. https://doi.org/10.3923/ijps.2010.1092.1096
Tuleun CD, Igba F (2008) Growth and carcass characteristics of broiler chickens fed water soaked and cooked velvet bean (*Mucuna utilis*) meal. Afr J Biotechnol 7(15):2676–2681
Tuleun CD, Igyem SY, Adenkola AY (2009) The feeding value of toasted mucuna seed meal diets for growing Japanese quail (Coturnix coturnix japonica). Int J Poult Sci 8(11):1042–1046. https://doi.org/10.3923/ijps.2009.1042.1046
Tusar MA, Ali MS, Das SC, Alam MS, Matin MA, Sufian MKNB, Paul RC (2015) Growth performance of broiler after inclusion of Pigeon pea (Cajanus cajan) seed as an unconventional feed ingredient in diets. Wayamba J Anim Sci 7:1223–1231
Ukachukwu SN, Obioha FC (2007) Effect of processing methods on the nutritional value of Mucuna cochinchinensis to broiler chicks. Aust J Exp Agric 47(2):125–131. https://doi.org/10.1071/ea03111
USDA (2000) Subject: egg-grading manual. Agriculture handbook number 75. http://www.ams.usda.gov/poultry/pdfs/EggGrading%20manual.pdf. Accessed Apr 2004
Uzun A, Gucer S, Acikgoz E (2011) Common vetch (Vicia sativa L.) germplasm: correlations of crude protein and mineral content to seed traits. Plant Foods Hum Nutr 66:254–260. https://doi.org/10.1007/s11130-011-0239-z
Vadivel V, Janardhanan K (2000) Nutritional and anti nutritional composition of velvet bean: an under-utilized food legume in South India. Int J Food Sci Nutr 51(4):279–287. https://doi.org/10.1080/09637480050077167
Vadivel V, Pugalenthi M, Doss M, Parimelazhagan T (2011) Evaluation of velvet bean meal as an alternative protein ingredient for poultry feed. Animal 5(1):67–73. https://doi.org/10.1017/S175173111000159X
Van der Poel AFB, Gravendeel S, Boer H (1991) Effect of different processing methods on tannin content and in vitro protein digestibility of faba bean (Vicia faba L.). Anim Feed Sci Technol 33:49–58. https://doi.org/10.1016/0377-8401(91)90045-T
Van der Poel AFB (1989) Effects of processing on antinutritional factors (ANF) and nutritional value of legume seeds for non-ruminant feeding. In: Huisman J, Van der Poel AFB, Liener IE (eds) Recent advances of research in antinutritional factors in legume seeds. Wageningen Academic Publishers, Pudoc Wageningen, p 213–229
Vinh NT, Tuan BQ, Hang NM (2013) The use of Mung bean (*Phaseolus aureus*) hulls in diets of laying hens. Livest Res Rural Dev 25:14. http://www.lrrd.org/lrrd25/1/vinh25014.htm
Viveros AA, Brenes R, Elices I, Arija CR (2001) Nutritional value of raw and autoclave kabuli and desi chickpeas (Cicer arietinum L.) for growing chickens. Br Poult Sci 42:242–251. https://doi.org/10.1080/00071660120048500
Von Sengbusch R (1942) Sweet lupins and oil lupins. The history of the origin of some new crop plants. Landwirtsch Jahrb 91:719–880
Waldroup PW (1982) Whole soybeans for poultry feeds. Worlds Poult Sci J 38:28–35. https://doi.org/10.1017/WPS19820003
Williams TO, Fernandez-Rivera S, Kelly TG (1997) The influence of socio-economic factors on the availability and utilization of crop residues as animal feeds. In: Renard C (ed) Crop residues in sustainable mixed crop/livestock farming systems. International Crops Research Institute for the Semi-Arid Tropics, Wallingford, pp 25–40
Wiryawan KG, Dingle JG (1995) Screening tests of the protein quality of grain legumes for poultry production. Br J Nutr 74(5):671–679. https://doi.org/10.1079/bjn19950170

Wiryawan KG, Dingle JG (1999) Recent research on improving the quality of grain legumes for chicken growth. Anim Feed Sci Technol 76:185–193

Wiryawan KG, Miller HM, Holmes JHG (1997) Mung beans (*Phaseolus aereus*) for finishing pigs. Anim Feed Sci Technol 66:297–303. https://doi.org/10.1016/S0377-8401(96)01061-9

Wisaniyasa NW, Suter IK, Marsono Y, Putra IK (2015) Germination effect on functional properties and antitrypsin activities of pigeon pea (Cajanus cajan (L.) Millsp.) sprout flour. Food Sci Qual Manag 43:79–83

Wondifraw Z (2018) Replacement of soybean grain with cowpea grain (*Vigna unguiculata*) as protein supplement in Sasso x Rir crossbred chicks diet. Poult Fish Wildl Sci 6:1–6. https://doi.org/10.4172/2375-446X.1000188

Woodworth JC, Tokach MD, Goodband RD, Nelssen JL, O'Quinn PR, Knabe DA, Said NW (2001) Apparent ileal digestibility of amino acids and digestible and metabolisable energy content of dry extruded-expelled soybean meal and its effect on growth performance of pigs. J Anim Sci 79:1280–1287. https://doi.org/10.2527/2001.7951280x

Woyengo TA, Nyachoti CM (2012) Ideal digestibility of amino acids for zero-tannin faba bean (Vicia faba L.) fed to broiler chicks. Poult Sci 91:439–443. https://doi.org/10.3382/ps.2011-01678

Yaklich RW (2001) Beta-Conglycinin and glycinin in high-protein soybean seeds. J Agric Food Chem 49:729–735. https://doi.org/10.1021/jf001110s

Yalcın S, Sehu A, Kaya I (1998) The effect of common vetch seed (*Vicia sativa* L.) added to the quail rations on growth, carcass yield and some blood parameters. Turk J Vet Anim Sci 22:37–42

Chapter 4
Potential Impact of Annual Forage Legumes on Sustainable Cropping Systems in Turkey

Süleyman AVCI

Abstract Including annual forage legumes (AFLs) in crop rotation systems is a leading application in efficient, sustainable agricultural production intended to preserve environmental resources. This review identifies some of the effects of AFLs' potential for growth in both irrigated and dry regions of Turkey on economic, efficient and sustainable agricultural production. AFLs are considered in terms of both hay production and their contributions to soil fertility. Among its findings, the review indicates that AFLs can readily be cultivated during bare periods between two primary summer and winter crop cycles instead of leaving fields fallow in the yearly agricultural cycle. Although AFLs are important sources of forage for livestock, their production as dry matter can be profitable only under irrigated conditions due to the high input of agriculture in Turkey. In addition to dry matter production, AFLs increase the yields of subsequent plants by facilitating nitrogen fixation and organic matter recovery. In particular, green manuring with AFLs introduces high volumes of organic matter and carbon into the soil, which improves nutrient availability, microbial activity and the physical and chemical properties of soil. Moreover, the mulch layer obtained from green manuring enhances weed and erosion control and the efficient use of water. In sum, including AFLs in cropping systems in Turkey can allow more profitable, more efficient and more sustainable production than using non-legume crops or bare soils.

Keywords Organic matter · Nitrogen fixation · Soil management · Productivity · Economics · Weed management · Reduce tillage · Organic carbon · Green manuring · Mulching

S. AVCI (✉)
Faculty of Agriculture, Department of Field Crops, Eskişehir Osmangazi University, Eskişehir, Turkey
e-mail: savci@ogu.edu.tr

P. Guleria et al. (eds.), *Sustainable Agriculture Reviews 51*, Sustainable Agriculture Reviews 51, https://doi.org/10.1007/978-3-030-68828-8_4

4.1 Introduction

The sustainable management of natural resources is important to respond to the current and future needs of humans and other living organisms. Worldwide, the food supply continues to vary amid global climate change in particular, although also for reasons of increased input costs, losses due to diseases and pests, the increased use of pesticides, soil pollution, reduced soil fertility, soil erosion, reduced biodiversity and desertification (Lichtfouse et al. 2009). Sustainable agriculture is thus an alternative, ecologically conscious solution for basic and applied problems related to food production (Lal 2008).

Including annual forage legumes (AFLs) in crop rotation systems can be pivotal for sustainable agriculture. AFLs not only provide high-quality roughage for livestock but also increase soil fertility by raising levels of soil organic matter as a result of green manuring and by fixing nitrogen (N_2) from the atmosphere to the soil by way of *Rhizobium* bacteria (Fageria et al. 2005). In turn, higher levels of soil organic matter and carbon benefit soil biota, nutrient availability and, ultimately, agricultural productivity (Lal 2004; Rengel 2015). AFLs also protect the soil by mitigating erosion and dispelling weeds, insects and diseases (Sheaffer and Seguin 2008). Thus, cultivating AFLs can enhance sustainable soil management as well as favour the economics of agriculture by reducing the need for chemicals (e.g. fertilisers, pesticides and herbicides) while increasing the availability of nutrients.

In agriculture, legumes are the most important source of N_2 fixation, defined as the reduction of molecular N_2 in the atmosphere to various forms of ammonium (Follett 2008). The biological potential of AFLs to facilitate N_2 fixation also involves the potential to reduce subsequent crops' demand for N fertilisers without limiting growing seasons in cropping systems (Parr et al. 2011; Dube et al. 2012; Liebmana et al. 2018). In parallel, limiting the use of N chemical fertilisers can reduce the accumulation of inorganic salt and increase the biological activity in soil for efficient and economic crop production (Fageria et al. 2005).

Including AFLs in cropping systems to maintain the sustainability of agricultural production has become increasingly popular due to intensive agricultural practices. Although have traditionally been used in rotation with cereals in arid and semi-arid conditions, they also show promise as cover crops for winter or second crops in areas of Turkey with sufficient annual precipitation or that are irrigated. The aim of this review is to describe the current and potential contributions of AFLs to agricultural production and productivity in sustainable cropping systems, primarily with examples from Turkey.

4.2 Cultivation of AFLs

Turkey has always had agricultural importance given its vast arable land, rich soil, biological diversity, favorable climate and geological conditions (Kan et al. 2015). Encompassing Continental, Mediterranean, Marmara, and Black Sea climates, Turkey is located at the intersection of Mediterranean, Euro–Siberian and Irano–Turanian floristic regions. It hosts more than 12,000 plant taxa, 3649 of which are endemic (Dönmez et al. 2018). Although members of the *Fabaceae* family in Turkey provide considerable agricultural biodiversity for forage crops in the *Lotus*, *Onobrychis*, *Trigonella*, *Astragalus*, *Coranilla*, *Lathyrus*, *Lupinus*, *Pisum*, *Medicago*, *Melilotus*, *Trifolium* and *Vicia* genera (TUBIVES 2019), too few existing crop species are used for agricultural purposes (Table 4.1).

Across Turkey, agricultural land occupies 37.8 million ha, 15.4 million ha of which is cultivated every year (Fig. 4.1, Anonymous 2018a). Although the sowing rate of forage crops in total utilized agricultural land is 5.28%, it should be closer to 25.0% as in other developed countries (Anonymous 2017; Anonymous 2018b) in order to meet the hay requirements of livestock and allow efficient, sustainable agricultural production in Turkey. The most cultivated forage crop in Turkey is alfalfa (31.7%), followed by corn (23.6%), vetch (19.3%), oat (10.7%) and sainfoin (9.0%), as shown in Table 4.1. Forage legumes with perennial and annual life cycles account

Table 4.1 Forage crops cultivation areas in Turkey

Families	Forages	Sowing areas (da)
Fabaceae	Alfalfa[a]	6,351,052
	Sainfoin[a]	1,817,338
	Vetch	3,869,465
	Clover	25
	Vetch (*Vicia ervilia*)	27,879
	Pea	104,377
	Gasspea	127,907
Gramineae	Corn (Silage+Hay)	4,726,428
	Wheat	196,798
	Barley	255,517
	Rye	40,717
	Oat	2,142,574
	Sorghum	17,922
	Triticale	135,397
	Ryegrass	103,410
Chenopodiaceae	Fodder beet	18,880
Crucifera	Feed Turnip	56,914
Total		19,992,600

[a]Indicates only perennial life cycle and the others are annual (Anonymous 2018a)

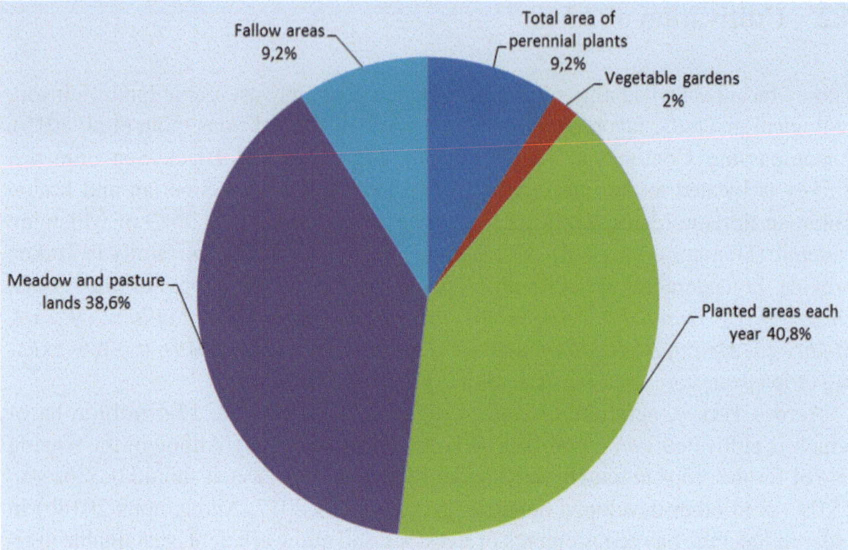

Fig. 4.1 Distribution of agricultural land in Turkey (Anonymous 2018a)

for approximately 61.5% of the total area where forage crops are cultivated, while the rate of AFLs in the total area where forage legumes are cultivated is 31.5%.

AFLs can play an important role in increasing sowing areas for forage crops in both dry and irrigated conditions in Turkey. Every year, periods between the cycles of two primary crops from harvest to sowing dates, whether short or long, can easily accommodate the cultivation of AFLs due to their annual life cycles without restricting the growing season of the subsequent crops. For more efficient and sustainable land use and agricultural production, suitable crops or crop mixtures should be planted during those periods instead of leaving the soil bare (Miguez and Bollero 2005).

In agriculture, fallowing is performed to improve the soil quality of arid lands that receive less than 400 mm of precipitation in uneven quantities throughout the year. In Turkey, fallow lands represent 9.2% of all agricultural land (Fig. 4.1, Anonymous 2018a), and in those areas, water and wind erosion increase soil degradation due to the soil's lying bare during the year (Peterson et al. 2006; Lal 2012). AFLs, which conserve soil and water resources, can be cultivated instead of leaving land fallow, provided that they are removed from the field or incorporated in the soil before the summer season. For that purpose, AFLs grown need to be able to survive in winter, mature early and be cultivated for hay or green manure. Although forage peas and narbon vetch are suitable for fertile soils, the winter forms of hairy and common vetches, Hungarian vetch and grass pea are more appropriate for less fertile soils prone to drought conditions (Ekiz et al. 2009; İleri et al. 2016).

In Turkey, AFLs are cultivated in both winter and summer in irrigated conditions in central regions and as winter crops before main summer crops in coastal regions.

AFLs should be harvested as soon as possible due to their effect on the growing season of subsequent crops. They must be sown a month before the first frost (i.e. approximately 1–15 October) so that the fields can be harvested in time and to avoid frost damage in central regions (Kalkan and Avcı 2020). The sowing date of summer forage legumes is affected by the harvest time of cereals in the second cropping, and their sowing should be performed between 5 and 15 July in central regions (Dereli 2015). In coastal regions, by contrast, sowing also be performed until mid-December (Cakmakcı and Cecen 1999).

The cultivars of Hungarian vetch, hairy vetch, forage peas and narbon vetch have the potential to be grown safely as winter crops in Turkey's central regions, which generally have a harsh continental climate (Kalkan and Avcı 2020). Common vetch, narbon vetch, hairy vetch, bitter vetch, grass peas, forage peas, fenugreek, berseem clover and Iranian clover should be cultivated as winter crops in coastal regions and as summer crops after the harvest of cereals in central regions (Cakmakcı and Cecen 1999; Dereli 2015).

4.3 Effects of AFLs on Plant Production

Many factors—soil fertility and quality, water, weeds, diseases and pests—that affect the yields and quality of crops can be optimized to benefit crop production by including forage legumes in crop rotations. Including AFLs in cropping systems for the purposes of producing hay or green manuring is one of the most practical methods of enriching soil with N and organic matter, loosening the soil and improving soil structure (Ozyazıcı and Manga 2000; Dabney et al. 2001). However, it can take more than four years to fully their affect some indicators of soil quality in cropping systems (Jokela et al. 2009).

If AFLs are used as green manure, then attention should be paid to their plant growth and rooting, the subsoil and decomposition process, the C:N ratio, nutrients left in the soil and the amount and quality of soil organic matter (Fageria et al. 2005; Reddy 2016). Due to their fast decomposition and release of N after termination (Gatiboni et al. 2011; Sieversa and Cook 2018), vetches and forage peas are the most commonly planted forage legumes for green manuring in Turkey. Ozyazıcı and Manga (2000) reported higher yields of corn and sunflower after common and narbon vetches had been incorporated into the soil than when they had been mowed; with common and narbon vetches, yields increased by 51.7% and 50.0% for corn and 36.8% and 36.4% for sunflower, respectively. Kuo and Jellum (2002) have also shown that though removing the top growth of hairy vetch limited the carbon inflow and reduced organic carbon in the soil, it also greatly reduced nitrate leaching (Gabriel et al. 2013). Ketterings et al. (2015) indicated that cover crop termination with herbicides compared to tillage decelerated degradation and the release of N. Liebmana et al. (2018) determined that the highest available N was obtained at 6–10 weeks after termination in the growth of hairy vetch and forage peas. They also found that disked hairy vetch provided the highest N level of 16 termination combinations.

In winter, rotating AFLs with silage or grain maize has shown great growth potential in irrigated areas in Turkey's coastal and central regions. Due to increased costs of N, farmers have experimented with new approaches to decreasing agriculture input in cropping systems. N is the most limiting nutrient in the yield potential of cereal crops such as corn (Pontes et al. 2018) for which the recommended amount of N ranges from 1.0 to 1.5 kg of N per 56 kg of product (Marsh 2014). When AFLs are the preceding crops, primary crops receive both a significant amount of biologically fixed N_2 and organic matter that decomposes easily due to the low C:N ratio (Fageria et al. 2005). In a meta-analytic study, corn yields were shown to have increased by 37% when legumes were used as cover crops in winter, whereas applying N fertiliser reduced that rate of growth (Miguez and Bollero 2005). Parr et al. (2011) indicated that the highest corn yield was obtained from hairy vetch and hairy vetch–rye biculture. The effects of hairy vetch on corn yields were confirmed by Liebmana et al. (2018), who found similar corn yields following the cultivation of forage peas such as hairy vetch. Most recently, Kalkan and Avcı (2020) determined that narbon vetch, when used as a preceding crop in winter, provided a yield increase of 30% relative to the control plot, and the highest biomass yield (13.2 t da^{-1}) in silage maize was obtained after planting narbon vetch compared to after planting forage peas and Hungarian vetch.

The efficient use of nutrients in the soil depends on the availability of water (Gheysari et al. 2009) Many crops, particularly maize, respond positively to increases in the amount of water and N applied until production is optimal (Fatih et al. 2008). The cultivation of AFLs in humid soils with high water-holding capacity and dry soils with low water-holding capacity does not prevent the growth of subsequent crops (Badaruddin and Meyer 1989; Pedrosa de azevedo et al. 1999). On the contrary, they increase surface water infiltration by providing soil organic matter and generate a mulch layer with green manuring, which prevents evaporation and, in turn, allows the more efficient use of water on the soil surface (Unger and Vigil 1998; Aldridge et al. 2019). In a study conducted in light-textured sandy soils, Skoufogianni et al. (2013) observed that cultivating peas and incorporating them into the soil improved soil fertility and the yields of subsequent crops (i.e. corn and sunflowers). That same year, Hubbard et al. (2013) determined that high residues from fall and winter legumes can add C to the soil, preserve available N in soil organic matter, improve the efficiency of using fertilisers and enrich the physical properties of highly sandy soils.

Weeds compete with crops for water, nutrients and light and can cause heavy losses in yields (Lukangila et al. 2016). A wide range of ecological processes and agricultural practices can be utilized to control weeds in sustainable ways (Liebmana et al. 2003). Including AFLs in crop rotations positively affect weed control in alternative systems, because AFLs suppress weeds both by competition and by allelopathic effects when used to fill gaps during short periods alone or in crop mixtures. In Mediterranean farming systems, hairy vetch is known to affect weeds allelopathically by accumulating cyanamide (Kamo et al. 2003). Hairy vetch and peas can also be grown as part of intercropping and for mulching for weed management purposes. Mohammadi (2010) reported that corn–hairy vetch intercropping reduced the dry weight of weeds by 50.9% and increased corn yields by 11%, depending on the planting rate of hairy vetch. Moreover, Cicek et al. (2014) recommended planting pea–oat mixtures and hairy vetch to enhance levels of dry matter production and weed competition. Similarly, Annicchiarico et al. (2017) and Mesbahab et al. (2019) found that weed content was lower than pure stands in mixtures of cereals (i.e. oats and triticale) and legumes (i.e. field peas, hairy vetch, narbon vetch and common vetch).

Although few researchers have furnished evidence of the positive effects of AFLs against diseases and pests, the effects of using the mulch of forage legumes such as *T. pratense*, *T. fragiferum* and *Desmodium* sp. on oviparity have been determined by Khan et al. (2000), Chabi-Olaye et al. (2005) and Bjorkman et al. (2007). Nevertheless, it remains unclear whether increased yields observed in those studies were due to pest suppression or the presence of legumes. In the context of organic farming, Jackson and Harrison (2008) found noteworthy results by using an oat–crimson clover mixture to ward off pests and diseases. In particular, they found that sweet potatoes grew successfully under oat–crimson clover mulch, because soil insect pests in the sweet potato roots were not significantly higher in that mulch than in conventionally tilled soil. In contrast to those results, however, Bashyala et al. (2019) reported that a rye–hairy vetch mixture did not affect pest suppression (i.e. corn earworm damage) or fungal ear rots. Last, in Prasifka et al.'s (2006) study, although the sole effect of living mulches did not provide sufficient pest suppression, their potential to improve biological control should be considered together with their other agronomic benefits.

4.4 Role of AFLs in N_2 Fixation

N is the most essential nutrient for the growth, development and increased yield potential of crops, especially cereals (N'Dayegamiye et al. 2015). The N uptake efficiency of crops is roughly 50% on average (Bergstrom and Kirchmann 2004), and the excess use of N, upon leaching into groundwater, threatens the environment and human health (Scharf et al. 2006). N is provided to crops in an organic, inorganic or organic-inorganic form to maximize production (Fleming et al. 1981). Currently, the most widely used source of N for plant nutrition is synthetic N fertiliser produced by the Haber–Bosch process with natural gas (Boddey et al. 2009). However, due to increased costs and environmental concerns, a highly efficient and sustainable alternative to that process is required to minimize the use of synthetic N.

The fixed N_2 contributions of AFLs by including them in cropping systems can play an important role in sustainably maintaining high yields and coping with increased costs of N fertiliser (Peoples et al. 1995; Fageria et al. 2005; Parr et al. 2011; Liebmana et al. 2018). As the gaseous form of N, N_2 diffuses into the soil from the atmosphere as a result of legume–rhizobia symbiosis and is converted into ammonia (NH_3) in the presence of nitrogenase (Sylvia et al. 2005). Symbiotic N_2 fixation is affected by legume species, bacterial activity and ecological factors such as soil moisture, soil acidity, salinity, soil N status and the presence of other available nutrients (Zahran 1999).

Depending on the species, AFLs with potential importance in Turkey can fix N_2 to the soil at various rates. As shown in Table 4.2, vetch, grass peas and forage peas fix N_2 more than clover and fenugreek. However, Briggs et al. (2005) have indicated that the amount of remaining N after harvest was greater following vetch than forage peas. Liebmana et al. (2018) confirmed those findings by observing that the highest biomass N was obtained by hairy vetch (226 kg ha^{-1}), followed by forage peas (189 kg ha^{-1}) and red clover (181 kg ha^{-1}). Table 4.2 also indicates that N provided to the soil can be higher if green manuring is performed with those species.

The presence of active and host-specific *Rhizobium* bacteria in the root zone is necessary for the efficient use of forage legumes in agricultural production. Host specificity between legumes and *Rhizobium* species is evident in the early stages of infection and mediates plant–bacterial signal change (Sharma et al. 1993; Wippel and Long 2019). Because Turkey is located in a diverse region where many forage legumes originated (TUBIVES 2019), host-specific *Rhizobium* strains belonging to cultivated AFLs are naturally distributed in the soil of Turkey (Table 4.3). However, new cultivated species introduced into cropping systems require bacterial inoculation for effective nodulation and symbiosis due to genetically incompatible legume–*Rhizobium* sp. pairs (Ozdemir et al. 1999).

Water stress adversely affects the nodule formation, growth, development and functions (González et al. 2015). In *Rhizobium*–legume symbiosis, because N_2

Table 4.2 The amount of N supplied to soil by AFL used in Turkey

Number	Annual legume species	[a]Nitrogen derived from atmospheric N_2, (kg ha^{-1} year^{-1})	Total N with green manuring (kg ha^{-1} year^{-1})	Sources
1	*V. pannonica*	74–78	–	
2	*V. sativa*	107–131	125 ± 13[b]	
3	*V. narbonensis*	–	100–200	Ozyazıcı and Manga (2000)
4	*V. villosa*	100–159	100–200; 143 ± 13[b]	Sato et al. (2011)
5	*V. ervilia*	–	–	
6	*L. sativus*	101–149	–	
7	*P. sativum* ssp. *arvense*	102–115	130 ± 13[b]	
8	*T. foenum-graecum*	7–62	–	
9	*T. alexandrinum*	35–59	151 ± 13[b]	
10	*T. resupinatum*	46–65	–	

[a]From Büchi et al. (2015)
[b]From Parr et al. (2011). Data, belong to only Piedmont in 2010. ± indicated standard error

Table 4.3 Effective *Rhizobium* strains for AFL used in Turkey

Number	Annual legume species	*Rhizobium* strains	Sources
1	*V. pannonica*	*R. leguminosarum* bv. *viciae*	Kumar et al. (2014)
2	*V. sativa*	*R. leguminosarum* bv. *viciae*	Mothapo et al. (2013), Guinet et al. (2018)
3	*V. narbonensis*	*R. leguminosarum* bv. *viciae*	Guinet et al. (2018)
4	*V. villosa*	*R. leguminosarum* bv. *viciae*	Yuan et al. (2016), Mothapo et al. (2013)
5	*V. ervilia*	*R. leguminosarum* bv. *viciae*	Siddique and Sykes (1997)
6	*L. sativus*	*R. leguminosarum* bv. *viciae*	Han et al. (2009)
7	*P. sativum* ssp. *arvense*	*R. leguminosarum* bv. *viciae*	Mothapo et al. (2013), Guinet et al. (2018)
8	*T. foenum-graecum*	*R. leguminosarum* bv. *trifolii*	Melino et al. (2012), Guinet et al. (2018)
9	*T. alexandrinum*	*R. leguminosarum* bv. *trifolii*	Mazur et al. (2013), Kumar et al. (2014)
10	*T. resupinatum*	*R. leguminosarum* bv. *trifolii*	Mazur et al. (2013), Kumar et al. (2014)

fixation is strongly influenced by the physiological state of the host plant, the *Rhizobium* strain is not expected to be expressed at maximum capacity under the effect of abiotic stress factors (Zahran 1999; Yuan et al. 2016). The stress-based effects on nodulation can be reduced by selecting appropriate genotypes for drought conditions, given the presence of genotypic variation in water stress among some AFLs (Haffani et al. 2013). In areas where annual precipitation is less than 300 mm, *L. sativus* and *V. narbonensis* are better adapted than *V. sativa*, *V. pannonica*, *V. ervilia*, *V. villosa* and *P. arvense* species, the latter group of which is better grown in areas with annual precipitation exceeding 300 mm. (Ates et al. 2014). Meanwhile, *T. alexandrinum* and *T. resupinatum* species are important annual forage legumes in semi-arid regions (Iannucci et al. 2002). Investigating the effect of arbuscular mycorrhizal symbiosis (AMS) on N_2 fixation under conditions of water stress in *T. alexandrinum*, Saia et al. (2014) determined that AMS increased N_2 fixation.

Worldwide, cultivated land amounts to roughly 1.5 billion hectares, 77 million hectares (5%) of which are unfavorable for plant growth due to high salinity (Selvakumar et al. 2014; Miransari 2017). Meanwhile, more than 20% of irrigated areas are adversely affected by salinity as well (Selvakumar et al. 2014). That latter rate is approximately 4% in Turkey, and salinity has increased in irrigated lands due to excessive and improper irrigation practices (Ekmekci et al. 2005). Both rhizobial strains and their hosts must be resistant to salinity in the cultivation of AFLs in salt-affected soils (Zahran 1999; Thrall et al. 2008). It is generally known that *Rhizobium* spp. tolerate salinity more than their hosts; however, their nodulation period (i.e.

colonization and infection) is particularly sensitive to salt stress (Plá and Cobos-Porras 2015). Among rhizobial strains obtained from wild legumes, *Rhizobium* and *Sinorhizobium* spp. are the most tolerant strains under conditions of salt stress (Brígido et al. 2012). Mandal (2014) observed that five *Rhizobium* strains belonging to berseem clover grown in different locations were resistant to 3% NaCl concentration. Ventorino et al. (2012) reported that common vetch increased N availability more than faba beans in soils with salinity. At the same time, Agarwal et al. (2010) identified genotypes resistant to saline soils in *Trifolium alexandrinum*, while others have found similar genotypes in *Trifolium fragiferum* (Rumbaugh et al. 1993) and *Melilotus* spp. (Rogers et al. 2008).

The soil acidity (pH <5.5) of arable land worldwide is about 30.0% on average (Lin et al. 2012), whereas that rate is approximately 9.5% in Turkey (Eyüpoğlu 1999). To support the tolerance of soil acidity, attention should be paid to the breeding and genetic modification of both host and bacterial strains, which might counteract major limitations in the N_2 fixation of legumes due to Al toxicity and P deficiency (Bordeleau and Prevost 1994; Valentine et al. 2011). Hairy vetch and grass pea are highly efficient in low-yield or acidic soils compared to other AFLs (IPGRI 2000; Sheaffer and Seguin 2008). Soil acidity and Al toxicity adversely affect the nodule development and function of clover species [(*Trifolium* spp.), (Wood et al. 1984)].

Leguminous crops usually require P, K, S and Ca^{2+} nutrition for biological N_2 fixation (Biswas et al. 2003; Sulieman et al. 2013; Divitoa and Sadras 2014). However, the presence of available mineral N in the soil can negatively affect N_2 fixation at various rates according to legume species and amounts of mineral N (Tamagno et al. 2018). Indeed, increasing NO_3^- inhibited the number of nodules and N_2 fixation in *P. sativum* and *L. sativus* in the studies of Tang et al. (1999) and Voisin et al. (2003). By contrast, Akbar et al. (2013) determined that low N and cobalt supply is initially required to increase nodulation and yields in the cultivation of peas. Phosphorus and potassium fertilisers also play important roles in root and nodule development in species of vetch (Karadag and Buyukburc 2001; Qing-yu et al. 2014), while sulfur affects N_2 fixation in *P. sativum* (Zhao et al. 1999). The effect of molybdenum on nodulation in hairy and common vetches was investigated by Alam et al. (2015) and Zolotarev (2016), who concluded that the number and weight of nodules increased with its application.

4.5 Effects of AFLs on Soil Management and Erosion

In agricultural production, intensive soil tillage makes organic matter and carbon decrease rapidly in soil (Lal 2001), which affects the physical, chemical and biological properties of soil (Melero et al. 2006). The loss of soil organic matter and carbon prompts the rapid deterioration of soil structure and decreases soil infiltration and water retention capacity (Dexter 2004). At the same time, it adversely affects soil nutrient cycling and the capacity of soil–cation exchange (Brady and

Weil 2005). Most soil-based organisms are heterotrophic, and their activity depends on the presence of organic carbon in the soil (Thies and Grossman 2006). Generally, increasing organic matter and carbon content in soil facilitates microbial activity (Shah et al. 2010).

Amounts of added soil organic matter is extremely low in both dry (e.g. cereal–fallow) and irrigated farming systems in central Turkey. The highest soil organic matter and carbon sequestration occurs when the highest amount of plant residue remains in the soil (Paustian 2000). Leguminous crops grown in rotation produce exceptionally high volumes of such residue, which protect soil organic carbon from mineralization by increasing the formation and stabilization of soil aggregates (Kapur et al. 2007).

Although perennial cropping systems sequester more soil carbon than annual ones (Müller-Stöver et al. 2012), AFLs are easily adapted to various annual cropping systems and produce a high amount of residue with green manuring (Fageria 2007). Masri and Ryan (2006) have indicated that crop rotation with legumes (i.e. *Medicago sativa* and *Vicia faba*) increased soil organic matter and the rate of soil infiltration compared to rotations with continuous wheat and wheat–fallow in fragile, semi-arid areas in the Mediterranean zone. Similarly, Armstrong et al. (2018) observed that soil organic carbon was sequestered more when rotations included legumes than when fields were left fallow. In their study on the effects of winter cover crop–maize rotations and fertiliser on soil organic carbon, Dube et al. (2012) found that although vetch sequestered soil organic carbon better than oats in low-input systems not using fertiliser, better sequestration was achieved at 0–20 cm when using oats with fertiliser. The effects of cover crops (i.e. barley and vetch) on organic C and N sequestration, aggregate stability, water infiltration and nitrate leaching in irrigated semi-arid areas have been reported by García-González et al. (2018). In particular, they revealed that replacing winter fallowing with barley cultivation improved soil characteristics better than with vetch cultivation, which exerted only a limited effect relative to fallowing. Although Mazzoncini et al. (2011) observed in their long-term study that forage legumes with high N fixation increased soil organic carbon more than low and non-legume crops did, in some studies legume crops in rotation adversely affected soil carbon or reduced its concentration (Sainju et al. 2002; Lange et al. 2015). At the same time, other findings have included that legume–grass mixtures do not significantly increase soil carbon stock compared to leguminous monoculture (Crème et al. 2016 and Li et al. 2019).

In soil, microbial activity is a potential indicator of soil quality, for it can mineralize organic nutrients that play an important role in the growth and development of plants (Schloter et al. 2018). Many researchers have shown that AFLs positively affect soil microorganisms. Among them, Tao et al. (2017) investigated the impact of four green manure crops (i.e. *Vicia villosa*, *Vicia sativa*, milk vetch and radish) on maize growth in variations of soil bacterial communities and soil properties. As a result, they revealed that milk vetch increased maize yield by 31.3% by positively influencing soil microbes and soil properties. Also that year, Finney et al. (2017) determined that hairy vetch and oat treatments had the highest bacteria:fungi ratio in spring. Using similar crops, Nevins et al. (2018) examined the effects of oats, hairy vetch and their biculture on soil biota and their relationship with the decomposition of their residue throughout the subsequent crop growing season. Their results indicated that soil biota included more taxa with larger effect sizes after termination at 21 days in the hairy vetch treatment, at 39 days in the oat treatment and at 53 and 75 days in the oat–hairy vetch mixture. Herencia (2018) confirmed that result by showing that the best chemical and microbiological properties of soil were obtained when vetch was used as a cover crop and its debris left on the soil surface.

Covering the soil surface with monocrops or intercrops prevents erosion and the loss of nutrients compared to leaving soil bare (Ali et al. 2007). Cover crops help to prevent soil erosion by decelerating the effects of water, improving soil's ability to absorb water and stabilizing soil particles around root systems (McCartney and Fraser 2010). In addition, mulches derived from cover crops with green manuring reduce organic carbon loss and increase the carbon storage capacity of soil (Poeplau and Don 2015). Including AFLs in cropping systems for the purposes of green manuring or hay production reduces soil erosion by increasing soil organic carbon, which improves the physical and chemical properties of soil, and by providing soil cover (McCartney and Fraser 2010; Blanco-Canqui et al. 2013).

4.6 Economics

For profitable agricultural production, AFLs should be included in crop rotations as forage or green manure crops, for they increase soil organic matter and N content as well as provide erosion control and weed management (McCartney and Fraser 2010). As the largest input for plant nutrition, N has risen in price in recent years due to the increased use of N fertiliser in intensive cropping system and the related costs (Boddey et al. 2009, WBB 2019). The N contribution of forage legumes to subsequent crops or intercropping can be invaluable to maintaining high yields and coping with the rising costs of N fertiliser (Frankenberger and Abdelmajid 1985; Meena and Lal 2018).

Duşa and Roman (2009) has reported that the rotation of forage and grain legumes provided 30 and 90 kg N ha^{-1} fertilisers compared to rotations of non-leguminous crops. N'Dayegamiye et al. (2015) confirmed that result by showing that legumes left N fertiliser to other plants to use, in amount of 51–77 kg ha^{-1} for corn and 5–37 kg ha^{-1} for wheat, compared with the recommended amount of fertiliser as 175 and 90 kg N ha^{-1} for corn and wheat, respectively. Most recently, Iqbal et al. (2019) indicated that the intercropping of sorghum–legumes provided 9–55% higher productivity than cultivating sorghum alone. According to those results, the costs of N fertiliser can be reduced for subsequent crops or intercropping by approximately a third depending on legume crops. In addition to the N that it contributes, including legumes in rotations with cereals in Europe can save US$22–67 per ha and more than US$56 ha^{-1} by reducing tillage before planting winter cereals and by using herbicides and fungicides, respectively (Luetke-Entrup et al. 2003; Reckling et al. 2014; Preissel et al. 2017). The gross margins of rotations including forage legumes are higher (US$5–115 ha^{-1} $year^{-1}$) than rotations with non-legume crops, and forage legumes also tend to generate equal or higher gross margins than grain legumes (Preissel et al. 2017).

Although the N contribution of AFLs to subsequent crops by mowing is less than that by incorporating crops into the soil (Kessel and Hartley 2000), AFLs are usually mowed due to the need for quality roughage and economic return in Turkey, where the production of dry matter ranges from 40–80 kg ha^{-1} and 20–25 kg ha^{-1} in irrigated and dry areas, respectively (Ekiz et al. (2009). While the produced dry matter is not profitable in drylands due to high input, it provides an average additional gross margin of US $300 ha^{-1} in irrigated lands during the bare period in Turkey. Preissel et al. (2017) have indicated that winter wheat grown following forage legumes in Europe generated an additional gross margin of US$130–337 ha^{-1}. Last, according to Scott Christiansen et al. (2015), including vetch in crop rotations can yield a profit of US$25.4 da^{-1} compared to practising continuous wheat production, as is common in dry areas of West Asia and North Africa.

4.7 Conclusion

At present, forage legumes are important to include in crop rotations in order to overcome increasing costs of agriculture and environmental concerns. Because the intensive use of chemicals and fertilisers, which adversely affect environmental resources, is expected to hinder profitable agricultural production, the effects of AFLs on sustainable agricultural productivity and environmental resources were evaluated. AFLs can easily adapt to different rotation systems as crops for forage or green manuring and do not negatively influence the growing season of subsequent crops. AFLs include important green manure crops whose incorporation into the soil provides organic matter and organic carbon recovery, both of which improve the physical and chemical properties of soil. In particular, increased organic matter and carbon positively affect soil's water-holding capacity, nutrient availability and erosion control. Cultivating AFLs also benefits subsequent crops by fixing N from the atmosphere at various rates depending on genotype. In addition to their positive effects on soil fertility, AFLs provide more profitable production as well by reducing the need for weed control and tillage. Although the effect of mowing AFLs on soil health and fertility is lower than incorporating crops into the soil, AFLs are generally mowed for quality forage in Turkey. Altogether, AFLs can better sustain efficient agricultural production than non-leguminous crops and should be included in cropping systems in Turkey.

References

Agarwal S, Singh B, Kumar A, Kumar P (2010) Salinity tolerance of Berseem (*Trifolium alexandrinum* L.) during germination and early seedling growth. Vegetos 23(1):63–82

Akbar FM, Zafar M, Hamid A, Ahmed M, Khaliq A, Khan MR, Rehman Z (2013) Interactive effect of cobalt and nitrogen on growth, nodulation, yield and protein content of field grown pea. Hort Environ Biotechnol 54(6):465–474. https://doi.org/10.1007/s13580-013-0001-6

Alam F, Kim TY, Kim SY, Alam SS, Pramanik P, Kim PJ, Lee YB (2015) Effect of molybdenum on nodulation, plant yield and nitrogen uptake in hairy vetch (*Vicia villosa* Roth). Soil Sci Plant Nutr 61:664–675. https://doi.org/10.1080/00380768.2015.1030690

Aldridge CA, Baker BH, Omer AR (2019) Investigation of short-term effects of winter cover crops on compaction and total soil carbon in a long-term no-till agricultural system. J Soil Water Conserv 74(1):77–84. https://doi.org/10.2489/jswc.74.1.77

Ali I, Khan F, Bhatti AU (2007) Soil and nutrient losses by water erosion under mono-cropping and legume inter-cropping on sloping land. Pak J Agric Sci 20(3–4):161–166

Annicchiarico P, Alami IT, Abbas K, Pecetti L, Melis RAM, Porqueddu C (2017) Performance of legume-based annual forage crops in three semi-arid Mediterranean environments. Crop Pasture Sci 68(11):932–941. https://doi.org/10.1071/CP17068

Anonymous (2017) Total U.S. cropland area 2012–2018. Available at https://www.statista.com/statistics/201762/projection-for-total-us-cropland-area-from-2010/. Accessed in 13 July 2019

Anonymous (2018a) Turkish Statistical Institute. Available at http://www.tuik.gov.tr/PreTablo.do?alt_id=1001. Accessed 13 July 2019

Anonymous (2018b) U.S. forage statistics. Available at https://www.progressivepublish.com/downloads/2019/general/2018-pf-stats-highres.pdf. Accessed 13 July 2019

Armstrong RD, Perris R, Munn M, Dunsford K, Robertson F, Hollaway GJ, O'Leary GJ (2018) Effects of long-term rotation and tillage practice on grain yield and protein of wheat and soil fertility on a Vertosol in a medium-rainfall temperate environment. Crop Pasture Sci 70(1):1–15. https://doi.org/10.1071/CP17437

Ates S, Feindel D, El Moneim A, Ryan J (2014) Annual forage legumes in dryland agricultural systems of the West Asia and North Africa regions: research achievements and future perspective. Grass Forage Sci 69:17–31. https://doi.org/10.1111/gfs.12074

Azevedo DMP, Landivar J, Vieira RM, Moseley D (1999) The effect of cover crop and crop rotation on soil water storage and on Sorghum yield. Pesqui Agropecu Bras 34(3):391–398. https://doi.org/10.1590/S0100-204X1999000300010

Badaruddin M, Meyer DW (1989) Water use by legumes and its effect on soil water status. Crop Sci 29(5):1212–1216. https://doi.org/10.2135/cropsci1989.0011183X002900050025x

Bashyal M, Ferguson JC, Perez-Hernandez O, Hoilett N (2019) Effect of cereal rye and hairy vetch on pest suppression and corn yield. Commun Soil Sci Plant Anal 50(9):1093–1105. https://doi.org/10.1080/00103624.2019.1603309

Bergström L, Kirtchmann H (2004) Leaching and crop uptake of nitrogen from nitrogen-15 labeled green manures and ammonium nitrate. J Environ Qual 33:1786–1792. https://doi.org/10.2134/jeq2004.1786

Biswas P, Hosain D, Ullah M, Akter N, Bhuiya MAA (2003) Performance of groundnut (Arachis hypogaea L.) under different levels of bradyrhizobial inoculums and nitrogen fertilizer. SAARC J Agr 1:61-68. Available at https://www.cabdirect.org/cabdirect/FullTextPDF/2010/20103304731.pdf. Accessed 06 Feb 2021

Bjorkman M, Hamback PA, Ramert B (2007) Neighbouring monocultures enhance the effect of intercropping on the turnip root fly (Delia floralis). Entomol Exp Appl 124(3):319–326. https://doi.org/10.1111/j.1570-7458.2007.00589.x

Blanco-Canqui H, Shapiro CA, Wortmann CS, Drijber RA, Mamo M, Shaver TM, Ferguson RB (2013) Soil organic carbon: the value to soil properties. J Soil Water Conserv 68(5):129–134. https://doi.org/10.2489/jswc.68.5.129A

Boddey RM, Alves B Jr, Henrique B, Soares L, Jantalia CP, Urqulaga S (2009) Biological nitrogen fixation and the mitigation of greenhouse gas emissions. In: Emerich DW, Krishnan HB (eds) Nitrogen fixation in crop production. Agronomy Monographs 52:387–314. https://doi.org/10.2134/agronmonogr52.c14

Bordeleau LM, Prevost D (1994) Nodulation and nitrogen fixation in extreme environments. Plant and Soil 161:115–124. https://doi.org/10.1007/BF02183092

Brady NC, Weil RR (2005) The nature and properties of soils, 13th edn. Macmillan Publishing Company, New York, pp 279–313

Briggs SR, Cuttle S, Goodlass G, Hatch D, King J, Roderick S, Shepherd M (2005) Soil nitrogen building crops in organic farming – Defra research project OF0316 project publication. Available at http://www.efrc.com/manage/authincludes/article_uploads/iota/technical-leaflets/soil-nitrogen-building-crops-in-organic-farming.pdf. Accessed 04 Aug 2019

Brígido C, Alexandre A, Oliveira S (2012) Transcriptional analysis of major chaperone genes in salt-tolerant and salt-sensitive mesorhizobia. Microbiol Res 167:623–629. https://doi.org/10.1016/j.micres.2012.01.006

Büchi L, Gebhard CA, Liebisch F, Sinaj S, Ramseier H, Charles R (2015) Accumulation of biologically fixed nitrogen by legumes cultivated as cover crops in Switzerland. Plant and Soil 393:163–175. https://doi.org/10.1007/s11104-015-2476-7

Cakmakcı S, Cecen S (1999) The possibilities at entering crop rotation system of certain annual legume plants in Antalya. Turk J Agric For 23:119–123. Available at http://journals.tubitak.gov.tr/agriculture/issues/tar-99-23-1/tar-23-1-14-96195.pdf. Accessed 04 August 2019. (In Turkish with English Summary)

Chabi-Olaye A, Nolte C, Schulthess F, Borgemeister C (2005) Abundance, dispersion and parasitism of the stem borer *Busseola fusca* (Lepidoptera: Noctuidae) in maize in the humid forest zone of southern Cameroon. Bull Entomol Res 95(2):169–177. https://doi.org/10.1079/BER2004347

Christiansen S, Ryan J, Singh M, Ates S, Bahhady F, Mohamed K, Youssef O, Loss S (2015) Potential legume alternatives to fallow and wheat monoculture for Mediterranean environments. Crop Pasture Sci 66(2):113–121. https://doi.org/10.1071/CP14063

Cicek H, Martens JRT, Bamford KC, Entz MH (2014) Forage potential of six leguminous green manures and effect of grazing on following grain crops. Renew Agr Food Syst 30(6):503–514. https://doi.org/10.1017/S1742170514000349

Crème A, Rumpel C, Gastal F, Gil MD, Chabbi A (2016) Effects of grasses and a legume grown in monoculture or mixture on soil organic matter and phosphorus forms. Plant and Soil 402:117–128. https://doi.org/10.1007/s11104-015-2740-x

Dabney SM, Delgado JA, Reeves DW (2001) Using winter cover crops to improve soil and water quality. Commun Soil Sci Plant Anal 32(7):1221–1250. https://doi.org/10.1081/CSS-100104110

Dereli DN (2015) Cultivation possibility of some annual legumes as a second crop under Eskisehir ecological conditions. Master thesis from Deparment of Field Crops, Eskisehir Osmangazi University, Eskişehir, Turkey. Available at https://tez.yok.gov.tr/UlusalTezMerkezi/tezSorguSonucYeni.jsp. Accessed 12 July 2020. (In Turkish with English Summary)

Dexter AR (2004) Soil physical quality. Part I. Theory, effects of soil texture, density and organic matter, and effects on root growth. Geoderma 120:201–214. https://doi.org/10.1016/j.geoderma.2003.09.004

Divitoa GA, Sadras VO (2014) How do phosphorus, potassium and sulphur affect plant growth and biological nitrogen fixation in crop and pasture legumes? A meta-analysis. Field Crop Res 156:161–171. https://doi.org/10.1016/j.fcr.2013.11.004

Dönmez Ş, Koçak N, Çakır M (2018) A new design approach for Süleyman Demirel University Botanical Garden. Turk J For 20(1):62–71. https://doi.org/10.18182/tjf.474236

Dube E, Chiduza C, Muchaonyerwa P (2012) Conservation agriculture effects on soil organic matter on a Haplic Cambisol after four years of maize–oat and maize–grazing vetch rotations in South Africa. Soil Tillage Res 123:21–28. https://doi.org/10.1016/j.still.2012.02.008

Duşa EM, Roman GV (2009) Researches regarding the productivity of oat-lentil intercropping in the organic agriculture system. Res J Agric Sci 41(1):22–26

Ekiz H, Kendir H, Şahin Demirbağ N (2009) Forage crop cultivation in fallow land. In: Avcıoğlu R, Hatipoğlu R, Karadağ Y (eds) Forage crops (General section) Volume 1, Republic Of Turkey Ministry Of Agriculture And Forestry, pp 113–120. (In Turkish, electronic version is not available)

Ekmekci E, Apan M, Kara T (2005) The effect of salinity on plant growtht. J Fac Agric OMU 20(3):118–125. Available at https://dergipark.org.tr/download/article-file/187589. Accessed 12 July 2020. (In Turkish with English Summary)

Eyüpoğlu F (1999) The soil fertility status of Turkey. Republic of Turkey Prime Ministry General Directorate of Rural Services, Soil and Fertiliser Research Institute publications. Publication No: 220, Technical Publication No: T–67 pp 122. (In Turkish, electronic version is not available)

Fageria NK (2007) Green manuring in crop production. J Plant Nutr 30(5):691–719. https://doi.org/10.1080/01904160701289529

Fageria NK, Baligar VC, Bailey BA (2005) Role of cover crops in improving soil and row crop productivity. Commun Soil Sci Plant Anal 36(19–20):2733–2757. https://doi.org/10.1080/00103620500303939

Fatih MK, Ustun S, Yasemin K, Talip T (2008) Determining water-yield relationship, water use efficiency, crop and pan coefficients for silage maize in a semi-arid region. Irrig Sci 16:112–123. https://doi.org/10.1007/s00271-008-0127-y

Finney DM, Buyer JS, Kaye JP (2017) Living cover crops have immediate impacts on soil microbial community structure and function. J Soil Water Conserv 72(4):361–373. https://doi.org/10.2489/jswc.72.4.361

Fleming AA, Giddens JE, Beaty ER (1981) Corn yields as related to legumes and inorganic nitrogen. Crop Sci 21:977–980. https://doi.org/10.2135/cropsci1981.0011183X002100060041x

Follett RF (2008) Transformation and transport processes of nitrogen in agricultural systems. In: Hatfield JL, Follett RF (eds) Nitrogen in the environment sources, problems, and management. Academic Press/Elsevier, Amsterdam, pp 19–50. https://doi.org/10.1016/B978-0-12-374347-3.00002-0

Frankenberger WT, Abdelmajid HM (1985) Kinetic parameters of nitrogen mineralization rate of leguminous crops incorporated into soil. Plant and Soil 87:257–271. https://doi.org/10.1007/BF02181865

Gabriel JL, Garrido A, Quemada M (2013) Cover crops effect on farm benefits and nitrate leaching: linking economic and environmental analysis. Agr Syst 121:23–32. https://doi.org/10.1016/j.agsy.2013.06.004

García-González I, Hontoria C, Gabriel JL, Alonso-Ayuso M, Quemada M (2018) Cover crops to mitigate soil degradation and enhance soil functionality in irrigated land. Geoderma 322:81–88. https://doi.org/10.1016/j.geoderma.2018.02.024

Gatiboni LC, Coimbra JLM, Denardin RBN, Wildner LP (2011) Microbial biomass and soil fauna during the decomposition of cover crops in no-tillage system. Rev Bras Ciênc Solo 35:1151–1157. https://doi.org/10.1590/S0100-06832011000400008

Gheysari M, Mirlatifi SM, Bannayan M, Homaee M, Hoogenboom G (2009) Interaction of water and nitrogen on maize grown for silage. Agric Water Manag 96:809–821. https://doi.org/10.1016/j.agwat.2008.11.003

González EM, Larrainzar E, Marino D, Wienkoop S, Gil-Quintana E, Arrese-Igor C (2015) Physiological responses of N_2-fixing legumes to water limitation. In: Sulieman S, Tran LS (eds) Legume nitrogen fixation in a changing environment-achievements and challenges. Springer, New York, pp 5–33. https://doi.org/10.1007/978-3-319-06212-9_2

Guinet M, Nicolardot B, Revellin C, Durey V, Carlsson G, Voisin AS (2018) Comparative effect of inorganic N on plant growth and N_2 fixation of ten legume crops: towards a better understanding of the differential response among species. Plant and Soil 432:207–227. https://doi.org/10.1007/s11104-018-3788-1

Haffani S, Mezni M, Slama I, Ksontini M, Chaibi W (2013) Plant growth, water relations and proline content of three vetch species under water-limited conditions. Grass and Forage Sci 69:323–333. https://doi.org/10.1111/gfs.12034

Han TX, Tian CF, Wang ET, Chen WX (2009) Associations among rhizobial chromosomal background, nodgenes, and host plants based on the analysis of symbiosis of indigenous rhizobia and wild legumes native to Xinjiang. Microb Ecol 59(2):311–323. https://doi.org/10.1007/s00248-009-9577-x

Herencia JF (2018) Soil quality indicators in response to long-term cover crop management in a Mediterranean organic olive system. Biol Agric Hortic 34(4):211–231. https://doi.org/10.1080/01448765.2017.1412836

Hubbard RK, Strickland TC, Phatak S (2013) Effects of cover crop systems on soil physical properties and carbon/nitrogen relationships in the coastal plain of southeastern USA. Soil Tillage Res 126:276–283. https://doi.org/10.1016/j.still.2012.07.009

Ileri O, Avci S, Koc A (2016) Seed and biological yields of narbon vetch genotypes under Central Anatolia condition, Turkey. J Int Sci Publ: Ecol Saf 10:430–435. Available at https://www.scientific-publications.net/get/1000017/1466405637361712.pdf. Accessed 06 Aug 2019

Iannucci A, Russo M, Arena L, Di Fonzo N, Martiniello P (2002) Water deficit effects on osmotic adjustment and solute accumulation in leaves of annual clovers. Eur J Agron 16:111–122. https://doi.org/10.1016/S1161-0301(01)00121-6

IPGRI (2000) Descriptors for *Lathyrus* spp. international plant genetic resources institute, Rome, Italy. Available at https://www.bioversityinternational.org/e-library/publications/detail/descriptors-for-lathyrus-spp/. Accessed 12 July 2020

Iqbal MA, Hamid A, Ahmad T, Siddiqui MH, Hussain I, Ali S, Ali A, Ahmad Z (2019) Forage sorghum-legumes intercropping: effect on growth, yields, nutritional quality and economic returns. Bragantia 78(1):82–95. https://doi.org/10.1590/1678-4499.2017363

Jackson DM, Harrison HF (2008) Effects of a killed-cover crop mulching system on sweet potato production, soil pests, and insect predators in South Carolina. J Econ Entomol 101(6):1871–1880. https://doi.org/10.1603/0022-0493-101.6.1871

Jokela WE, Grabberb JH, Karlenc DL, Balserd TC, Palmquiste DE (2009) Cover crop and liquid manure effects on soil quality indicators in a corn silage system. Agron J 101(4):727–737. https://doi.org/10.2134/agronj2008.0191

Kalkan F, Avcı S (2020) Effects of applying nitrogen on yield of silage maize grown after forage legumes. KSU J Agric Nat 23(2):336–342. https://doi.org/10.18016/ksutarimdoga.vi.646221

Kamo T, Hiradate S, Fujii Y (2003) First isolation of natural cyanamide as a possible allelochemical from hairy vetch (*Vicia villosa*). J Chem Ecol 29:275–283. https://doi.org/10.1023/A:1022621709486

Ketterings QM, Swinka SN, Duikerb SW, Czymmeka KJ, Beegleb DB, Coxc WJ (2015) Integrating cover crops for nitrogen management in corn systems on Northeastern U.S. dairies. Agron J 107(4):1365–1376. https://doi.org/10.2134/agronj14.0385

Khan ZR, Pickett JA, Van Den Berg J, Wadhams LJ, Woodcock CM (2000) Exploiting chemical ecology and species diversity: stem borer and striga control for maize and sorghum in Africa. Pest Manag Sci 56(11):957–962. https://doi.org/10.1002/1526-4998(200011)56:11<957::AID-PS236>3.0.CO;2-T

Kumar N, Lad G, Giuntini E, Kaye ME, Udomwong P, Shamsani NJ, Young JPW, Bailly X (2014) Bacterial genospecies that are not ecologically coherent: population genomics of Rhizobium leguminosarum. Open Biol 5:140133. https://doi.org/10.1098/rsob.140133

Kuo S, Jellum EJ (2002) Influence of winter cover crop and residue management on soil nitrogen availability and corn. Agron J 94(3):501–508. https://doi.org/10.2134/agronj2002.5010

Kan M, Küçükçongar M, Keser M, Morgounov A, Muminjanow H, Özdemir F, Qualset C (2015) Wheat landraces in farmers' fields in Turkey national survey, collection and conservation, 2009–2014. FAO, Ankara, 178 pages. Available at http://www.fao.org/3/a-i5316e.pdf. Accessed 12 July 2020

Kapur S, Ryan J, Akça E, Çelik İ, Pagliai M, Tülün Y (2007) Influence of mediterranean cereal-based rotations on soil micromorphological characteristics. Geoderma 142(3–4):318–324. https://doi.org/10.1016/j.geoderma.2007.08.024

Karadag Y, Buyukburc U (2001) Research on the effects of different phosphorus doses on root, nodule and plant growth in some vetch species. Turk J Agric For 25(6):359–368. Available at https://dergipark.org.tr/tbtkagriculture/issue/11648/138685. Accessed 12 July 2020. (In Turkish with English Summary)

Kessel VC, Hartley C (2000) Agricultural management of grain legumes: has it led to an increase in nitrogen fixation? Field Crop Res 65:165–181. https://doi.org/10.1016/S0378-4290(99)00085-4

Lal R (2001) World cropland soils as a source or sink for atmospheric carbon. Adv Agron 71:141–191. https://doi.org/10.1016/S0065-2113(01)71014-0

Lal R (2004) Soil carbon sequestration impacts on global climate change and food security. Science 304(5677):1623–1627. https://doi.org/10.1126/science.1097396

Lal R (2008) Soils and sustainable agriculture. A review. Agron Sustain Dev 28:57–64. https://doi.org/10.1051/agro:2007025

Lal R (2012) Climate change and soil degradation mitigation by sustainable management of soils and other natural resources. Agric Res 1(3):199–212. https://doi.org/10.1007/s40003-012-0031-9

Lange M, Eisenhauer N, Sierra CA, Bessler H, Engels C, Griffiths RI, PG M-V'z, Malik AA, Roy J, Scheu S, Steinbeiss S, Thomson BC, Trumbore SE, Gleixner G (2015) Plant diversity increases soil microbial activity and soil carbon storage. Nat Commun 6:6707. https://doi.org/10.1038/ncomms7707

Luetke-Entrup N, Pahl H, Albrecht R (2003) Fruchtfolgewert von Körnerleguminosen. UFOP-Praxisinformation. Union zur Förderung von Oel- und Proteinpflanzen (UFOP), Berlin, Germany. Available at https://www.ufop.de/files/3513/4080/9712/Koernerleguminosen.pdf. Accessed 12 July 2020. (In German)

Lukangila MA (2016) Response of weeds and crops to fertilization alone or in combination with herbicides: a review. Am J Plant Nutr Fertil Tech 6:1–7. https://doi.org/10.3923/ajpnft.2016.1.7

Li Q, Zhoua D, Denton MD, Conga S (2019) Alfalfa monocultures promote soil organic carbon accumulation to a greater extent than perennial grass monocultures or grass-alfalfa mixtures. Ecol Eng 131:53–62. https://doi.org/10.1016/j.ecoleng.2019.03.002

Lichtfouse E, Navarrete M, Debaeke P, Souchere V, Alberola C, Menassieu J (2009) Agronomy for sustainable agriculture. A review. Agron Sustain 29(1):1–6. https://doi.org/10.1051/agro:2008054

Liebmana M, Mohler CL, Staver CP (2003) Ecological management of agricultural weeds. Cambridge University Press, Cambridge, p 532. https://doi.org/10.1017/CBO9780511541810

Liebmana AM, Grossman J, Brownb M, Wellsc MS, Reberg-Hortond SC, Shie W (2018) Legume cover crops and tillage ımpact nitrogen dynamics in organic corn production. Agron J 110(3):1046–1057. https://doi.org/10.2134/agronj2017.08.0474

Lin M, Gresshoff PM, Ferguson BJ (2012) Systemic regulation of soybean nodulation by acidic growth conditions. Plant Physiol 160:2028–2039. https://doi.org/10.1104/pp.112.204149

Mandal HK (2014) Isolation of salt tolerant strains of *Rhizobium Trifolii*. Int J Agric Food Sci Tech 5(4):325–332

Marsh MC (2014) Winter field pea as a leguminous cover crop in corn production. Master thesis from Crop, Soil & Environmental Sciences, Arkansas University, US, 2095, 47 p. Available at http://scholarworks.uark.edu/etd/2095. Accessed 12 July 2020

Masri Z, Ryan J (2006) Soil organic matter and related physical properties in a Mediterranean wheat-based rotation trial. Soil Tillage Res 87(2):146–154. https://doi.org/10.1016/j.still.2005.03.003

Mazur A, Stasiak G, Wielbo J, Koper P, Kubik-Komar A, Skorupska A (2013) Phenotype profiling of *Rhizobium leguminosarum* bv. *trifolii* clover nodule isolates reveal their both versatile and specialized metabolic capabilities. Arch Microbiol 195(4):255–267. https://doi.org/10.1007/s00203-013-0874-x

Mazzoncini M, Sapkota TB, Bàrberi P, Antichi D, Risaliti R (2011) Long-term effect of tillage, nitrogen fertilization and cover crops on soil organic carbon and total nitrogen content. Soil Tillage Res 114(2):165–174. https://doi.org/10.1016/j.still.2011.05.001

McCartney D, Fraser J (2010) The potential role of annual forage legumes in Canada: a review. Can J Plant Sci 90(4):403–420. https://doi.org/10.4141/CJPS07182

Meena RS, Lal R (2018) Legumes and sustainable use of soils. In: Meena R, Das A, Yadav G, Lal R (eds) Legumes for soil health and sustainable management. Springer, Singapore. https://doi.org/10.1007/978-981-13-0253-4_1

Melero S, Ruiz Porras JC, Herencia JF, Madejón E (2006) Chemical and biochemical properties in a silty loam soil under conventional and organic management. Soil Tillage Res 90:162–170. https://doi.org/10.1016/j.still.2005.08.016

Melino VJ, Drew EA, Ballard RA, Reeve WG, Thomson G, White RG, O'Hara GW (2012) Identifying abnormalities in symbiotic development between *Trifolium* spp. and *Rhizobium leguminosarum* bv. *trifolii* leading to sub-optimal and ineffective nodule phenotypes. Ann Bot 110(8):1559–1572. https://doi.org/10.1093/aob/mcs206

Mesbahab A, Nilahyanea A, Ghimireb B, Beckbc L, Ghimire R (2019) Efficacy of cover crops on weed suppression, wheat yield, and water conservation in winter wheat-sorghum-fallow. Crop Ecol Manag Qual 59(4):1745–1752. https://doi.org/10.2135/cropsci2018.12.0753

Miguez FE, Bollero GA (2005) Review of corn yield response under winter cover cropping systems using meta-analytic methods. Crop Sci 45:2318–2329. https://doi.org/10.2135/cropsci2005.0014

Miransari M (2017) Arbuscular mycorrhizal fungi and soil salinity. In: Johnson NC et al (eds) Mycorrhizal mediation of soil. Elsevier, pp 263–277. https://doi.org/10.1016/B978-0-12-804312-7.00015-2

Mohammadi GR (2010) Weed control in corn (*Zea mays* L.) by hairy vetch (*Vicia villosa* L.) interseeded at different rates and times. Weed Biol Manag 10(1):25–32. https://doi.org/10.1111/j.1445-6664.2010.00363.x

Mothapo NV, Grossman JM, Sooksa-Nguan T, Maul J, Bräuer SL, Shi W (2013) Cropping history affects nodulation and symbiotic efficiency of distinct hairy vetch (*Vicia villosa* Roth.) genotypes with resident soil rhizobia. Biol Fertil Soils 49:871–879. https://doi.org/10.1007/s00374-013-0781-y

Müller-Stöver D, Hauggaard-Nielsen H, Eriksen J, Ambus P, Johansen A (2012) Microbial biomass, microbial diversity, soil C storage and stability after incubation of soil from grass–clover pastures of different age. Biol Fertil Soils 48:371–383. https://doi.org/10.1007/s00374-011-0633-6

N'Dayegamiye A, Whalen JK, Tremblay G, Nyiraneza J, Grenier M, Drapeau A, Bipfubusa M (2015) The benefits of legume crops on corn and wheat yield, nitrogen nutrition, and soil properties improvement. Agron J 107(5):1653–1665. https://doi.org/10.2134/agronj14.0416

Nevins CJ, Nakatsu C, Armstrong S (2018) Characterization of microbial community response to cover crop residue decomposition. Soil Biol Biochem 127:39–49. https://doi.org/10.1016/j.soilbio.2018.09.015

Ozdemir S, Karadavut U, Erdogan C (1999) Influence of rhizobium inoculation and fertilization on nodulation and yield of pea (*Pisum sativum* L.). Turk Agric J For 23(4):869–874. Available at https://journals.tubitak.gov.tr/agriculture/abstract.htm?id=2872. Accessed 12 July 2020. (In Turkish with English Summary)

Ozyazici MA, Manga I (2000) The effects of some leguminous forage crops used as green manure and plant residues on yield and quality of maize and sunflower under irrigated conditions of Çarşamba Plain. Turk J Agric For 24:95–103. Available at http://journals.tubitak.gov.tr/agriculture/issues/tar-00-24-1/tar-24-1-13-98217.pdf. Accessed 12 July 2020. (In Turkish with English Summary)

Parr M, Grossman JM, Reberg-Horton SC, Brinton C, Crozier C (2011) Nitrogen delivery from legume cover crops in no-till organic corn production. Agron J 103(6):1578–1590. https://doi.org/10.2134/agronj2011.0007

Paustian K (2000) Modelling soil organic matter dynamics-global challenges. In: Rees et al (eds) Sustainable management of soil organic matter. CABI Press, Oxon, pp 45–53

Poeplau C, Don A (2015) Carbon sequestration in agricultural soils via cultivation of cover crops – a meta-analysis. Agric Ecosyst Environ 200:33–41. https://doi.org/10.1016/j.agee.2014.10.024

Peoples MB, Herridge DF, Ladha JK (1995) Biological nitrogen fixation: an efficient source of nitrogen for sustainable agricultural production? In: Ladha JK, Peoples MB (eds) Management of biological nitrogen fixation for the development of more productive and sustainable agricultural systems. Developments in Plant and Soil Sciences, Springer, Dordrecht. https://doi.org/10.1007/BF00032239

Peterson GA, Unger PW, Payne WA (2006) Dryland agriculture. Agronomy monographs 23. ASA, CSSA, SSSA, Madison. https://doi.org/10.2134/agronmonogr23.2ed.frontmatter

Pla CL, Cobos-Porras L (2015) Salanity: physiological impacts on legume nitrogen fixation. In: Saad S, Lam-Son PT (eds) Legume nitrogen fixation in a changing environment. Springer, Cham, pp 35–65. https://doi.org/10.1007/978-3-319-06212-9_3

Pontes LS, Tullio GF, Martins AS, Moletta JL, Silva VP (2018) Corn yield for silage and grains in different integrated crop-livestock systems. Rev Cienc Agron 49(2):315–323. https://doi.org/10.5935/1806-6690.20180036

Prasifka JR, Schmidt NP, Kohler KA, O'neal ME, Hellmich RL, Singer JW (2006) Effects of living mulches on predator abundance and sentinel prey in a corn-soybean-forage rotation. Environ Entomol 35(5):1423–1431. https://doi.org/10.1603/0046-225X(2006)35[1423:EOLMOP]2.0.CO;2

Preissel S, Reckling M, Bachinger J, Zander P (2017) Introducing legumes into european cropping systems: farm-level economic effects. In: D M-B, Stoddard FL, Watson C. (eds) Legumes in Cropping Systems. CABI Publishing, pp 209–225. https://doi.org/10.1079/9781780644981.0209

Qing-yu Z, Yu-min Y, Liang-yu P, Ji Z, Li T, Jian-rong L, Wen-qiang F (2014) Nodule and root growth of *Vicia villosa* in purple hilly area of Sichuan ender different fertilizer treatments. Southwest China J Agricultur Sci 27(5):2000–2005. (In China with English Summary)

Reckling M, Preissel S, Zander P, Topp CFE, Watson CA, Murphy-Bokern D, Stoddard FL (2014) Effects of legume cropping on farming and food systems. Legume Futures Report 1.6. Available at www.legumefutures.de/results/agriculturaleffects.html. Accessed 12 July 2020

Reddy PP (2016) Cover/Green manure crops. In: Reddy PP (ed) Sustainable intensification of crop production. Springer, Singapore, pp 55–67. https://doi.org/10.1007/978-981-10-2702-4_4

Rengel Z (2015) Availability of Mn, Zn and Fe in the rhizosphere. J Soil Sci Plant Nutr 15(2):397–409. https://doi.org/10.4067/S0718-95162015005000036

Rogers ME, Colmer TD, Frost K, Henry D, Cornwall D, Hulm E, Deretic J, Hughes SR, Craig AD (2008) Diversity in the genus *Melilotus* for tolerance to salinity and waterlogging. Plant and Soil 304:89–101. https://doi.org/10.1007/s11104-007-9523-y

Rumbaugh MD, Pendery BM, James DW (1993) Variation in the salinity tolerance of strawberry clover (*Trifolium fragiferum* L.). Plant and Soil 153:265–271. https://doi.org/10.1007/BF00012999

Saia S, Amato G, Frenda AS, Giambalvo D, Ruisi P (2014) Influence of arbuscular mycorrhizae on biomass production and nitrogen fixation of berseem clover plants subjected to water stress. Plos One 9(3):Article Number: e90738. https://doi.org/10.1371/journal.pone.0090738

Sainju UM, Singh BP, Whitehead WF (2002) Long-term effects of tillage, cover crops, and nitrogen fertilization on organic carbon and nitrogen concentrations in sandy loam soils in Georgia, USA. Soil Tillage Res 63(3–4):167–179. https://doi.org/10.1016/S0167-1987(01)00244-6

Sato T, Sato E, Takakai F, Yokoyama T, Kaneta Y (2011) Effects of hairy vetch foliage application on nodulation and nitrogen fixation in soybean cultivated in three soil types. Soil Sci Plant Nutr 57:313–319. https://doi.org/10.1080/00380768.2011.569375

Scharf PC, Brouder SM, Hoeft RG (2006) Chlorophyll meter readings can predict nitrogen need and yield response of corn in the North-Central USA. Agron J 98(3):655–665. https://doi.org/10.2134/agronj2005.0070

Schloter M, Nannipieri P, Sørensen SJ, van Elsas JD (2018) Microbial indicators for soil quality. Biol Fertil Soils 54(1):1–10. https://doi.org/10.1007/s00374-017-1248-3
Selvakumar G, Kim K, Hu S, Sa T (2014) Effect of salinity on plants and the role of arbuscular mycorrhizal fungi and plant growth-promoting rhizobacteria in alleviation of salt stress. In: Ahmad P, Wani MR (eds) Physiological mechanisms and adaptation strategies in plants under changing environment. Springer, New York, pp 115–144. https://doi.org/10.1007/978-1-4614-8591-9_6
Siddique KHM, Sykes J (1997) Pulse production in Australia past, present and future. Aust J Exp Agric 37:103–111. https://doi.org/10.1071/Ea96068
Sieversa T, Cook RL (2018) Aboveground and root decomposition of cereal rye and hairy vetch cover crops. Soil Fertil Plant Nutr 82(1):147–155. https://doi.org/10.2136/sssaj2017.05.0139
Shah Z, Ahmad SR, Rahman H (2010) Soil microbial biomass and activities as influenced by green manure legumes and N fertilizer in rice-wheat system. Pak J Bot 42(4):2589–2598
Sharma PK, Kundu BS, Dogra RC (1993) Molecular mechanism of host specificity in legume-rhizobium symbiosis. Biotechnol Adv 11(4):741–779. https://doi.org/10.1016/0734-9750(93)90002-5
Sheaffer CC, Seguin P (2008) Forage legumes for sustainable cropping systems. J Crop Prod 8(1–2):187–216. https://doi.org/10.1300/J144v08n01_08
Skoufogianni E, Danalatos NG, Dimoyiannis D, Efthimiadis P (2013) Effects of pea cultivation as cover crop on nitrogen-use efficiency and nitrogen uptake by subsequent maize and sunflower crops in a sandy soil in Central Greece. Commun Soil Sci Plant Anal 44:861–868. https://doi.org/10.1080/00103624.2013.749446
Sulieman S, Schulze J, Tran LS (2013) Comparative analysis of the symbiotic efficiency of Medicago truncatula and Medicago sativa under phosphorus deficiency. Int J Mol Sci 14:5198–5213. https://doi.org/10.3390/ijms14035198
Sylvia DM, Furhmann JJ, Hartel PG, Zuberer DA (2005) Principles and application of soil microbiology, 2nd edn. Pearson Prentice Hall, Upper Saddle River, p 640
Tamagno S, Sadras VO, Haegele JW, Armstrong PR, Ciampitti IA (2018) Interplay between nitrogen fertilizer and biological nitrogen fxation in soybean: implications on seed yield and biomass allocation. Sci Rep 8:17502. https://doi.org/10.1038/s41598-018-35672-1
Tang C, Unkovich MJ, Bowden JW (1999) Factors affecting soil acidification under legumes. III. Acid production by N-fixing legumes as influenced by nitrate supply. New Phytol 143:513–521. https://doi.org/10.1046/j.1469-8137.1999.00475.x
Tao J, Liu X, Liang Y, Niu J, Xiao Y, Gu Y, Ma L, Meng D, Zhang Y, Huang W, Peng D, Yin H (2017) Maize growth responses to soil microbes and soil properties after fertilization with different green manures. Appl Microbiol Biotechnol 101(3):1289–1299. https://doi.org/10.1007/s00253-016-7938-1
Thies JE, Grossman JM (2006) The soil habitat and soil ecology. In: Uphoff et al (eds) Biological approaches to sustainable soil systems. CRC Press, Boca Raton, pp 59–78
Thrall PH, Bever JD, Slattery JF (2008) Rhizobial mediation of Acacia adaptation to soil salinity: evidence of underlying trade-offs and tests of expected plants. J Ecol 96:746–755. https://doi.org/10.1111/j.1365-2745.2008.01381.x
TUBIVES (2019) Turkish Plants Data Service. Available at http://194.27.225.161/yasin/tubives/index.php. Accessed 12 July 2020
Unger PW, Vigil MF (1998) Cover crop effects on soil water relationships. J Soil Water Conserv 53(3):200–207
Valentine AJ, Benedito VA, Kang Y (2011) Legume nitrogen fixation and soil abiotic stress: from physiology to genomics and beyond. Annu Plant Rev 42:207–248. https://doi.org/10.1002/9781444328608.ch9
Ventorino V, Caputo R, De Pascale S, Fagnano M, Pepe O, Moschetti G (2012) Response to salinity stress of *Rhizobium leguminosarum* bv. *viciae* strains in the presence of different legume host plants. Ann Microbiol 62(2):811–823. https://doi.org/10.1007/s13213-011-0322-6

Voisin AS, Salon C, Jeudy C, Warembourg FR (2003) Root and nodule growth in *Pisum sativum* L. in relation to photosynthesis: analysis using ^{13}C-labelling. Ann Bot 92(4):557–563. https://doi.org/10.1093/aob/mcg174

WBB 2019. Worl Bank Blogs 2019. Fertilizer prices to rise in 2019 on supportive fundamentals. Available at https://blogs.worldbank.org/developmenttalk/fertilizer-prices-rise-2019-supportive-fundamentals. Accessed 04 Aug 2019

Wippel K, Long SR (2019) Symbiotic performance of Sinorhizobium meliloti lacking ppGpp depends on the *Medicago* host species. Mol Plant Microbe Interact 32(6):717–728. https://doi.org/10.1094/MPMI-11-18-0306-R

Wood M, Cooper JE, Holding AJ (1984) Aluminum toxicity and nodulation of *Trifolium repens*. Plant and Soil 78:381–391. https://doi.org/10.1007/BF02450371

Yuan K, Miwa H, Iizuka M, Yokoyama T, Fujii Y, Okazaki S (2016) Genetic diversity and symbiotic phenotype of hairy vetch rhizobia in Japan. Microbes Environ 31(2):121–126. https://doi.org/10.1264/jsme2.ME15184

Zahran HH (1999) Rhizobium-legume Symbiosis and nitrogen fixation under severe conditions and in an arid climate. Microbiol Mol Biol Rev 63(4):968–989

Zhao FJ, Wood AP, McGrath SP (1999) Effects of sulphur nutrition on growth and nitrogen fixation of pea (*Pisum sativum* L.). Plant and Soil 212:209–219. https://doi.org/10.1023/A:1004618303445

Zolotarev VN (2016) Responsiveness to inoculation and application of micronutrient fertilizers in common vetch (*Vicia sativa* L.) new varieties at seed production. Russ Agric Sci 42(1):46–49. https://doi.org/10.3103/S1068367416010249

Chapter 5
Alternative RNA Splicing and Editing: A Functional Molecular Tool Directed to Successful Protein Synthesis in Plants

Regina Sharmila Dass, Pooja Thorat, and Rathijit Mallick

Abstract Intervening sequences interrupt most plant genes. The corresponding mRNA transcripts are required to go through the splicing pathway in order to get rid of the non-functional sequences. Alternative splicing (AS) has come in handy as an important post transcriptional tool that makes multiple mRNA products from single copy of precursor-mRNA molecule in eukaryotic organisms. Computational biology analyses have finally revealed that alternative splicing is not as less frequent in plants as it was thought originally. This mechanism is universal in almost all eukaryotic system for successful gene expression (e.g., plants, mammals, invertebrates etc.) that enhances or improves their coding capacity. The genes that encode functional proteins in a cell, are all spliced by several pathways, amongst which maximum are alternatively spliced. Here we present a brief discussion about the characteristics of various mRNA processing pathways with special reference to alternative splicing in leguminous plants, one of the elaborate and crucial plant families.

Keywords Alternative splicing · Genome complexity · Legumes · mRNA · RNA splicing · Protein synthesis

5.1 Introduction

Chloroplasts, the cyanobacterial endosymbiotic descendants, exhibit salient genomic similarity, metabolic activities and biogenesis with prokaryotes. During their co-evolution alongside the eukaryotic counterpart, it has led to prominent

R. S. Dass (✉) · P. Thorat · R. Mallick
Fungal Genetics and Mycotoxicology Laboratory, Department of Microbiology, School of Life Sciences, Pondicherry University, Pondicherry, India

P. Guleria et al. (eds.), *Sustainable Agriculture Reviews 51*, Sustainable Agriculture Reviews 51, https://doi.org/10.1007/978-3-030-68828-8_5

modifications of typical prokaryotic characteristics and has resulted in the acquisition of some novel characteristics by the way of losing common pathogenic attributes, that are absent in their ancestral counterparts. Two features of chloroplasts demonstrate the integration of this organelle into the regulatory circuits of plants. First, the abundance of intervening sequences in chloroplasts, which otherwise, is scarce in bacteria. Even more remarkable feature that chloroplasts exhibit is, an RNA processing event called RNA splicing and editing which is completely absent in prokaryotes Schmitz-Linneweber et al. (2006). In this process, the protein non-coding genes (*introns*) in eukaryotic transcript are spliced away from primary mRNA transcript and protein coding genes (*exons*) are joined to generate mature functional mRNAs (Wang and Brendel 2006). Alternative splicing or differential splicing process is a post transcriptional modification which leads to vast diversification of proteins. Alternative Splicing is a highly regulated process that influences the destiny of the mRNA and increases the intricate complexities of gene expression. It has a vital role in cellular differentiation and development of the organism (Wang et al. 2014a, b). Higher the levels of Alternative Splicing, higher will be the complexity of the organism.

The occurrence of Alternative Splicing in plants was reckoned as rare. Nevertheless, a number of emerging computer-based studies have now exhibited a high occurrence of alternatively spliced genes in plants than those that were predicted earlier. The splice sites and the types of Alternative Splicing in plants are generally conserved compared with mammals (Carninci et al. 2005). Over 54 plant species with 400 genes are proven to be alternatively spliced (Wang et al. 2008).

5.2 Introns

Genome of a eukaryotic cell harbors intron or the intervening sequence and their propagation in the cell is mysterious. These sequences are excised in a mature RNA and do not code for proteins. Nevertheless, some introns also exhibit direct and indirect roles in the cell, among which regulation of alternative splicing is one of the roles (Jo and Choi 2015). Introns were initially identified by a group of researchers working with adenoviruses (Berget et al. 1977; Chow et al. 1977). In 1978, Darnell discovered introns in the gene associated with hemoglobin and immunoglobulin. Subsequently, Knapp and co-workers observed splicing of RNA transcripts and removal of introns from t-RNA. Group I introns or self splicing introns were first discovered in the protozoan *Tetrahymena thermophila* (Clancy 2008). Group II self splicing introns are found in mitochondrial genes. The self-splicing activity of Group I and Group II self splicing introns is facilitated by distinct RNA structures. Both groups contain ORF's that assist the removal and propagation of introns from the RNA transcripts (Roy and Gilbert 2006). On the contrast, spliceosomal introns are the one which are present in the nuclear genomes of eukaryotes and absent in prokaryotes. Unlike self splicing introns they lack ORF's. Their length varies between species from ten bases to hundreds of kilo bases. Their numbers also widely vary from hundreds to thousands in different species (Roy and Gilbert 2006). Spliceosomal introns are used as tools for the genomic and evolutionary analysis. The two theories, intron-early and intron-late, are a topic of extensive debate in the origin and evolution of spliceosomal introns. The intron-early theory proposes that introns were abundant in the deep ancestral prokaryotes and eukaryotes, but eventually were lost from the prokaryotic genomes (Gilbert 1987). Intron-early theory also suggests that the spliceosomal introns evolved from the Group II self splicing introns as they share similarities in their excision mechanisms (Clancy 2008).

5.3 Alternative Fates of mRNA by Diverse Processing Pathways

The life cycle of eukaryotic mRNA is more extended than that of prokaryotic mRNA. In eukaryotes, the transcription of mRNA is carried out in a usual way by initiating with a 5′ triphosphate end and immediate capping of 5′ end with guanine nucleotide which is methylated on the seventh position. RNA polymerase, the main enzyme conducting transcription is not released from the DNA template strand and continues transcription of the entire gene. The protein coding regions of eukaryotic genome are usually flanked by junk regions. Thus, the transcript produced (pre-mRNA) includes exons as well as introns. The 3′ end is generated since termination of transcription is executed by cleaving the transcript rather than with termination at specific site. Immediately, after cleavage of the transcript, addition of a series of polyadenylates or poly (A)s at the 3′ end is carried out. Such transcripts generated from interrupted genes necessitates for further processing in order to remove the introns (protein non-coding genes).

5.3.1 RNA Splicing

RNA splicing is a form of RNA processing in which the protein non-coding genes are excised and coding genes are joined together. Splicing of the mRNA at specific sites is executed by a complex of ribonucleic proteins (RNPs) or also called as *Spliceosome* (molecular machine called as "spliceosome" or "spliceosomal complex" which comprises about 150 proteins, five RNAs and catalyzes the splicing reaction by hydrolyzing several ATP molecules). There are two key steps in the splicing process; first the assembly of small nuclear ribonucleic proteins (snRNPs) to form a complex called spliceosome followed by the splicing of pre-mRNA. The small nuclear ribonucleic proteins (snRNPs); U1, U2 and the U4/U6 construct a spliceosome multimolecular complex. The eight evolutionarily conserved DExD/H-type RNA-dependent ATPases/helicases and specific spliceosomal complexes (E, A, B and others) together perform a cascade of splicing steps that result in exonic ligation by cutting out the intronic sequences (Wang et al. 2014a, b) and the transcript is transformed into a mature messenger RNA. This mature mRNA is then transported from the nucleus to the cytoplasm for successful protein translation.

5.3.2 Alternative Splicing

The process of RNA splicing performed in differential forms is known as '*Alternative splicing*' (AS). The outcome of such splicing mechanism is the combination of distinct splice junctions which are available in precursor-mRNA transcripts. This leads to the production of a multifarious range of mRNA and causes an escalation of the protein products for a certain genetic unit, thereby elaborating the potential informational content of the so-called eukaryotic genetic storehouses. This process offers diversification of ultimate protein products that are encoded by the genetic make-up of an organism. Synthesis of the various types takes place in an unpredictable manner. Higher eukaryotic organisms produce tremendous amounts of mRNAs. These mRNAs, sometimes can be alternatively spliced in several alternate ways to produce two or more processed mRNA transcripts and in turn different functional protein products or isoforms. A more strict and well-coordinated regulation process has to be maintained in a cell, all the time, in order to make sure that different alternative forms are being produced in different cell types and in response to distinct conditions and/or stimulus (Watson et al. 2014).

5.4 Differences Between RNA Splicing and Alternative Splicing (Table 5.1)

5.5 Similarities Between RNA Splicing and Alternative Splicing

1. RNA splicing and Alternative Splicing occur within the nucleus
2. Both the biological processes are examples of post-transcriptional modification systems at the time of gene expression in eukaryotes.
3. The starting molecule in RNA splicing and AS is the primary RNA transcript.
4. Both the processes are involved in splicing of exons by the removal of introns.
5. Both RNA splicing and Alternative Splicing are responsible in the synthesis of the mature mRNA molecule which could be translated into proteins.

Table 5.1 Differences between RNA splicing and alternative splicing (Wang et al. 2014a, b)

Sl. no.	RNA splicing	Alternative splicing
1	RNA splicing comprises of the modification of the newly synthesized precursor-mRNA in which intruding sequences are withdrawn/eliminated and coding regions are combined preceding protein synthesis.	Alternative splicing constitutes a mechanism that allows the mRNA itself to govern concoction of a variety/assortment of different protein isoforms that could supposedly have cellular function/altered function and behave differently.
2	The splicing away of coding sequences from the immature RNA transcript through RNA splicing results in the formation of a mature mRNA.	Alternative splicing splices the exons of the primary transcript resulting in the formation of differential combination of exons.
3	The mature mRNA usually consists of all the coding regions in the single stranded primary transcript.	The specific exonic sequences of a gene may be incorporated within or eliminated from being included in the ultimate mature mRNA produced from the gene.
4	RNA splicing links up the protein coding regions by the removal of the non-coding regions in the mRNA molecule.	The focal point of alternative splicing is to emphatically enhance the diversity of the data or information and the cell's proteome.

5.6 Alternative Splicing (AS) in Plants: Definition

The end result of different Alternative RNA processing (or splicing) pathways just like mammals, leads to the combination of different splice junctions in plant systems as well, which are accessible in the so-called raw mRNA materials not been processed yet. This culminates in the heterogeneity of mRNA and protein products of a genetic unit, thereby elaborating the data encoded in the eukaryotic DNAs of different origin. These kinds of researches would impart knowledge about the progression of alternative splicing as an evolutionary phenomenon. This in turn would reveal the biological consequences it holds for an organism or species at the genomic level. On the other hand, at the genetic realm, the conservation of alternative splicing can enlighten us about gene functions and gradual changes during the course of evolution regarding specific genes (Kan et al. 2005; Lareau et al. 2004; Sorek and Ast 2003; Modrek and Lee 2003). Alternative splicing is yet another "alternative" method, as the name suggests, allowing the cell to generate multiple polypeptide products through mature mRNAs from one single gene (rather from one pre-mRNA). This is a way to make one polypeptide with multiple subunits (e.g. *Haemoglobin*) or different sets of proteins altogether (Watson et al. 2014).

This process (Alternative Splicing) was long been thought of as an intermittent event in plants, until the intervention of computational biology to prove that the existence of alternatively spliced genes is comparatively higher than previous approximation (Wang and Brendel 2006; Campbell et al. 2006). Alternative splicing is a process where bringing close and fusing various 5′ and 3′ regions take place from an individual gene to produce multiple mRNAs, encoding proteins with high diversity and at times with antagonistic functionality (Lander 2001).

5.7 Molecular Mechanism of Alternative Splicing

Alternative splicing bestows a cell with abilities to functionally select/choose different exonic sequences to produce a final, mature mRNA. Although, a diverse range of proteins (isoforms) can be made by mix and matches, the initial DNA sequence to start with, remains the same in each case. Almost 75% of the splicing reactions that take place in exons, have been found to be protein coding (Blencowe 2006). Generation of longer or shorter exons is dependent on the recognition of different splice sites by the spliceosome which can sometimes retain the entire introns through this '*exonization*' process. These events completely rely on the 'strength' of the competing splice sites. Meaning that the extent of matching or similarity between the consensus sequence and the frequently occurring splice sites will determine the firm binding of the spliceosome machinery subunits and in turn use or discard them. Apart from this, certain auxiliary *cis*-acting elements present both in exons and/or introns can influence employment of certain spliceosome factors, which regulate the introduction of a specific exon either favourably or in an abortive manner (Smith and Valcárcel 2000; Grabowski 1998). Splice site strength also seems to depend on them. These sequences are known as splicing *enhancers/silencers*, which are usually present nearby the exons and/or introns. SR proteins define a family of Serine-Arginine (SR) proteins which include some protein factors that in turn bind to the enhancers. Negative factors, on the other hand, bind to a family of nuclear riboproteins that are heterogenous in nature (hnRNPs). However, this is not always the scenario. There are evidences where these basic amino acids (eg: Serine, Arginine) are found to down-regulate the splicing reaction from occurring. During adenovirus mediated infection, SF2/ASF, a group of SR proteins that bind to a repressor element in the intron close to the branch point of the viral pre-mRNA (Kanopka et al. 1996). This turn of event inhibits the U2 snRNP deployment, preventing the access of 3′ splice site. In a retrospective manner, the hnRNPs can also help activate than repress splicing (Martinez-Contreras et al. 2006). Clearly, these activating and repressing abilities coincide in synchrony (Charlet et al. 2002) in a cell and it remains a mystery till date as to how and why one cell dominates activation and the other, repression. SR proteins and hnRNPs, both possess RNA and protein binding domains through which they bind the regulatory sequences and to each other with relatively low specificity. It is this interactive measure between the pre-mRNA and the proteins that forms the basis of the so-called '*splicing code*' (Hertel 2008). The splicing reactions depend enormously on the *enhancers* and *silencers* and their variability is quite higher than the actual splice site sequences. But very little is known so far as to why changes in the sequence of these elements affect the splicing factor binding to them. Mutations that upset this otherwise natural balance in a cell can give rise to anomalous alternative splicing regulation. This catastrophic outcome can up- or down-regulate or cause unnecessary protein expression in a particular cell at any developmental stage (Fig. 5.1).

Versatility has been observed among enhancer and silencer sequences compared to splice site sequences. Although very little information is documented about them,

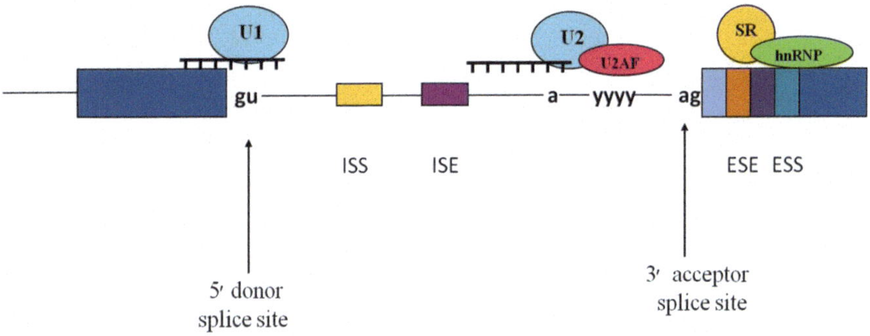

Fig. 5.1 Regulatory elements controlling the splicing mechanism. U1 and U2 snRNPs bind to the loose consensus sequence which takes place through complementary base pairing at 5′ splice site and branch point respectively. U2 snRNPs act as auxiliary factor (U2AF), recognizing and firmly binding to the Poly (Pyrimidine) tract. This helps in accurate interaction of U2. SR proteins and hnRNPs bind to ESEs (Douglas and Wood 2011), although each has a completely separate set of outcomes. The SR proteins increase the splice site use while hnRNPs impose an adverse effect on using splice site. Many other proteins involved in splicing bind to intronic splicing enhancers (ISEs) and silencers (ISSs)

it is yet to be known why change in these sequences alters binding of various splicing factors involved. Splicing factors are examples of *trans*-acting factors. These factors act in trans, i.e., they can regulate function of other target sequences as well, except the gene that they are made from. The variability of these sequences is a proof that the underlying interactions between proteins and RNA that intimate the splicing factor binding process are not strong enough to impart a higher rapport. Even though the identification of enhancers and silencers becomes difficult for this incapacitated binding, it is this exact mechanism that helps in accurately directing the regulation and control phase (Fig. 5.2).

5.7.1 Types of Alternative Splicing Observed in Plant Species

There are five fundamental means of Alternative splicing (AS) namely, Exon skipping or Cassette-type alternative exon, Mutually exclusive exons, Alternative 3′ splice site, Alternative 5′ splice site and Intron retention.

Exon skipping is a major and widespread motif of alternative splicing in both vertebrates and non-vertebrates. And intron retention (IntronR) has continued to remain the rarest type of alternative splicing observed in lower metazoans (Kim et al. 2007). From an evolutionary perspective, comparison of RNA sequences between cross species within a particular family has unveiled the preservation of intron sequences, which are known to be conserved sequences in order to govern a biological function (Kalyna et al. 2006; Iida and Go 2006).

Tools like expressed sequence tags (EST's) or cDNA-based analysis has helped in extensively studying AS in mammalian systems. In the case of humans, the naive

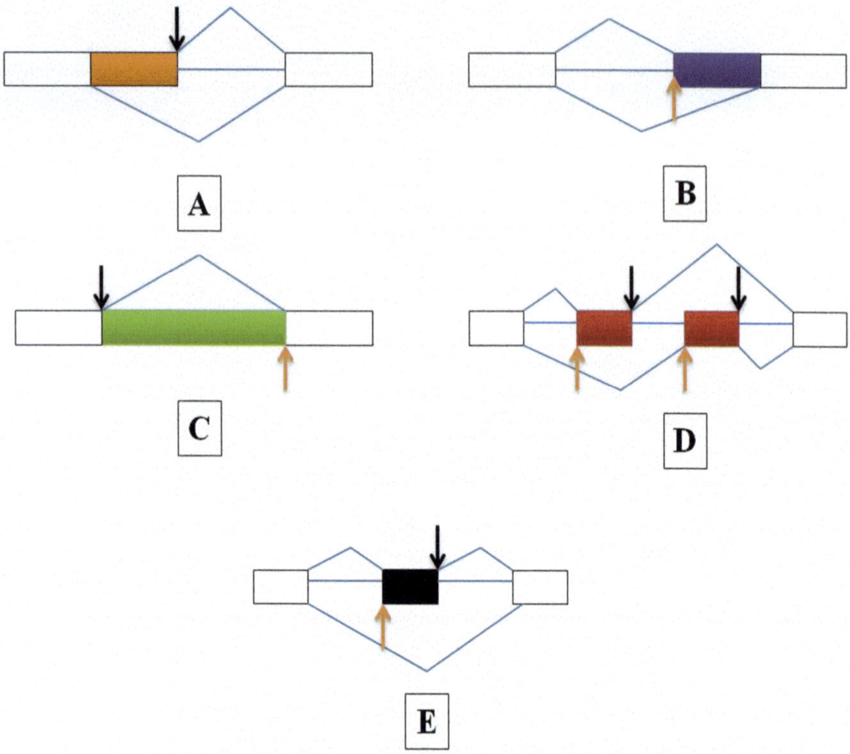

Fig. 5.2 Various alternative splicing models: white boxes represent constitutive exons and the vertical blue lines denote introns. Light orange arrows show the 3′ splice site and the black arrows show the 5′ splice sites. The black box in (E) indicates the cassette type alternative exon which is the most common form of alternative splicing. The colored boxes (orange in A, purple in B, green in C and magenta in D) show other types of splicing methods (Tazi et al. 2009) [A: Alternative 5′ splice site; B: Alternative 3′ splice site; C: Retained intron; D: Mutually exclusive; E: Cassette]

mRNA is maximally processed by alternative splicing pathway (Gupta et al. 2004; Brett et al. 2002; Kan and Gish 2002; Modrek et al. 2001; Croft et al. 2000; Mironov et al. 1999). Among the types of alternative splicing, Exon skipping (Exon S) is the most common type, amounting to nearly 58% of all the alternative splicing events which involve a single exon, and about 11% where in multiple exons are skipped in tandem (Gupta et al. 2004); ≈70–80% of the alternative splicing episodes occur in functional regions (Modrek and Lee 2002) and approximately 1/3rd produce premature termination codons (PTCs) (Lewis et al. 2003). The PTC-containing sequences act as targets for nonsense-mediated mRNA decay (Lewis et al. 2003). Of all the predicted alternative splicing events, there may exist some, which are not real and functional, mainly because of the existence of the possible sources of false positives (Modrek and Lee 2002). Sorek and team (2004), have stated that, a functional splicing event is only then, when it is required at any point in the life cycle of an organism, which is been activated and regulated in a precise manner.

Zhou and co-workers (2003) reported that over four hundred (400) genes in about fifty-four (54) species of plants, which are known to be spliced by alternative pathway. Only a few alternative splicing events have been studied and reported in legumes (Fabaceae). Two isoforms produced from a phytochelatin synthase gene (LjPCS2), namely LjPCS2-7N from the nodules and LjPCS2-7R from the roots have been studied in *Lotus japonicas*. The proteins encoded by these two isoforms differ only in five amino acids, where one protein (LjPCS2-7R) confers tolerance to Cadmium (Cd) while the other namely LjPCS2-7R does not, when studied in yeast models (Ramos et al. 2007).

A nodulin gene (LjN0D70), which belongs to a small family of closely related *Lotus japonicas* genes, responsible for nodule development, shows an intron retention event in *Lj*, where the spliced isoforms are not so plenteous in nodules. Intron retention event in *Lj* was shown in a nodule specific gene (LjN0D70) wherein the spliced isoforms is less abundant in nodules (Szczyglowski et al. 1998). Expressed Sequence Tag (EST) data mining has been used to identify a nodule specific gene (GmPGN) in soyabean (*Glycine max*). Jeong and team (2006) have confirmed with experiments that *Gm* tissue specificity and alternative splicing events for this particular gene (GmPGN). On the other hand, in *Phaseolus vulgaris* commonly known as kidney beans, possesses a particular gene (PvSBE2) which can be modified by alternative splice mechanism to produce two starch-branching enzyme isoforms, each of which manifest distinctly marked characteristics and localization (Hamada et al. 2002).

Large-scale genome sequencing projects are in progress for two plants namely *Medicago truncatula* (*Mt*) and *Lotus japonicas* (*Lj*) (Young et al. 2005). Both the legume models have relatively mammoth EST collections (over 150,000 sequences). Transcript sequences from other legume species especially Soybean are available in large numbers. Features such as these make *Mt* and *Lj* ideal plant tools for computational comparison of alternative splicing episodes in legumes and other plants.

In a particular investigation, Wang et al. (2008) have used all the available transcript sequences from model legumes namely *Mt* and *Lj*. They reported that alternative splicing events are widespread in *Mt* and *Lj*, as in *At* (*Arabidopsis thaliana*) and *Os* (*Oryza sativa*). A multitude of alternative splicing events were identified in *Mt* through a combinational examination of same and cross-species EST alignments (Chen et al. 2007). Intron R was found to be the most abundant in different subsets of genes, as previously observed and studied in *At* and *Os*. EST alignments within cross species has revealed several novel and conserved alternative splicing events. Detailed information on the methods adopted for analyses are available in particular studies (Wang et al. 2008).

During the developmental transitions in plants, the differential expression of the genes encoding alternatively spliced transcripts is identified as optional (Srinivasan et al. 2016). This concept is validated by sequencing studies extensively performed by tracking changes in the expression and alternative events throughout during a plant's embryonic development (Narsai et al. 2017; Thatcher et al. 2016; Sun and Xiao 2015; Aghamirzaie et al. 2013). The reproduction in plants initiates with the double fertilization process wherein, one female gamete fuses with two male gametes. During fertilization, one male gamete fuses with the female gamete and forms a diploid (2n) embryo and diploid central cell is fertilized by a second sperm cell which grows into an endosperm (Szakonyi and Duque 2018; Berger et al. 2008; Raghavan 2003). In embryogenesis, the one-cell zygote develops into a mature embryo through firmly regulated developmental stages. Following the embryo morphogenesis, the next phase i.e., seed maturation begins in which, nutrients required for growth are accumulated, tolerance to desiccation is gained, reduction in metabolic activities occur and dormancy stage emerges making embryo competent until commending and congenial environmental conditions allocate germination (Graeber et al. 2012).

The process of germination commences with imbibition (uptake of water by membrane) which restores the key biochemical and molecular processes in plants. The growth of embryo results in the bursting of seed coat and appearance of the radical. Under darkness, the seedlings endure skotomorphogenesis, giving rise to physical characters namely root, elongated hypocotyls, apical hook, etc. However, after exposure to light, photomorphogenesis gets activated which prevent elongation of hypocotyls, and allows the cotyledons to open and expand, and initiate development of photosynthetic pigments (Wu 2014).

The stages of early plant development are regulated by the plant hormones namely auxins, cytokinins, abscisic acid (ABA), gibberellic acid (GA) and brassinosteroids. Auxins and cytokinins play a vital role in embryo formation, while ABA is responsible for the activities such as seed maturation, development of dormancy, and inhibition of seed germination. Whereas, this inhibition effect of ABA is counteracted by GA, ethylene, and brassinosteroids (Palovaara et al. 2016). The genes involved in these stages of development and maturation processes could be regulated by alternative splicing.

The studies carried out on time points involved in the early developmental time course of soybean seed maturation revealed that, the genome of soybean contains 54,175 genes and 73,320 transcripts. The average number of transcripts encoded by a single gene is 1.35 whereas this number in *Arabidopsis* is 1.29. A number of 217,371 different novel transcripts are generated by alternative splicing of 47,331 genes with variations in levels (Aghamirzaie et al. 2013). A total of 1395 genes are reported to have more than one differentially expressed transcript. The later stages or the last time point in embryogenesis, where the embryo is completely matured and heading into, establishment of dormancy, demonstrate maximum alternative splicing events. The produced isoforms exhibit differential expression in relation to each other since one isoform poses changes in its transcription levels, whereas the other isoform(s) does not change.

Genes encoding central carbon and nitrogen metabolism (CCNM) enzymes are insignificant in proportion of the total genes in plant genomes. The enzymes encoding for carbon metabolism and hormone mediated signaling pathways were also differentially spliced. Some of the alternative splicing events are conserved between the two process in seed maturation; one, prevention and repair of cells during rehydration and second, response to abiotic stress in plants.

Similar to alternative splicing regulation in plant early development, it is also significant in response to environmental factors such as incidence of light on photoreceptors. Light perceived by plants pose a strong impact on the seed germination and their conversion to autotrophic growth. Recently, studies on response of seedlings on exposure to light and its effects on alternative splicing were done using RNA-seq techniques (Mancini et al. 2016; Hartmann et al. 2016; Shikata et al. 2014) which discovered several genes that are phytochrome-mediated and are differentially expressed or with changed alternative splicing pattern, and later transcription acts as a dominant regulatory mechanism. The data observed for phytochrome-dependent AS represents that the splicing related genes namely Serine-Arginine proteins, and two spliceosomal subunits (U1 and U2) are overrepresented, while the gene sequences coding for proteins regulating transcription were the major differentially expressed genes. The PIF6 mRNA sequence variants have influence on response to light in seedlings, and only the short isoform is reported to have markable functions in the stages of seed germination (Penfield et al. 2010). After extensive studies on alternative splicing in developing soyabean embryos, researchers have come to the conclusion that some genes only encode a single transcript, while others encode multiple transcripts. And the tests performed on similar types of tissues of *Arabidopsis* and *G. max* suggest significant similarity in the frequencies of alternative splicing.

5.8 Alternative Splicing in Legumes

Legumes are the second most important crops and constitute one of the most important and largest plant families. They contain a high protein content (ranging from 15–18%) largely being used as food or fodder to support human and animal health. Legumes possess symbiotic relationship with nitrogen-fixing bacteria and therefore have the ability to thrive in low or modest-nitrogen concentrations. They serve as ideal tools for genomic studies as proven by their evolutionary history. However, studies reported on alternative splicing events in legumes are few compared to other plant species. Some of the important legumes being consumed worldwide include common beans/kidney beans (*Phaseolus vulgaris*), chick peas (*Cicer arietinum*), green peas (*Pisum sativum*) and lentils (*Lens culinaris*). Common bean is a good source of proteins and micronutrients (iron, zinc, calcium, magnesium, phosphorous, and copper) while soybean is a superior source of protein commonly used in the production of cooking oil, animal feed and poultry feed mixtures. The annotated genomes of both *Phaseolus vulgaris* and *Glycine max* include ~10% and ~23% of gene isoforms produced by alternative splicing, respectively. However, these are found in lower proportions in comparison with alternative splicing genes in other plant species.

In legumes, alternative acceptor (AA) is the most frequent type of alternative splicing event reported, followed by alternative donor (AD) and exon skipping (ES) events. The order of frequencies for the alternative splicing events in *P. vulgaris* subsequent to AA, AD, ES are, intron retention (IR), retained exon (RE) and new intron (NI), whereas, intron retention (IR), retained exon (RE) and new intron (NI) occur less frequent. In *P. vulgaris*, events subsequent to the AA and ES are IR, RE and NI while in *G. max* the order is RE, NI and IR While in *G. max*, the order of frequencies following the first three events is retained exon (RE), new intron (NI) and intron retention (IR). The alternative splicing site event is the least reported for both the plants. However, these results are contradictory compared with the alternative splicing profiles of various other plants where the most prevalent AS event is IR. On the other hand, studies done using a different methodology for RNA seq and EST's analysis, Iniguez and his co-workers (2017), discovered a significant resemblance in recognition of alternative splicing event where 2343 and 115,881 new alternative splicing events in *P. vulgaris* and *G. max* were identified. Though there is similarity in percentage of genes affected by AS reported in other plants.

In legumes, the size of intron through wide range of variety is similar and is significantly longer than other plant species while exon size tends to be slightly short. The intron retention (IntronR) type of AS in model legumes, *M. truncatula* and *L. japonicus*, is 40–50% which is lower than other plants. Nevertheless, these levels are higher than levels in mammals (Wang et al. 2008). The introns with size shorter than 70 nt (nucleotides) are liable to be retained in legumes since the structure formation and assembly of splicing complex will be inhibited. Therefore, this may lead to incompetent splicing and IntronR. Alternative splicing pathway is a mixture of functional splicing events and splicing errors, where the functional splicing events may not be conserved.

5.9 Conclusion

RNA processing is the most crucial step in a cell for its efficient and diverse protein synthesis as a mode of survival strategy. An mRNA's fate is changed in accordance to the pathway it undergoes to make a particular type of protein, in a specific cell under precise and stringent environmental conditions. Explanation of the fundamental details of the entire molecular process is beyond the scope of this chapter. However, on a daily basis, new and emerging studies have decoded these underlying unanswered questions with better understanding. The highly organized biochemical pathways, numerous enzymes intertwined with subtle chemical reactions are really challenging to decipher the multilayered complexity of this entire process. Here, only the fruitful outcome of processing pathways has been described. The other side of the coin has not been outlined in the current work. The consequences of flawed mRNA processing can be severe and fatal, ultimately leading to cell death. The destiny of a cell depends completely on how, why and when a cell chooses a set of reactions which directly is linked to their genetic makeup and their ecological surroundings. As our understanding of this topic unfolds, sooner or later, we might be able to control the over or under-expression of certain genes in order to prevent cell aging or achieve a heightened state of immunity against certain pathogens, increase product yield of economically important crops, so on and so forth. Also, modification to use a certain splicing pathway by phosphorylation and not the others can also change the pattern of gene expression enormously.

References

Aghamirzaie D, Nabiyouni M, Fang Y, Klumas C, Heath L, Grene R, Collakova E (2013) Changes in RNA splicing in developing soybean (*Glycine max*) embryos. Biology (Basel) 2(4):1311–1337

Berger F, Hamamura Y, Ingouff M, Higashiyama T (2008) Double fertilization–caught in the act. Trends Plant Sci 13(8):437–443

Berget SM, Moore C, Sharp PA (1977) Spliced segments at the 5′ terminus of adenovirus 2 late mRNA. Proc Natl Acad Sci 74(8):3171–3175

Blencowe BJ (2006) Alternative splicing: new insights from global analyses. Cell 126(1):37–47

Brett D, Pospisil H, Valcárcel J, Reich J, Bork P (2002) Alternative splicing and genome complexity. Nat Genet 30(1):29–30

Campbell MA, Haas BJ, Hamilton JP, Mount SM, Buell CR (2006) Comprehensive analysis of alternative splicing in rice and comparative analyses with *Arabidopsis*. BMC Genomics 7(1):327

Carninci P, Kasukawa T, Katayama S, Gough J, Frith MC, Maeda N, Kodzius R (2005) The transcriptional landscape of the mammalian genome. Science 309(5740):1559–1563

Charlet BN, Singh G, Cooper TA, Logan P (2002) Dynamic antagonism between ETR-3 and PTB regulates cell type-specific alternative splicing. Mol Cell 9(3):649–658

Chen FC, Wang SS, Chaw SM, Huang YT, Chuang TJ (2007) Plant gene and alternatively spliced variant annotator. A plant genome annotation pipeline for rice gene and alternatively spliced variant identification with cross-species expressed sequence tag conservation from seven plant species. Plant Physiol 143(3):1086–1095

Chow LT, Gelinas RE, Broker TR, Roberts RJ (1977) An amazing sequence arrangement at the 5′ ends of adenovirus 2 messenger RNA. Cell 12(1):1–8

Clancy S (2008) RNA splicing: introns, exons and spliceosome. Nat Educ 1(1):31

Croft L, Schandorff S, Clark F, Burrage K, Arctander P, Mattick JS (2000) ISIS, the intron information system, reveals the high frequency of alternative splicing in the human genome. Nat Genet 24(4):340–341

Darnell JE (1978) Implications of RNA-RNA splicing in evolution of eukaryotic cells. Science 202(4374):1257–1260

Douglas AG, Wood MJ (2011) RNA splicing: disease and therapy. Brief Funct Genomics 10(3):151–164

Gilbert W (1987) The exon theory of genes. Cold Spring Harbor Symp Quantitative Biol 52:901–905

Grabowski PJ (1998) Splicing regulation in neurons: tinkering with cell-specific control. Cell 92(6):709–712

Graeber K, Nakabayashi K, Miatton E, Leubner-Metzger G, Soppe WJ (2012) Molecular mechanisms of seed dormancy. Plant Cell Environ 35(10):1769–1786

Gupta S, Zink D, Korn B, Vingron M, Haas SA (2004) Genome wide identification and classification of alternative splicing based on EST data. Bioinformatics 20(16):2579–2585

Hamada S, Ito H, Hiraga S, Inagaki K, Nozaki K, Isono N, Matsui H (2002) Differential characteristics and subcellular localization of two starch-branching enzyme isoforms encoded by a single gene in *Phaseolus vulgaris* L. J Biol Chem 277(19):16538–16546

Hartmann L, Drewe-Boß P, Wießner T, Wagner G, Geue S, Lee HC, Rätsch G (2016) Alternative splicing substantially diversifies the transcriptome during early photomorphogenesis and correlates with the energy availability in *Arabidopsis*. Plant Cell 28(11):2715–2734

Hertel KJ (2008) Combinatorial control of exon recognition. J Biol Chem 283(3):1211–1215
Iida K, Go M (2006) Survey of conserved alternative splicing events of mRNAs encoding SR proteins in land plants. Mol Biol Evol 23(5):1085–1094
Iñiguez LP, Ramírez M, Barbazuk WB, Hernández G (2017) Identification and analysis of alternative splicing events in *Phaseolus vulgaris* and *Glycine max*. BMC Genomics 18(1):650
Jeong SC, Yang K, Park JY, Han KS, Yu S, Hwang TY, Park YI (2006) Structure, expression, and mapping of two nodule-specific genes identified by mining public soybean EST databases. Gene 383:71–80
Jo BS, Choi SS (2015) Introns: the functional benefits of introns in genomes. Genomics Informatics 13(4):112
Kalyna M, Lopato S, Voronin V, Barta A (2006) Evolutionary conservation and regulation of particular alternative splicing events in plant SR proteins. Nucleic Acids Res 34(16):4395–4405
Kan Z, Gish W (2002) Selecting for functional alternative splices in ESTs. Genome Res 12(12):1837–1845
Kan Z, Garrett-Engele PW, Johnson JM, Castle JC (2005) Evolutionarily conserved and diverged alternative splicing events show different expression and functional profiles. Nucleic Acids Res 33(17):5659–5666
Kanopka A, Mühlemann O, Akusjärvi G (1996) Inhibition by SR proteins of splicing of a regulated adenovirus pre-mRNA. Nature 381(6582):535–538
Kim E, Magen A, Ast G (2007) Different levels of alternative splicing among eukaryotes. Nucleic Acids Res 35(1):125–131
Lander ES (2001) Initial sequencing and analysis of the human genome. International human genome sequencing consortium. Nature 409(6822):860–921
Lareau LF, Green RE, Bhatnagar RS, Brenner SE (2004) The evolving roles of alternative splicing. Curr Opin Struct Biol 14(3):273–282
Lewis BP, Green RE, Brenner SE (2003) Evidence for the widespread coupling of alternative splicing and nonsense-mediated mRNA decay in humans. Proc Natl Acad Sci U S A 100(1):189–192
Mancini E, Sanchez SE, Romanowski A, Schlaen RG, Sanchez-Lamas M, Cerdan PD, Yanovsky MJ (2016) Acute effects of light on alternative splicing in light-grown plants. Photochem Photobiol 92(1):126–133
Martinez-Contreras R, Fisette JF, Nasim FUH, Madden R, Cordeau M, Chabot B (2006) Intronic binding sites for hnRNP A/B and hnRNP F/H proteins stimulate pre-mRNA splicing. PLoS Biol 4(2):e21
Mironov AA, Fickett JW, Gelfand MS (1999) Frequent alternative splicing of human genes. Genome Res 9(12):1288–1293
Modrek B, Lee C (2002) A genomic view of alternative splicing. Nat Genet 30(1):13–19
Modrek B, Lee CJ (2003) Alternative splicing in the human, mouse and rat genomes is associated with an increased frequency of exon creation and/or loss. Nat Genet 34(2):177–180
Modrek B, Resch A, Grasso C, Lee C (2001) Genome-wide detection of alternative splicing in expressed sequences of human genes. Nucleic Acids Res 29(13):2850–2859
Narsai R, Gouil Q, Secco D, Srivastava A, Karpievitch YV, Liew LC, Whelan J (2017) Extensive transcriptomic and epigenomic remodelling occurs during *Arabidopsis thaliana* germination. Genome Biol 18(1):172
Palovaara J, de Zeeuw T, Weijers D (2016) Tissue and organ initiation in the plant embryo: a first time for everything. Annu Rev Cell Dev Biol 32:47–75
Penfield S, Josse EM, Halliday KJ (2010) A role for an alternative splice variant of PIF6 in the control of *Arabidopsis* primary seed dormancy. Plant Mol Biol 73(1–2):89–95
Raghavan V (2003) Some reflections on double fertilization, from its discovery to the present. New Phytol 159(3):565–583
Ramos J, Clemente MR, Naya L, Loscos J, Pérez-Rontomé C, Sato S, Becana M (2007) Phytochelatin synthases of the model legume *Lotus japonicas*: a small multigene family with differential response to cadmium and alternatively spliced variants. Plant Physiol 143(3):1110–1118

Roy SW, Gilbert W (2006) The evolution of spliceosomal introns: patterns, puzzles and progress. Nat Rev Genet 7(3):211–221

Schmitz-Linneweber C, Williams-Carrier RE, Williams-Voelker PM, Kroeger TS, Vichas A, Barkan A (2006) A pentatricopeptide repeat protein facilitates the trans-splicing of the maize chloroplast rps12 pre-mRNA. Plant Cell 18(10):2650–2663

Shikata H, Hanada K, Ushijima T, Nakashima M, Suzuki Y, Matsushita T (2014) Phytochrome controls alternative splicing to mediate light responses in Arabidopsis. Proc Natl Acad Sci 111(52):18781–18786

Smith CW, Valcárcel J (2000) Alternative pre-mRNA splicing: the logic of combinatorial control. Trends Biochem Sci 25(8):381–388

Sorek R, Ast G (2003) Intronic sequences flanking alternatively spliced exons are conserved between human and mouse. Genome Res 13(7):1631–1637

Sorek R, Shamir R, Ast G (2004) How prevalent is functional alternative splicing in the human genome? Trends Genet 20(2):68–71

Srinivasan A, Jiménez-Gómez JM, Fornara F, Soppe WJ, Brambilla V (2016) Alternative splicing enhances transcriptome complexity in desiccating seeds. J Integr Plant Biol 58(12):947–958

Sun Y, Xiao H (2015) Identification of alternative splicing events by RNA sequencing in early growth tomato fruits. BMC Genomics 16(948):1–13

Szakonyi D, Duque P (2018) Alternative splicing as a regulator of early plant development. Front Plant Sci 9:1174

Szczyglowski K, Kapranov P, Hamburger D, de Bruijn FJ (1998) The *Lotus japonicus* LjNOD70 nodulin gene encodes a protein with similarities to transporters. Plant Mol Biol 37(4):651–661

Tazi J, Bakkour N, Stamm S (2009) Alternative splicing and disease. Biochim Biophys Acta 1792(1):14–26

Thatcher SR, Danilevskaya ON, Meng X, Beatty M, Zastrow-Hayes G, Harris C, Li B (2016) Genome-wide analysis of alternative splicing during development and drought stress in maize. Plant Physiol 170(1):586–599

Wang BB, Brendel V (2006) Genomewide comparative analysis of alternative splicing in plants. Proc Natl Acad Sci 103(18):7175–7180

Wang BB, O'Toole M, Brendel V, Young ND (2008) Cross-species EST alignments reveal novel and conserved alternative splicing events in legumes. BMC Plant Biol 8:17

Wang PL, Bao Y, Yee MC, Barrett SP, Hogan GJ, Olsen MN, Salzman J (2014a) Circular RNA is expressed across the eukaryotic tree of life. PLoS One 9(6):e90859

Wang Y, Liu J, Huang BO, Xu YM, Li J, Huang LF, Wang XZ (2014b) Mechanism of alternative splicing and its regulation. Biomed Rep 3(2):152–158

Watson JD, Baker TA, Bell SP, Gann A, Levine M, Losick R (2014) Molecular biology of the gene, 7th edn. Cold Spring Harbour Laboratory Press, New York

Wu SH (2014) Gene expression regulation in photomorphogenesis from the perspective of the central dogma. Annu Rev Plant Biol 65:311–333

Young ND, Cannon SB, Sato S, Kim D, Cook DR, Town CD, Tabata S (2005) Sequencing the genespaces of Medicago truncatula and Lotus japonicus. Plant Physiol 137(4):1174–1181

Zhou Y, Zhou C, Ye L, Dong J, Xu H, Cai L, Wei L (2003) Database and analyses of known alternatively spliced genes in plants. Genomics 82(6):584–595

Keywords Abiotic stress · Breeding · Crop yield · Drought · Legumes · Metal(loid)s · Omics · Salinity · Tolerance · Transgenics

6.1 Introduction

Legumes are the second main crop in the world, after cereals, being extensively used as a food source, due to their high content of proteins, complex carbohydrates, micronutrients, bioactive phytochemicals and dietary fiber (Maphosa and Jideani 2017). Food legumes can be divided in two groups, the first one are oil seeds, in which soybean and peanut have great importance due to their high oil content, and the second one are pulses, which include dry seeds used as traditional food (Annor et al. 2014). Agronomically, legumes enhance phosphorus availability in soil (Jensen et al. 2012) and mitigate greenhouse gases emissions (Foyer et al. 2016). One of the particularities of Leguminosae family is the ability to form root nodules in a symbiotic relationship established with soil bacteria, collectively named "rhizobia". This phenomenon allows bacteria to fix atmospheric nitrogen that is used by the plant as nitrogen source while the plant provides, via phloem, photo-assimilates to the nodule. This symbiotic interaction results not only in an improvement of yield but also, and most important to the biosphere, in the substitution or minimization of chemical nitrogen fertilizers application. The nutritional properties of legumes have increased their demand in the world resulting in a higher production of these crops.

Plants are constantly rearranging their metabolism in order to respond to external stimuli. Some are natural causes, such as daily variation in light intensities or relative humidity, canopy or development.While others are sporadic events such as episodes of drought and/or salinity, herbivory, xenobiotics, extreme temperatures, among others. On the other hand, some conditions depend on the implantation site, so, plants develop all their life cycle in stressful conditions such as salinity or metal(loid)s presence. Despite the differences in the nature of stressors, at the whole plant level, plants respond inducing a cascade of several conserved physiological and molecular events. In particular, drought, salinity and cold trigger similar responses since they all cause physiological dehydration at the cellular level (Mitra 2018).

Salinity is defined as the presence of an excessive concentration of soluble salts in soil and water. The dissolved mineral salts can form electrolytes of cations and anions due to the weathering of rocks, constituting primary salinity. Other natural process can occur due to the salt transport from oceans to lands by wind and rain. On the other hand, secondary salinity is caused by anthropogenic practices that alter water incomes and outcomes in soils between water applied by irrigation or rainfall and water consumed by crops by transpiration. In this sense, deforestation and plantation of annual crops instead of perennial vegetation, irrigation with salt-rich water, or, soils with poor drainage are the main causes of secondary salinity (Chhabra 2017).

Drought stress is a consequence of combined soil and atmospheric low water potentials and elevated temperature, creating higher evapotranspiration flux than water income (Lipiec et al. 2013). An alarming phenomenon occurring nowadays is the changing climatic condition affecting worldwide crops; but more severely, the soils and productivity of many developing regions. Increased temperatures and consequent elevated evapotranspiration losses combined with decreasing rainfalls are of enormous concern (Rijsberman 2006; Lobell et al. 2008). The temperature raise and increased frequency of drought, flood and heatwaves predicted by climatic models will continue affecting the agroecosystems in the future (Mittler and Blumwald 2010).

Heavy metals and metalloids (metal(loid)s) are widely distributed in nature. The majority are naturally associated with organic and inorganic ligands (Carpena et al. 2006). The natural release of metal(loid)s can be attributed to igneous, sedimentary or metamorphic rocks due to volcanic activity, sedimentation and transformation by intense heat and pressure, respectively. Also, anthropogenic activities (mining, modern agricultural practices and industrialization) have had an additional impact on the contribution of metal(loid)s to nature (Tiwari and Lata 2018). The presence of these metal(loid)s in agricultural ecosystems seriously affects not only the crop productivity but also animal and human health by transfer through the trophic chain.

Plants respond at morphological, physiological and biochemical levels trying to cope unpredictable adverse environmental conditions (Huber and Bauerle 2016). These responses can be classified in two groups: avoidance mechanisms which are morphological and physiological adjustments that help to decrease evapotranspiration such as lower leaf area with increased thickness, leaf folding or rolling, decreased stomatal conductance and deeper root systems (Sicher et al. 2012; Goufo et al. 2017). The second group comprises tolerance traits commanded mainly by cellular and biochemical modifications that help maintain tissue water potentials specifically through osmotic adjustment (Khan et al. 2015a; Blum 2017). At the cellular level, the signal perception occurs through cell-surface receptors which activate signaling cascades. Plant hormones are messengers involved in the propagation of the signal and also communicate distant organs. Among them, abscisic acid (ABA) may be the most important molecule involved in the modulation of tolerance mechanisms in a variety of ways and in response to multiple stressful challenges (Cho et al. 2018). Second messengers such as Ca^{2+} increase their intracellular levels also by several signal molecules such as: inositol phosphates, diacylglycerol and reactive oxygen species (ROS) (Hirayama and Shinozaki 2010).

ROS accumulation is a conspicuous response to abiotic stresses. These molecules have a double role, as signaling molecules (Mittler 2017) or leading to protein, lipid, carbohydrate or DNA damage and ultimately oxidative stress (Zlatev and Lidon 2012). Upon the imposition of ROS accumulation, plant cells respond activating an intricate detoxifying system comprised of scavenging enzymes and non-enzymatic antioxidants (Mittler 2002; Gill and Tuteja 2010). Afterwards, calcium binding proteins, sense the elevated Ca^{2+} levels (Kudla et al. 2010), and activate calcium dependent protein kinases. These proteins can bind phosphate reversibly to transcription factors (TFs) and hence, modulate the induction or repression of stress-responsive genes (Reddy et al. 2011). Finally, the signaling cascade ends in the production of heat shock proteins (HSPs) or enzymes involved in antioxidant response or in the synthesis of compatible osmolytes.

Osmotic adjustment can be achieved by the production of low molecular mass metabolites being soluble carbohydrates (i.e., glucose and fructose) and amino acids the most important ones (Vinocur and Altman 2005; Szabados and Savouré 2010; Arbona et al. 2013; Morales et al. 2013). The accumulation of them confers stress tolerance by maintaining cell turgor, protecting protein structures and stabilizing cellular membranes (Arbona et al. 2008, 2013; Zlatev and Lidon 2012). In the amino acids group, proline accumulation has been reported ubiquitously, which contributes to osmotic adjustment and the redox balance of plant cells (Campos et al. 2011; Morales et al. 2013; Sinay and Karuwal 2014; Zandalinas et al. 2017).

Plant acclimation and persistence under fluctuating environmental conditions can be achieved also by the increased production of secondary metabolites such as coumarins, anthocyanins, flavonoids, lignin, and tannins (Fraser and Chapple 2011). Recently, new retrograde signals, this means signals from organelles to nucleus, were reported. An example is 3′-phosphoadenosine 5′-phosphate which accumulates during drought and high light regulating ABA signaling and stomatal closure in oxidized environments (Pornsiriwong et al. 2017).

The Fig. 6.1 represents the impact of diverse abiotic stresses (salinity, drought, metal(loid)s) on legumes and their symbiotic interaction with rhizobial strains occupying nodules. The physiological, biochemical and molecular effects triggered upon stress imposition as well as the adaptive responses evoked by plants, are summarized.

Human efforts to improve legume tolerance to abiotic stresses comprise a wide number of possibilities ranging from agronomic or breeding strategies to transgenics or genome editing. The agronomic approaches most implemented are related with crop management practices including soil and irrigation management, crop residues and mulching treatments, and the selection of best crop varieties for determined regions. The genetic and genomics biotechnological tools can help counteract the complexity associated with traits controlled by multiple quantitative trait loci (QTL) (Parmar et al. 2017). In this sense, marker-assisted recurrent selection and backcrossing strategies, using QTLs, showed beneficial results in several breeding programmes (Mir et al. 2012). Other strategies involve the study of extremely drought tolerant species used as models, such as resurrection plants, in searching of tolerant traits applicable to agronomic crops (Giarola et al. 2017). In the past decade, epigenetics began to be considered as a source for variable traits searching in order to be applied in plant breeding programmes (Mirouze and Paszkowski 2011). Then, the introduction of desirable genes through transgenic-based approaches is the most widespread tool used to confer tolerance traits into crop varieties. However, transgenics lack of public acceptation so, a new approach had emerged as a plausible solution, the Genome Editing (GEd). The advantage of GEd over transgenics is that the introduced genetic modifications produce minor genome changes. In GEd technologies, double-strand breaks are created by sequence-targeted DNA nucleases in loci selected for editing. The current reported methods of GEd are clustered

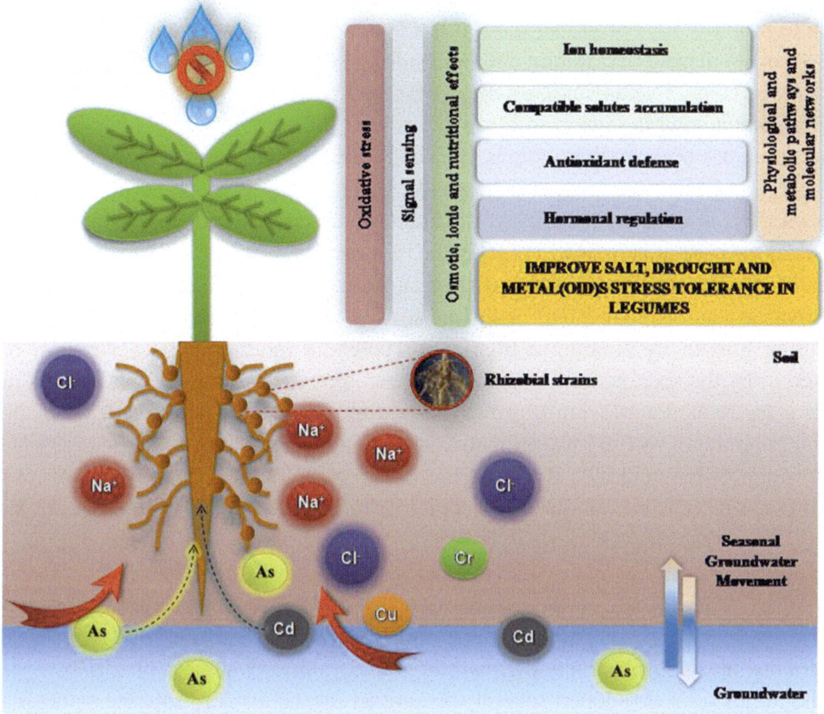

Fig. 6.1 Overview of abiotic stresses effects on legume crops and their response in agroecosystems. Legumes can be affected by salinity (e.g. high concentrations of Na^+ and Cl^-), drought and metal(oid)s (e.g. high concentrations of Cu, Cd, Cr and As present in groundwater and/or soil). These stresses can generate the production of reactive oxygen and nitrogen species (ROS/RNS) with a known double role, both as part of a cell signaling process as well as damage agents to biomolecules and the consequent oxidative stress. Thus, sensing abiotic stresses can generate osmotic, ionic and nutritional effects that trigger ionic homeostasis maintenance, compatible solutes accumulation, antioxidant defense and hormonal regulation. Therefore, understanding physiological and metabolic pathways and molecular networks can help to establish strategies to improve legume tolerance to unfavorable environmental conditions

regularly interspaced, short palindromic repeats (CRISPR), transcription activator-like effector nucleases (TALENs) and Zinc Finger (ZF) nucleases (Voytas 2013).

6.2 Salt Stress

Increased salinity is a major limiting factor for plant production around the world. One of the main causes of saline soils is the irrigation of arable lands with saline water in semi-arid and arid climates with elevated evapotranspiration (Munns and Gilliham 2015, Kaashyap et al. 2017). The irrigation systems are used in almost 300 million hectares in the world, so agriculture uses about 90% of the total water available to produce 36% of global food. It has been estimated that the increase in salinized soils will have devastating effects, resulting in a loss of 30% of soil productivity within the next 25 years, and 50% by 2050 (FAO 2015).

The major soluble salts in soils are the cations Na^+ (sodium), Ca^{2+} (calcium) and Mg^{2+} (magnesium), and the anions Cl^- (chloride), SO_4^{2-} (sulfate), HCO^{3-} (bicarbonate) and CO^{3-} (carbonate) (Chhabra 2017). There are several classification systems for salt-affected soils in the world, but all share an electrical conductivity (EC) value of the saturation extract (ECe) in the root zone that exceed 4 dS m^{-1} (approximately 40 mM NaCl) at 25 °C and have an exchangeable sodium content of 15% (Shrivastava and Kumar 2015). Salinized soils can inhibit the seed germination and seedling growth by creating a high water potential inhibiting the absorption of water, or by the specific ion toxicity. Moreover, the most common effect of salts in soils is the reduction of legume growth and development due to: I) a decrease in the water potential of soil; II) a specific toxicity, normally associated with the excessive absorption of Na^+ and Cl^-; III) a nutritional imbalance by the interference of ions with essential nutrients and IV) the combination of these effects (He et al. 2015). Salinity increases leaves necrosis and chlorosis with concomitant reduction in photosynthesis and further leaf senescence in legumes. Also disturbs the grain composition and yield (Khan et al. 2015b, 2017). Therefore, salt accumulation has detrimental effects on legumes by affecting germination and reproductive stages, biological nitrogen fixation and growth in grain legumes (Kumar et al. 2019).

Legumes have developed several specialized mechanisms to tolerate salts (Jha et al. 2019). However, salt tolerance of plants depends on diverse aspects such as nature of the salts species, variety, growth stage, and environmental factors (Shabala and Munns 2017). Several model food and forage legumes are adapted for growth in varying soil salt concentrations. Legumes such as chickpea, pea, cowpea, common bean, soybean and forage legumes as alfalfa plants, are less tolerant to salinity as compared to other grain legumes like black gram and mungbean. However, there is sparse availability of information on abiotic stress responses and tolerance mechanisms of these legumes (Araújo et al. 2015) so, their study may contribute to understanding different responses or/and mechanisms related to salinity tolerance.

Salt tolerance is a multifactorial complex pathway including molecular, physiological and metabolic networks. An exhaustive study of response mechanisms to salinity at all the mentioned levels and integrated methodological approaches, are imperative for developing salt tolerant legumes in salinized soils. Recent researches have demonstrated several response mechanisms to salinity at molecular, physiological and metabolic levels, although the entire response underlying salt tolerance is not yet completely understood. Legume mechanisms involve ion uptake, its transport and compartmentalization; the synthesis of compatible solutes or osmoprotectants; the synthesis and regulation of antioxidants and phytohormones, among others.

Ion homeostasis is an essential mechanism of plant cells modulated by ions flux regulation to keep low the toxic ion concentration such as Na^+, and elevated K^+. This balance is essential for the functioning of several proteins in the cytoplasm and for the maintenance and regulation of membrane potential and cell volume (Farooq et al. 2015). Mechanisms associated to ion homeostasis include compartmentation, exclusion and secretion of Na^{+}/H^{+} ions into vacuoles (Ahanger et al. 2017). The unique structural feature of plant cells is the presence of vacuoles. Plant Na^{+}/H^{+} exchangers (NHXs) are intracellular membrane proteins that regulate pH and K^{+} and Na^{+} homeostasis of plant cells. The Na^{+}/H^{+} antiporters localized in the tonoplast are involved in sequestration of Na^{+} within the vacuole under salinity conditions (Jia et al. 2017). These antiporters regulate the exchange of Na^{+}/H^{+} across the vacuole to decrease Na^{+} concentration in the cytoplasm (Arzani and Ashraf 2016). Thus, these antiporters facilitate Na^{+} uptake into vacuoles, which is mediated by the vacuolar proton gradient generated by the vacuolar (V-type) proton ATPase that acidifies the vacuolar lumen. The Na^{+} sequestrated in the vacuole protects important enzymatic reactions in the cytosol from Na^{+} excess, while maintaining water potential in the cell. Several SOS (salt overly sensitive) homologues, a group of proteins wich regulates the Na^{+}/H^{+} antiporters, have been identified and functionally analyzed in soybean plants (*Glycine max* and *Glycine soja*) (Nie et al. 2015; Zhou et al. 2015). Studies with transgenic plants showed that SOS1, a Na^{+} efflux transporter, could improve the tolerance of salinity in different legume species.

Under salinity conditions, osmotic adjustment occurs through ion uptake (Farooq et al. 2015) and by increasing the synthesis and accumulation of specific compatible solutes such as amino acids (proline), carbohydrates (trehalose, fructans), polyols (mannitol, sorbitol, pinnitol), amides and quaternary ammonium compounds (glycine-betaine), among others (Masood et al. 2016). These compatible solutes are characterized by their low molecular weight, hydrophilic nature and high polarity that allows vegetative cells to protect plasma membranes, enzymes and cell structures (Roychoudhury et al. 2015; Per et al. 2017). Thus, in order to maintain the photosynthetic pigments and improve plant growth under salinity stress, the accumulation of soluble sugars, proline, glycine-betaine and choline in chickpea results essential (Qurashi and Sabri 2013). Similarly, the synthesis of reducing sugars and proline in legumes, such as soybean plants (Sabagh et al. 2018), indicated that these solutes were involved in osmotic adjustment under salinity conditions. Some of these compounds also serve as osmoprotectants by stabilizing the membranes and the quaternary structures of proteins, scavenging free radicals and maintaining the cellular redox potential in plants under the mentioned stress (Kaur and Asthir 2015; Blum 2017). For instance, Kang et al. (2018) demonstrated that salt-tolerant *Medicago truncatula* genotypes accumulated more proline in the roots but less proline in leaves than sensitive genotypes, suggesting different roles of proline in both organs. The negative correlation between proline contents in leaves and overall salt tolerance in *M. truncatula* indicated that proline in the leaves was not a primary tolerance mechanism helping plants to avoid salinity stress, but rather a secondary line of defense to help plants survive when primary mechanisms failed (Kang et al. 2018). Taken together, the solutes accumulation helps to reduce the harmful effects of salinity on leguminous plants.

Reactive oxygen species (ROS) increases in plants growing under salinized soils. These species alter the normal cellular metabolic activities via oxidative damage to nucleic acids, proteins and lipids. The degree of damage results from the increased levels of ROS, which depends on the balance among the ROS production and ROS removal via antioxidant scavenging mechanisms (Hanin et al. 2016). Therefore, plants have developed a complex defensive antioxidant system for mitigating the oxidative stress caused by increased levels of ROS (You and Chan 2015). This system includes low molecular mass antioxidants and antioxidant enzymes such as superoxide dismutase (SOD), peroxidase (POX), ascorbate peroxidase (APX), glutathione reductase (GR) and the non-enzymatic antioxidants as ascorbate (ASC), glutathione (GSH) and phenolic compounds (Choi et al. 2017). Under salinity conditions, it has been observed that the antioxidant activities and the non-enzymatic antioxidant contents increased in several legumes. For instance, *Phaseolus vulgaris* plants showed an increase in APX, GR and CAT activities, which were related to an increase in the fresh and dry weights of plants in response to salinity (Osmon and Salim 2016). Similarly, Chatterjee et al. (2018) determined the toxic effects of sodium chloride on oxidative stress markers and GSH metabolism of three legumes (*Cajanus cajan*, *Vigna mungo* and *Lathyrus sativus*) and its possible reversal by NaCl pretreatment in seeds. They found that the degree of damage in GSH metabolism imposed by salinity in plants was ameliorated by priming of seeds with NaCl, which caused an increase in the antioxidant defense systems (Chatterjee et al. 2018). These studies suggest that the increased antioxidant activities in legumes help to improve salt tolerance by protecting them from oxidative stress.

It is well reported that phytohormones regulate numerous responses during the whole plant life cycle under salinity conditions (Llanes et al. 2016a). Abscisic acid (ABA) is among the most important plant hormones modulating different plant responses under normal conditions and upon the occurrence of abiotic stresses. Such responses include seed growth and development, delayed seed germination and dormancy, stomatal activities, protein and lipid synthesis, and leaf senescence. In addition, ABA content increases in stressed plants, and it has been indicated that ABA is able to regulate the enzyme activities and gene expression under salinity. In legumes exposed to salinity, rapid *de novo* synthesis of ABA was observed (Llanes et al. 2014; Zhang et al. 2015; Rothwell et al. 2015). This response leads to stomata closure, which helped in protecting the plants against water evaporation. Furthermore, the accumulation of ABA varied depending species, organ analyzed and type and concentration of salts present in the medium. This feature was reported in the halophytic legume *Prosopis strombulifera* (Llanes et al. 2014). The authors showed that ABA content in roots was lower than in leaves showing the rapid transport and biosynthesis from roots. In addition, the increase of ABA-glucose ester (ABA-GE), an ABA conjugate, was registered in both organs of salt-treated plants, suggesting that ABA-GE could be a reservoir of free ABA in roots as well as an ABA-transport form from roots to leaves in this species. Therefore, these results suggested that free ABA and ABA-GE may act together to perform a specific stress signal. This mechanism was not reported before in legumes.

Other plant hormones involved in salt responses include indole-3-acetic acid (IAA), jasmonates (JAs), gibberellins (GAs), ethylene (ET), strigolactones (SLs) and brassinosteroids (BRs). All of these can positively or adversely affect the salt responses in legumes. Recently, updates on the roles of ABA and GAs (Llanes et al.

2016a), ET (Khalid et al. 2017), BRs (Vazquez et al. 2019), SLs (Banerjee and Roychoudhury 2018), JAs (Sadeghipour 2017; Srivastava et al. 2018) in legume salt responses have been reported. However, the proposed mechanisms involving hormones in plant growth regulation under salinity vary widely and their regulation is not yet clear.

6.3 Drought Stress

Drought stress episodes can affect plant development in different growth stages. During germination, drought stress limits seed imbibition and reduces seedling vigor (Petrović et al. 2016; Berger et al. 2017). The limited water supply disturbs osmotic balance, metabolic activity and enhances ROS production leading to biomolecules damage, reduced respiration and less ATP production. In vegetative stages, drought reduces plant height, leaf area, stem diameter, and biomass (Apel and Hirt 2004; Farooq et al. 2009; Zheng et al. 2016). Nevertheless, the effect on root growth is different. Genotypes with large root systems showed high water use efficiency under drought condition as observed in peanut crops (Songsri et al. 2009). Besides, deep root systems were reported in drought-tolerant accessions of cowpea (Matsui and Singh 2003), chickpea (Kashiwagi et al. 2005), field pea (*Pisum sativum* L.) and soybean (*Glycine max* (L.) Merr.) (Benjamin and Nielsen 2006). In general, legumes varieties presenting prolific and deep root systems are desirable in agronomic crops challenged by water-deficit episodes (Belachew et al. 2018). This trait is directly related with sustained plant growth and consequent yield. A positive association between the increase in root length and ABA content was observed in several root systems (Manivannan et al. 2007). Metabolic changes induced by the stressful condition culminate in decreased crop yields ranging from 20 to 50%, as reported in several legume crops, such as mashbean (*Vigna mungo* L.) (Baroowa and Gogoi 2013, 2014), common bean (*Phaseolus vulgaris* L.) (Ghanbari et al. 2013), lentil (*Lens culinaris* L.) (Allahmoradi et al. 2013) and soybean (Maleki et al. 2013). Specifically in legumes, drought reduces the seed yield in a big extent by imbalances in the flowering process such as increased flower abortion, limited flower production, and consequent reduced seed size (Fang et al. 2010).

The main effects exerted in plants by drought stress imposition are related to imbalances in water relations. Thus, it is documented that drought stress dramatically decreases the values of several variables indicative of water status, such as relative water content, osmotic potential and water potential in legumes (Nayyar and Gupta 2006; Furlan et al. 2017). The declination in water content can be observed in almost all plant organs, as reported by Liu et al. (2004) in soybean roots, leaves, and pods. Besides, the severity of drought stress imposition is associated linearly with tissue water contents (Reddy et al. 2004).

Soil moisture has an important role in membrane permeability and cation active transport, as is the case of Ca^{2+}, K^{+} and Mg^{2+}. Thus, low moisture under drought stress can disturb the rate of nutrient inflow (Hu and Schmidhalter 2005; Farooq et al. 2009). Besides, an indirect effect of drought stress on nutrient status can be due to the impairment of enzyme activities involved in nutrient assimilation. In this sense, Ashraf and Iram (2005) reported that nitrate reductase activity decreased in leaves and nodules of *Sesbania aculeata* and common bean (*Phaseolus vulgaris* L.). In addition, drought can also induce deficiencies in some mineral micronutrients acquisition, i.e., ferrum (Fe), manganese (Mn), and molybdenum (Mo) (Hu and Schmidhalter 2005) in part due to the formation of insoluble and less oxidized ions (Havlin et al. 1999).

The ABA synthesis in roots and its transport to leaves lead to stomatal closure (Mittler and Blumwald 2015). But, ABA is synthesized in leaf cells and translocated around the plant too (Andrésa et al. 2014). ABA regulates many stress-responses such as osmotic adjustment, ion compartmentalization, root hydraulic conductivity, shoot and root growth and stomatal closure (Verslues and Zhu 2005; Pospíšilová et al. 2009). Also ABA modulates the gene expression through activation of drought-responsive genes in ABA-dependent and ABA-independent mechanisms. The difference is the required presence of ABA for the induction of the first group of genes (Kim 2014). The ABA-dependent TFs are: myelocytomatosis oncogene (MYC) and myeloblastosis oncogene (MYB), which binds to their respective cis-elements (MYCRS, MYBRS, ABRE), and zinc finger proteins (bZIPs). Also the C-repeat/drought-responsive element binding factor (CBF4), a member of CBF/DREB1 family, is included in the group of ABA-dependent regulators. DREBs (dehydration responsive element binding) are plant TFs involved in the regulation of gene expression mainly in an ABA-independent manner, while NAC (NAM, ATAF, and CUC) are exclusively ABA-independent TF´s. Both DREBs and NAC improve abiotic stress tolerance by interacting with a DRE/CRT and NACRS cis-elements, respectively. These elements are part of the promoter regions of stress-responsive genes.

The imbalances in plant water content affect the photosynthesis and gaseous exchange by decreasing not only leaf area but also the photosynthesis rate (Wahid and Rasul 2005; Farooq et al. 2009). The carbon fixation rate is slowed down due to restricted metabolism or limited carbon dioxide (CO_2) diffusion into leaves (Apel and Hirt 2004; Farooq et al. 2009). Besides, a big number of photosynthesis-associated mechanisms has been reported. These include decreased Rubisco activity (Bota et al. 2004), imbalances between systems involved in light capture and/or using (Foyer and Noctor 2005), damage to photosynthetic apparatus (Fu and Huang 2001) and oxidative damage of photosynthetic pigments (Anjum et al. 2003). It is essential to consider that the drought-induced limitations of carbon fixation, related to a reduction in photosynthetic pigment synthesis, are more complex than stomatal limitations (Reddy et al. 2004). The overreduction of the photosynthetic electron transport components, due to the limited intercellular CO_2 concentration, can reduce the molecular oxygen originating ROS, which is also deleterious to photosynthetic apparatus (Basu et al. 2016). In plants, ROS production increases linearly with the severity of stress conditions. Hydrogen peroxide (H_2O_2), hydroxyl radicals ($OH^{\cdot}$), singlet oxygen ($^1O_2$) and superoxide radicals ($O_2^{\cdot-}$) are the most studied ROS (Apel and Hirt 2004). ROS overproduction is primarily due to both decreased CO_2 fixation and elevated photorespiration rates with consequent O_2 over-reduction (Cruz de

Carvalho 2008). Thus, ROS accumulation and increased oxidative damage upon drought stress imposition have been extensively reported (Mano 2002; Blokhina et al. 2003). To counteract ROS overproduction, plants have evolved an increasing recognized network of antioxidants comprised of enzymatic and non-enzymatic compounds, namely the antioxidant system (Gill and Tuteja 2010).

Reactive nitrogen species (RNS) are also important molecules produced upon drought stress imposition. Among RNS molecules, nitric oxide (NO) stands out. NO is a gaseous signaling molecule involved in germination, development, hormone regulation, and stress management (Laxa et al. 2019). Several studies have suggested a positive correlation between NO accumulation and stress exposure in plants. Taking into account the RNS dual role as a damaging and signaling molecules, the alleviation of stress symptoms by the application of exogenous NO and improvement in antioxidant activity in plant species was reported by Nabi et al. (2019). In legumes, the application of an NO donor or the phytohormones ABA and an ethylene precursor to *Lotus japonicus* nodules, repressed nitrogenase activity, induced the expression of senescence-associated genes, and caused cytological changes evidencing nodule senescence (Fukudome et al. 2018). In the same species, drought-stressed roots showed an active metabolism of ROS and RNS characterized by an increase of protein nitration and accumulation of NO and S-nitrosothiols, suggesting that RNS had a negative impact on protein functioning in drought stressed roots (Signorelli et al. 2019).

Another important response in stressful conditions is the osmotic balance of plant cells. Through this process, plants can favor the water income or retention by inducing the decrease of osmotic potential, owing to solute accumulation and yielding a process known as osmotic adjustment (Zhang et al. 2018). The solutes accumulated can be very diverse such as ions (K^+, Ca^+, Cl^-), soluble sugars (trehalose), organic acids (glutamate), amino acids and derivatives (proline and glycine betaine). Therefore, the accumulation of compatible solutes in plant cells was associated with stress tolerance by inducing sustained growth, increased the photosynthetic rate and proper assimilate partitioning (Subbarao et al. 2000). In soybean, proline content and osmotic adjustment were associated to physiological characteristics, such as relative water content, and thus proline content was recommended as good indicator of water stress (Filová & Krivosudská 2017). Osmotic adjustment, the solutes involved in the mechanism, and its contribution to increased yields, were reported for chickpea, pea, pigeon pea and soybean (Blum 2017).

All the described responses are orchestrated in plants facing drought stress and finally end with a phenotype that expresses an escape or avoidance mechanism. In the first one, legumes tend to ensure the completion of their life cycle before reach extreme drought. In contrast, drought avoidance implies the retention of water within the cells to avoid plant dehydration (Babar et al. 2015).

When studying crops, and particularly legumes, a special consideration is highlighted when water deficit occurs during reproduction (the beginning of flowering and subsequent stages) and grain filling, a process known as terminal drought. The special emphasis in studying this growth stage is due to the devastating effects exerted in grain yield (Pushpavalli et al. 2015b; Farooq et al. 2017). The reduction in grain yield is associated with almost all the physiological and biochemical responses described above, such as reduced CO_2 fixation and translocation of C-assimilated compounds (Mondal et al. 2011, Zlatev and Lidon 2012) and impairment of photosynthesis due to oxidative damage (Farooq et al. 2009), all mediated by ABA, ROS and RNS accumulation. Besides, some specific traits related to reproductive organs are adversely affected by stress imposition during terminal drought. These include reduced flower number, fewer pollen grains, and impaired grain development and sink capacity (Andersen et al. 2002; Croser et al. 2003; Liu et al. 2004; Al-Ghzawi et al. 2009; Samarah et al. 2009; Vadez et al. 2012).

6.4 Metal(loid)s

Heavy metals and metalloids, or "Metal(oid)s", can be naturally found distributed in the Earth's crust. Some of them, like ferrum (Fe), cobalt (Co), copper (Cu), magnesium (Mg), manganese (Mn), molybdenum (Mb) and zinc (Zn) are essential for living organisms, participating in metabolic reaction in cells (Lane and Morel 2000). Others, such as mercury (Hg), cadmium (Cd), arsenic (As), do not have biological functions and can be extremely harmful for cells, even at low concentrations.

Metal(loid)s exposure of plants cause a wide range of physiological, biochemical and genotypic modifications that, considering time of exposure and concentration, could lead to a programmed cell death or an adaptation to the adverse growth condition (Singh et al. 2016). Therefore, it is important to understand the dynamics of these elements on ecosystem reservoirs by studying their uptake, translocation, concentration/speciation, metabolic processes involved, toxicity, ionic competition and interaction (Janadeleh et al. 2016). The main tissue affected by metal(loid)s in plants are roots (and in the case of legumionus plants also nodules), since is the first organ that interact with the contaminant element and could be easily accumulated as occur in soybean, peanut and other non-hyperaccumulator legume plants (Hernández et al. 2015; Bianucci et al. 2017a, 2018 ; Peralta et al. 2019).

Growth reduction, chlorosis and brownish of the root system are the first observable changes of plants exposed to metal(loid)s (Yadav 2010; Finnegan and Chen 2012) as occurred in peanut and soybean plants when exposed to As or Cd (Bianucci et al. 2012, 2017b, 2018, 2019). Alteration of metabolic processes such as decrease in the photosynthetic rate can also be observed in peanut and *Pisum sativum* plants growing with As, Cd, Cr, Cu, Ni and Zn (Sandalio et al. 2001; Stoeva and Bineva 2003; Paunov et al. 2018). In addition, Cd produces an imbalance of the chloroplast metabolism, inhibiting chlorophyll content and therefore reducing CO_2 fixation. Moreover, metal(loid)s cause disturbances on the mineral metabolism of plants by depletion of Fe (co-factor related to chlorophyll biosynthesis) by inhibiting the enzyme Fe (III) reductase or, the uptake and translocation of Fe to aerial parts (Fodor et al. 1996; Chang et al. 2003).

Plant production and accumulation of ROS in response to metal(loid)s exposure, can lead to oxidative stress by oxidative modifications of proteins and lipids. A significant source of ROS burst is the NADPH oxidase enzyme. This enzyme is located in the plasma membrane and produces $O_2^{\cdot -}$ (Mittler et al. 2004). This radical can be rapidly disrupted to H_2O_2, spontaneously or via enzymatic reactions (Torres and Dangl 2005), and it is known to be activated under metal stress (Hernández et al. 2015). Peanut plants exposed to As, significantly enhance the activity of NADPH oxidase when exposed to increasing As concentration. Moreover, histochemical staining of $O_2^{\cdot -}$ was detected at the cortex of root cells and in the epidermis (Bianucci et al. 2017b ; Peralta et al. 2020). In addition, in *Arabidopsis* and alfalfa plants grown under Cd and Hg treatment, an accumulation of root H_2O_2 at apoplastic level was associated to an activation of NADPH oxidase enzyme (Ortega-Villasante et al. 2005, 2007; Montero-Palmero et al. 2014).

The protein oxidative modifications increase carbonyls groups and originate modified amino acids. Proteins rich in sulfur and thiol groups are highly sensitive to ROS attack (Hancock et al. 2006; Oracz et al. 2007). Those damaged proteins show alterations of their structures being able to suffer spontaneous fragmentation or proteolysis (Matamoros et al. 2006; Møller et al. 2007; Rajjou et al. 2008; Flora 2011; Parkhey et al. 2014). Carbonylation of proteins was observed in plants exposed to metals as Co, Zn and Cu however, until now, scarce information is available on metalloids like As as observed in peanut plants (Karuppanapandian and Kim 2013; Ramakrishna and Rao 2013; Mustafa and Komatsu 2016; Abbas et al. 2018 ; Peralta et al. 2020).

One of the most severe processes triggered by metal(loid)s is lipid peroxidation. Plant cells present a wide variety of lipid components (fatty acids, phospholipids and isoprenoids) that act not only as a metabolic reserve but also have structural functions (Matamoros et al. 2006). Polyunsaturated fatty acids are particularly sensitive to be attacked by ROS; the peroxidation of these lipids entails important alterations in the structure and function of the membranes, causing loss of ions and membrane potential, inhibition of metabolite transport and modification of hormone receptors. Products of lipid peroxidation can form conjugates with lipids and their estimation, is considered a reliable index of oxidative stress (Groppa et al. 2001; Matamoros et al. 2006; Gill et al. 2011; Clemens and Ma 2016). Lipid peroxidation was observed in peanut and soybean plants exposed to As (Bianucci et al. 2017b, 2018 ; Peralta et al. 2020) and in *Arabidopsis thaliana* exposed to Cd and Cu (Skórzyńska-Polit et al. 2010).

The toxicity of metal(loid)s on plants can be attributed to: (a) negative interaction with the antioxidant defense system, as is the case of Cd that behaves as a sulphur-seeking heavy metal ion, disrupting their structure and function; (b) competition with cations at the root surface, for example due to the chemical similarity of As with P, the metalloid can be easily incorporated into plant cells through the high-affinity Pi transport system, and (c) production and/or accumulation of ROS, that could be directly achieved through Haber-Weiss reactions or indirectly by the interaction of metal(loid)s with the antioxidant defense system, the disruption of the electron transport chain or by disturbing the metabolism of essential elements (Dong et al. 2006; Qadir et al. 2004; Srivastava et al. 2004; Sharma and Dietz 2009; Verbruggen et al. 2009; Hernández et al. 2015; Bianucci et al. 2017a). Nevertheless, plants possess an integrated defense system that includes enzymes and metabolites to cope with metal toxicity.

Detoxification of ROS and mechanisms that avoid ROS production or accumulation and removal of oxidation products is achieved by the antioxidant defense system (Gupta et al. 2016; Bianucci et al. 2017a, 2018, 2019; Tiwari and Sarangi 2017). Moreover, plants are able to reduce metal(loid)s toxicity by synthesizing metal binding proteins (Piechalak et al. 2002; Gupta et al. 2004; Loscos et al. 2006; Degola et al. 2015), and also by eliminating the toxic elements from cells through specific transporters or compartmentalization (Dubey et al. 2018). Thus, the balance between the activation of the defense system and ROS production will define the initiation of an oxidative cascade signal and/or the programmed cell death (Møller et al. 2007). Enzymes such as CAT, PX and glutathione peroxidase-peroxiredoxin (GPX/PRX, able to use GSH and thioredoxins (TRX) as a substrate) are involved in the reduction of H_2O_2 to H_2O (Mittler 2002). In addition, enzymes of the ASC-GSH cycle, APX, GR, dehydroascorbate reductase (DHAR) and monodehydroascorbate reductase (MDHAR) efficiently remove H_2O_2 radicals. Glutathione related enzymes play a central role in the antioxidant activity in cells. GR allow maintaining the GSH/GSSG ratio in cells by the reduction of GSSG to GSH and, Glutathione S-transferases (GSTs) constitutes a large family of isoenzymes that uses GSH as a substrate, able not only to reduce organic and radicals, but also xenobiotics (Schröder et al. 2009; Bianucci et al. 2017a).

Non-enzymatic antioxidant defense involve low molecular weight compounds that remove ROS by their oxidation, maintaining the cellular redox state. GSH is a tripeptide with a sulfhydryl group (−SH) in the cysteine residue which can directly interact with ROS, producing GSSG, or act as an enzymatic co-factor (Duan et al. 2007; Das and Roychoudhury 2014; Choudhury et al. 2017). Moreover, GSH is the precursor of phytochelatins (PCs) that are nonprotein thiols synthesized by phytochelatin synthase (PCS) highly induced by metal(loid)s (Hernández et al. 2015; Bianucci et al. 2017a). In addition, many plant species can also detoxify metal(oid)s-induced toxicity by the production and/or accumulation of compatible solutes or osmolites such as glycinebetaine (Garg and Singla 2011; Jabeen et al. 2016) and proline (Mishra and Dubey 2006; Alyemeni et al. 2016; Chandrakar et al. 2018).

The antioxidant response of legume plant to metal(loid)s exposure depends on several factors such as plant species, organs sampled, stage of growth, metal(oid) dose, and exposure time (Sanitá di Toppi and Gabbrielli 1999). In this regards, peanut plants exposed to a low Cd concentration showed a reduction on GSH levels in roots and leaves and, a significant increment of PCs in roots, suggesting a crucial participation of GSH in order to avoid oxidative stress (Bianucci et al. 2012). In some legumes as *Lotus* and *Glycyne max*, the major thiol in cells is homoglutathione (hGSH), formed by an alanine substitution of glycine in the GSH biosynthesis process. Thus, these plants can accumulate homophytochelatins (hPCs) mitigating the toxic effects of metal(loid)s (Oven et al. 2001; Vazquez et al. 2009; Vazquez-Reina et al. 2005). Roots of *Lotus* exposed to high Cd concentration revealed a rapid increase of GSH, PCs and homophytochelatins (hPCs) (Ramos et al. 2007). In alfalfa plants exposed at a low Hg concentration, an increased GSH level was observed in shoots, but no significant changes were observed in roots which accumulated PCs. Moreover, GSH, hGSH and PCs were induced by a low Cd concentration in alfalfa roots, without changes in the levels of this thiol in shoots (Sobrino-Plata et al. 2009). Soybean plants exposed to Cd showed no evidence of oxidative stress in roots and leaves; however, nodules were affected showing a significant induction of PCs (Balestrasse et al. 2001, 2003, 2006). Regarding GSH related enzymes, the activity of GR was increased in shoots of alfalfa plants exposed to Cd and Hg, but was severely inhibited in roots exposed to Hg (Sobrino-Plata et al. 2009). An increase of GR activity was also observed in shoots of *Vigna unguiculata* and *M. sativa* exposed to As (Panigrahi et al. 2013) and, in soybean plants exposed to Cd, an increase of the relative gene expression of *GR-5* was observed (Pérez-Chaca et al. 2014). Nevertheless, in peanut and *Brassica juncea* plants, the enzyme was drastically reduced after As exposure (Srivastava et al. 2010; Bianucci et al. 2017a). A similar response was observed in soybean exposed to a mixture of Al and Cd (Shamsi et al. 2008). Regarding GPX/PRX enzyme, the exposure of peanut and *Brassica juncea* to As, and *Pissum sativum* to Cd, resulted in a significant decrease of the activity (Srivastava et al. 2010; Bianucci et al. 2017b). On the opposite, the activity of GST was significantly enhanced by As in peanut roots and in *Prosopis* and soybean leaves, becoming an interestingly biomarker of metalloid contamination (Mokgalaka-Matlala et al. 2009; Bianucci et al. 2017b, 2018). Also, an induction on the *GST* transcripts was found in *Medicago truncatula* exposed to As (Lafuente et al. 2015).

6.5 Achieving Legume Tolerance Through Biotechnology, Breeding or Agronomic Tools

The development of stress-tolerant genotypes and agronomic practices avoiding the detrimental effects of salinity, drought and metal(oid)s are required to sustain and increase the production of legumes under harsh environments. Therefore, scientists are looking for strategies to achieve an increase on legume productivity to manage the necessity of feeding an increasing population. Among the strategies for improving legume stress-tolerance under challenging environments, we will focus on reports of the newest biotechnological, genomic and agronomic approaches.

6.5.1 Biotechnological Approaches

6.5.1.1 Transgenics

Numerous events of plant transformation related to increases in stress tolerance have been reported in the major legume crops such as *Arachis* spp., *Cajanus cajan*, *Cicer arietinum*, *Glycine max*, *Lupinus* spp., *Medicago truncatula*, *Pisum sativum*,

Phaseolus spp., *Vicia* spp., and *Vigna* spp. (Reddy et al. 2012; Latef and Ahmad 2015). Nadeem et al. (2019) summarized a number of successful cases of legume plants that were transformed with single genes encoding enzymes that participate in water transport, biosynthesis of osmoprotectants, and late embryogenesis proteins. Some of the reported cases showed both salt and drought stress tolerance, since the responses triggered by one single gene can affect the metabolism in multiple ways. For example, An et al. (2017) reported that a tonoplast intrinsic protein (TIP), a type of aquaporin, from *Panax ginseng* (PgTIP1) expressed in transgenic soybean lines promoted both salt and drought tolerance. The tolerance mechanisms were related to the PgTIP1 primary functions in water status regulation but, remarkably, also to its secondary roles in the up-regulation of stress-related genes such as antioxidant enzymes, ABA synthesis, proline synthesis and ion transport.

Many efforts are necessary to develop salt tolerance in legume genotypes using transgenic approaches. However, some transgenic legumes for better salt tolerance have been developed. For instance, Zhang et al. (2015) reported the improvement in salt tolerance of transgenic soybeans resulting from overexpression of D1-pyrroline-5-carboxylate synthetase (P5CS) gene from *Solanum torvum*. This gene increased the synthesis and accumulation of proline in transgenic soybeans. Similarly, P5CS gene was overexpressed in chickpea and *Cajanus cajan* (Ghanti et al. 2011; Surekha et al. 2014). In another revision, Bao et al. (2016) demonstrated that transgenic alfalfa lines expressing both tonoplast NXH and H^+-PPase genes (*ZxNHX* and *ZxVP1-1* from the xerophyte *Zygophyllum xanthoxylum*) were able not only to increase growth but also to enhance tolerance to high salinity (Bao et al. 2016). Furthermore, an enhanced salt tolerance in transgenic mungbean overexpressing an *Arabidopsis* antiporter (*NHX1*) gene was reported by Sahoo et al. (2016). Under salinity, transgenic lines showed less damage and higher growth than wild type. Moreover, transgenic lines accumulated high K^+/Na^+ content in shoots and high Na^+ in roots and reduced membrane lipid peroxidation and H_2O_2 and $O_2^{\cdot-}$ levels. These responses were related to high levels of antioxidant enzyme activity and increased accumulation of proline and ascorbate registered in transgenic lines under salinity. Similar results were found by Stritzler et al. (2018) in *Medicago sativa* plants transformed with the AtNHX1 Na^+/H^+ antiporter and inoculated with the stress-resistant nitrogen-fixing strain *Sinorhizobium meliloti* B401. Plants showed high salt tolerance and nitrogen accumulation under salinity conditions. These studies demonstrated that heterologous *AtNHX1* contributed to enhanced salt tolerance in legumes. Moreover, the salt overly sensitive (SOS) pathway has been proposed to mediate cellular signaling under salinity, maintaining ion homeostasis and conferring salt tolerance (Martínez-Alcántara et al. 2015). Particularly, the SOS1 protein from *Arabidopsis thaliana* (AtSOS1) was the first plasma membrane-localized Na^+/H^+ antiporter described, mediating Na^+ efflux and controlling long-distance Na^+ transport from roots to shoots (Shi et al. 2002).

Recently, Karthik et al. (2019) transformed soybean plants via direct organogenesis with marker free construct of *p68* gene by *Agrobacterium*-mediated genetic transformation. Protein p68 is a prototype constituent of DEAD-box protein family, which is involved in RNA metabolism, induced during abiotic conditions. The transgenic soybean plants showed a higher accumulation of chlorophyll, proline, CAT, APX, SOD and relative water content than the wild type plants under salinity. These transgenic plants also showed high K^+ and low Na^+ accumulation. This study reported the p68 role against salinity, by enhancing the tolerance in legume plants. Taken together, these studies demonstrate that transgenic approaches using the overexpression of the activity of heterologous proteins are contributing to enhance the salt tolerance in legumes.

A special consideration of single-gene transgenic plants must be done when the targeted overexpressed genes are TFs. Thus, a single modification can have multiple effects on metabolic pathways activated to overcome the stressful conditions. A homeodomain-leucine zipper (HD-Zip) transcription factor, subfamily 1, from *Heliantus annus (HaHB11)* was introduced to alfalfa plants and conferred enhanced drought tolerance. The proposed mechanistic response were deciphered form *Arabidopsis thaliana* transgenic plants exposed to drought and salt stress and included physiological and molecular changes, mediated mainly by ABA and morphological responses such as leaf rolling, wider stems and deep root system (Cabello et al. 2017).

Dehydration responsive element binding (DREB) proteins are upstream a complex system involved in the up regulation of stress-responsive genes, including *kin1, kin2, rd29A, erd10* and *cor6.6*, involved in protecting plants from drought stress and low temperatures (Ray et al. 2011; Mizoi et al. 2012). Thus, the activation of *DREB2* transcription revealed increased capability to survive in water-limiting conditions in chickpea and pea plants (Acharjee and Sarmah 2013; Jovanovic et al. 2013). New alternative target genes to be used in legumes may be the TFs related to root architecture development that were identified in *Oryza sativa* (OsNAC9) and *Triticum aestivum* (TaWRKY1). Their overexpression in rice and transgenic *Arabidopsis* promoted root growth and enhanced water use efficiency (WUE) (Redillas et al. 2012; He et al. 2016). Also, other TFs linked to water use efficiency (WUE), such as those stimulating wax depositions in cuticle and suberin deposition were identified in other species (Legay et al. 2016) and may be good candidates to be used in legumes. The glyoxalase pathway detoxifies methylglyoxal, a very reactive cytotoxic compound produced upon stress imposition, and was also studied to enhance tolerance to abiotic stress in different plant species (Hasanuzzaman et al. 2017). The upregulation of glyoxalase I and II and their overexpression improved tolerance to various abiotic stresses (Lin et al. 2010; Tuomainen et al. 2011; Wu et al. 2013; Alvarez-Gerding et al. 2015). However, no such studies have been made in legumes, but, interest in these mechanism has been increased as the genome analysis and expression profiling of glyoxalase gene families was studied in *Glycine max* and *Medicago truncatula* (Ghosh 2017; Ghosh and Islam 2016) providing a basis for develop further tools aimed to improve stress tolerance.

The stomatal closure induced upon stress conditions leads to carbon shortage and excess of photochemical capacity, exceeding the ability of chloroplasts for its use and may induce photooxidative damage in chloroplast components. Plants respond accumulating the Early-light Inducible Proteins (ELIPs), a type of nuclear encoded proteins which belong to the chlorophyll a/b-binding protein (cab) family

(Araújo et al. 2015). *Medicago truncatula* plants overexpressing the *Dsp22* gene from *Craterostigma plantagineum,* which encodes an ELIP-like protein, display higher amounts of total chlorophyll, chlorophyll b and efficiency in energy conversion of PSII after the exposition to a drought stress and rehydration cycle. Thus, ELIPs engineering constitutes a useful approach to improve abiotic stress tolerance in legumes (Araújo et al. 2013).

Transgenic plants have been used also in the development of phytoremediation technologies (Krämer 2005). These strategies provide new insights into metal(oid) s-hyperaccumulation in plants. They also indicated that multiple genes may be needed in the engineering of transgenic plants for soil cleanup or for increasing crop resistance when growing under unfavorable environments. However, transforming plants with multiple genes and optimizing their expression is costly and time-consuming (Chen et al. 2013). Nevertheless, there are preliminary studies that seek to detect putative genes, whose cloning could give resistance to crops in metal(loid) s contaminated environments. Rahman et al. (2016) identified six genes involved in the homeostasis of abiotic stress, antioxidant activity and plant defense, in alfalfa roots exposed to As, and proposed them to be useful for increasing As tolerance, initially in alfalfa plants. On the other hand, Talukdar (2015) studied two pea genotypes (*Lathyrus sativus* L.), a glutathione deficient mutant (*gshL-1*) and the wild-type BioL-212 under (a) 40 μM As and (b) As + 1 mM BSO (L-buthionine-sulfoximin, an inhibitor of GSH synthesis). The authors reported that a major metabolic reorganization upstream and downstream the thiol cascade can confer As tolerance through stimulation of the GSH independent antioxidant system and maintenance of the Cys homeostasis in pea. Therefore, the engineering of GSH metabolism is a suitable strategy to increase metal(loid) tolerance in legumes. Pérez-Palacios et al. (2017) designed a double symbiotic system for Cu phytostabilization: *Medicago truncatula* plants were transformed with *mt4a* (metallothionein) from *Arabidopsis thaliana*. Besides, its symbiont, *Ensifer medicae,* was transformed with a copper resistance gene of *Pseudomonas fluorescens*. Thus, the *mt4a* expression in transgenic plants improved growth and nodulation and increased Cu tolerance. Authors reported lower levels of ROS-scavenger enzymes and thiobarbituric acid reactive substances (TBARs) content, which is a marker of lipid peroxidation, suggesting a reduction in oxidative stress. In addition, the inoculation of *M. truncatula* with the genetically modified *Ensifer medicae* strain further enhanced the Cu accumulation in roots, which led to low translocation factor (TF).

6.5.1.2 RNA-Based Applications

The discovery of functional RNA molecules, produced by non-coding RNA (ncRNA) genes, revolutionized the comprehension of gene expression regulation. The microRNAs (miRNAs) are ncRNAs involved in plant development and stress responses affecting target gene expression negatively in post-transcriptional events. As expected, in the last years, an increasing interest raised for the identification of miRNA related to drought stress response (Mantri et al. 2013). Then, 44 drought-associated miRNAs (30 upregulated and 14 downregulated upon stress imposition) were identified in cowpea (Barrera-Figueroa et al. 2011). Similarly, Wang et al. (2011) identified 22 miRNA that were upregulated and 10 miRNA that were downregulated in *M. truncatula* subjected to drought stress. Furthermore, Nageshbabu and Jyothi (2013) reported that miR395 was upregulated in a very sensitive manner in *Phaseolus vulgaris* L.exposed to 0.4 M NaCl and drought stress. Further, miR396 and miR172 were upregulated after exposure to similar conditions (Mantri et al. 2013; Nageshbabu and Jyothi 2013). Besides, Huang et al. (2017) explored the expression of a long non-coding miRNA gene that controls b-diketone wax formation, which could contribute to drought tolerance minimizing water loss. Then, these ncRNAs identified are a valuable source for exploring tolerance mechanisms in legumes exposed to drought and salt stresses.

6.5.1.3 Epigenetics

Increased interest has been developed in the study of heritable changes in gene expression that can occur due to methylation of DNA, histone modification, and RNA interference (RNAi), rather than changes in gene sequences, constituting a whole new research area, the epigenetics (Dwivedi et al. 2018). The identification of differentially methylated regions (DMRs) serves as a tool for comparing different growth conditions and its association with the phenotype in a determined species. In legumes, the last advances were reported in faba bean (*Vicia faba L.*), where five of the six drought-stressed DMRs reported had higher expression in drought tolerant (Bachar) than sensitive (F177) cultivars, suggesting its contribution to drought tolerance (Abid et al. 2017). Therefore, the polymorphism identification in DNA methylation can be considered to develop crops with increased drought tolerance. Recent methodological advances have made possible to describe methylomes at the very detailed level of a single base resolution (BS-seq) in soybean and chickpea (Shen et al. 2018; Piedra-Aguilera et al. 2019; Bhatia et al. 2018). Consequently, such tool can be used to decipher differential methylomes and the identification of candidate DMRs as markers of stress tolerance, as reported in *M. truncatula* salinized plants (Yaish et al. 2018). On the other hand, genome-wide cytosine methylation analysis was used for describing developmental changes in vegetative and reproductive tissues in soybean and common bean (Li et al. 2017; An et al. 2017; Kim et al. 2015). A breeding strategy was investigated for inducing epigenetic variation in soybean that may be of agronomical value (Raju et al. 2018). These authors used the MutS HOMOLOG1 (MSH1) system to induce epigenetic variation for agronomic traits and found that genes related to cell cycle, ABA biosynthesis, and auxin-response, were differentially expressed in lines that also showed increased yield. Thus, supporting the proposition that heritable phenotypic plasticity can be induced environmentally by epigenetic variation which may contribute to adaptation to challenging environments (Raju et al. 2018; Kumar et al. 2019).

6.5.1.4 Omics

Recent advances in omics technologies are providing opportunities leading to the identification of mechanisms involved in plant salt tolerance. Molecular technologies have made major advances over the last decade and genome sequences are now available for several legume species (Schmutz et al. 2014; Bertioli et al. 2016; Hirakawa et al. 2016; Vlasova et al. 2016; Hane et al. 2017). Moreover, microarrays and high-throughput RNA sequencing (RNA-seq) technologies have accelerated the global expression profiling of the relevant genes that respond to salinity (Vu et al. 2015; Conesa et al. 2016). These technologies improve the possibilities to decipher genes or transcripts that play essential role in the regulation of transcriptional machineries in legumes under salt conditions. Hiz et al. (2014) performed a comparative analysis of transcriptomes from salt-tolerant common bean genotypes grown under salinity conditions. The analysis revealed the up-regulation of several genes encoding Na^+/H^+ antiporters and vacuolar H^+ATPase. Besides, the transcription level of the gene coding for the non-selective cation channel increased in root from salt-tolerant genotypes exposed to stress. This channel could play a role in K^+ influx into root cells, helping them to maintain a high K^+/Na^+ ratio and to tolerate the salinity conditions. Recently, Zhang et al. (2018) performed the transcriptome analyses of two genotypes of *Strophostyles helvola* (halophytic vs glycophytic legume) stressed with NaCl. In this study, authors could demonstrated several factors as the analytic methods to generate the raw sequence, sequence alignment, quality control and differential expression studies. Therefore, this comparative expression profiling provides valuable resources of genomic data for understanding the salt tolerance in this species and other legumes. Furthermore, metabolomics comparisons provided new insights into salt tolerance mechanisms. For instance, metabolome profiling showed an accumulation of carbohydrates, fatty acids, phenylalanine, proline, and palmitic acid in the salt-tolerant *Glycine max* and *G. soja* plants as compared to salt-sensitive plants under salinity conditions (Zhang et al. 2018; Yang et al. 2017). Similar results were observed in the primary and secondary metabolite profiles of the halophytic legume *Prosopis strombulifera* (Llanes et al. 2016b). Several compounds (amino acids, carbohydrates, and organic acids, among others) were induced by salt in roots and leaves. Specific amino acids, such as leucine in roots, or peptides and cysteinyl-glycine in leaves, could be a response induced to maintain cell homeostasis during salinity. In addition, the elevated accumulation of carbohydrates reflected a strict regulation of several metabolites from glycolysis/gluconeogenesis pathways and some of these compounds (maltose, maltotriitol, sucrose, and cellobiose) could be involved in salt tolerance of this legume (Llanes et al. 2016b). Finally, the advances in omics approaches have been raised up together with the development of cost-efficient and precise phenotyping tools, necessary to dissect the molecular mechanisms associated with salt tolerance (Fernandez et al. 2017). The combined analysis of genotypic and phenotypic data will facilitate the identification and selection of genetic loci associated with essential features for the obtention of salt tolerant legumes.

6.5.2 Breeding Approaches

Legume survival under abiotic stress conditions can be enhanced through several strategies, from classical breeding to physiological genetic. Therefore, it is vital to grasp the germplasm collections that contain large amounts of genetic diversity, including landraces, wild relatives and varieties not used nowadays. It is known that landraces constitute a genetic pool bank as resulting from natural and artificial selection over centennials and constitute an important reservoir for traits to be used in stress-adaptation. Dwivedi et al. (2016) revealed that allelic variation between rice and wheat landraces allowed identifying agronomically helpful traits for abiotic stress tolerance. On the other hand, the knowledge of molecular markers, DNA regions tightly linked to agronomic traits in crops, will facilitate breeding strategies for crop improvement. Molecular markers have enabled researchers to characterize genetic diversity sooner and exactly, determine improved germplasms based on selected traits, choose genes that explain agronomic traits and develop molecular trials needed for scaling and breeding utilization. It is very important to develop platforms of cost-effective and superior genotyping combined with automation methodologies for phenotyping. Hence, it could encourage the genomic tool application in breeding programs and mark the continuity of an era of molecular improvements in legume genomics (Varshney et al. 2009).

It is well known that the breeding programmes in legumes consist of four stages: (a) making variations with hybridizations and induced mutations; (b) selection in early generations; (c) evaluation of selected lines; and (d) release of varieties (Toker and Mutlu 2011). Mutagenesis raises genetic variability and the probability of appearance of abiotic stresses resistance. Hybridization is widely used in breeding in order to obtain specific desirable traits within or between species. In several grain legumes, interspecific crosses posses variable success (Singh et al. 2018). For example, *Phaseolus vulgaris* crossed with its wild relative *Phaseolus acutifolius*, which has an elevated osmotic adjustment, allowed its transfer to cultivated beans (Basu et al. 2007). Another interesting germplasm accession was developed in a drought-related background. This was the case of chickpea (e.g., ICC 4958), which exhibited desired root traits, such as increased depth and density and also enhanced tolerance to terminal drought and elevated canopy temperature (Kashiwagi et al. 2005).

In vitro selection is an important step in breeding programmes, and can be used for selection of putative stress-resistant lines derived from both conventional breeding or transgenic approaches (Švábová and Lebeda 2005). Segregating populations of common beans tolerant to different abiotic stresses were studied; afterwards, superior genotypes were selected and used in conventional breeding. Furthermore, traits associated with drought resistance (avoidance and tolerance) were selected indirectly, taking into account the complexity of these traits.This profound analysis showed that the phenotypic differences exhibited in drought-stressed plants were associated with vigorous root system (Rubiales and Mikic, 2015; Graham and Vance 2003), effective use of water (Micheletto et al. 2007), efficient dry matter redistribution to grain (Fang et al. 2010; Farooq et al. 2016), biomass production at mid-pod filling (Siddique et al. 1999), phenological plasticity (Mittal et al. 2014) and earliness (Delmer 2005; Pushpavalli et al. 2015a). It was shown that the strategy of drought avoidance, i.e., early maturity, correlates with yield under stressful conditions; however, it has also disadvantages such as reduced yield potential. Taking together, the results showed that some phenotypic characters as days to physiological maturity, days to flowering and pod harvest index are more dependant on the genotype than on the environmental condition (Farooq et al. 2016).

Nowadays, there is a huge amount of sequences and expression data housed by the DNA Data Bank of Japan (DDBJ) (Mashima et al. 2016), GenBank (Benson et al. 2012) and Molecular Biology Laboratory (EMBL) (Kanz et al. 2005). Despite, the improvement of economically important crops depends on the integrated analysis of such data with other sequences variants and phenotypes. Nevertheless, the storage and management of that rising amount of public and private data is challenging (Batley and Edwards 2009; Lee et al. 2012). Therefore, it creates a strong demand on powerful computational platforms to be used for data storage, editing and processing. New database sources were developed by the United States National Plant Germplasm System (NPGS) and the European Cooperative Programme for Plant Genetic Resources Networks (ECPGR). Therefore, the availability of increasing germplasm information can be connected to large collections of genetic and phenotypic data. Thus, it could guarantee the accessibility of germplasm and its use in experimentation.

Soybean (Charlson et al. 2009; Du et al. 2009) and *M. truncatula* (Badri et al. 2011; Vandecasteele et al. 2011) are the most used legumes, among others, for studies of genetic and molecular basis of drought resistance via quantitative trait loci (QTL) or gene linkage and association mapping. Several research have been carried out to understand the molecular basis of salt tolerance in plants and some QTLs for traits associated with salt tolerance have been identified. Pushpavalli et al. (2015b) showed two key genomic regions which contain QTLs for salt tolerance in ICCV 2 (sensitive) × JG 11 (tolerant) derived chickpea (*Cicer arietinum*) recombinant native lines. Several putative candidate genes with roles in salt tolerance were identified. For instance, putative genes involved in ion transport encoding proteins of potassium channels and genes involved in ABA biosynthesis, metabolism and ABA-dependent signalling pathways. Genes involved in the biosynthesis of methionine and compatible solutes like glycine betaine were also identified. These genes could play a direct or indirect role in osmoregulation and help plants to cope with salinity and other abiotic stresses (Pushpavalli et al. 2015b). Similarly, studies combining QTL mapping and molecular specific gene expression in chickpea plants reported genes encoding for chlorophyll a/b binding proteins that regulate the photosynthetic efficiency and seed yield under abiotic stress conditions (Basu et al. 2018). In addition, genes encoding enzymes involved in photosynthesis, in the biosynthesis of carbohydrates, xyloglucans and amino acids, and several transcription factors of APETALA2/ethylene-responsive element binding protein (AP2/EREBP), WRKY and NAC families were up-regulated in stressed chickpea cultivars exhibiting increased tolerance to salinity (Garg et al. 2016).

Studies in the wild ancestor of *Glycine max*, *Glycine soja*, allowed to identify several genotypes with increased tolerance to salt and the genes coding for those traits (Qi et al. 2014; Jia et al. 2017). Particularly, *GmCHX1*, was detected by analyzing data from high-density-marker QTL mapping, whole genome *de novo* sequencing and genome resequencing. Jia et al. (2017) reported that this gene is a member of cation/H^+ exchanger superfamily from soybean and promotes high salinity and carbonate alkaline tolerance. However, it remains unclear if this gene will result in tradeoffs if used to develop salt-tolerant legumes. Finally, it is important to mention that the QTL results of several studies in legumes are environmentally dependent, and thus, to identify the trait associated with tolerance, it is required to integrate and compare QTL information from different assays.

The identification and mapping of QTLs for drought tolerance in common bean revealed twenty-two QTLs for chlorophyll, leaf temperature, leaf and stem biomass, days to flowering, number of seeds and pods per plant, seed and pod weight, and yield in well-watered and drought stress conditions. These detected QTLs were located in different chromosomes (Briñez et al. 2017). On the other hand, Diaz et al. (2018) identified a stable yield QTL on chromosome Pv04, in several locations tested and under different stress treatments and they suggested that it could serve as a good candidate for marker-assisted selection.

The increasing available data of germplasm resources from legumes and their wild relatives to gene discovery is of interest to elite-breeding germplasms. In this way, crops can be improved for increment and stabilize potential yields. Now, researchers have the challenge of integrating all the information contained in open databases, as a result of the rapid data generation, in order to accelerate the success of breeding programmes worldwide. In this context, genome editing raises as an opportunity that deserves analysis and consensus on its merits. It would help to satisfy the demands driven by global climate change and an increase growing population on plant breeders to supply high-yielding and climate-resilient cultivars. Genome editing is a robust tool that allows developing diversity on crop germplasm in a rapid and efficient manner. Then, yield-interesting traits associated with known genes can be selected for improving crops (Dwivedi et al. 2018). It is noteworthy that all the strategies mentioned rely on the knowledge of diverse germplasm pools under the influence of biotechnology advances.Thus, it can make safe global food in the future.

6.5.3 Agronomic Approaches to Improving Abiotic Stress Tolerance

6.5.3.1 Chemical Fertilizers

Legume yield in stress-prone areas can be improved at the very beginning by the adequate cultivar selection. Other simple strategies are adjustment in time of sowing, plant geometry, and fertilizer management (Jha et al. 2015). Besides, mulching, zero tillage, and deep ploughing in the rainy season confer drought stress tolerance to crops (Johnsona et al. 2018). The exogenous application of biological compounds arises as a mitigation strategy to abiotic stresses, although with limitations in terms of economic costs. The addition of exogenous compounds that confer plant tolerance, in particular proline, reversed the oxidative stress generated by As (Mishra and Dubey 2006; Chandrakar et al. 2018), and osmotic stresses (Hayat et al. 2012). Other molecules that provide As tolerance are sodium nitroprusside, a source of NO (Singh et al. 2013, 2017; Tripathi et al. 2017; Rodríguez-Ruiz et al. 2019), salicylic acid (SA) (Saidi et al. 2017; Odjegba 2012; Kang et al. 2015; Hemantaranjan et al. 2018; Guo 2019) and phosphate (Pi) (Niazi et al. 2017; Bustingorri et al. 2017). Talukdar (2016) investigated the effect of thiourea (TU) (a sulfhydryl bioregulator) application on lentils exposed to As. TU was able not only to sequester the metalloid at the root, but also prevented its translocation to shoots. In addition, TU coordinated the glyoxalase system (Gly) I and II (Gly I and II) and the redox state of ASC-GSH, and antioxidant enzymes in leaves and roots from the four lentil genotypes analyzed. The activation of the Gly system prevented the toxic over-accumulation of methylglyoxal, while ASC-GSH cycle enzymes, GST and catalase enzymes removed H_2O_2 and prevented the onset of oxidative damage. On the other hand, Xu et al. (2015) studied the interactive effects of salicylic acid (SA) and nitric oxide (NO) in the relief of Cd toxicity in two peanut cultivars (*Arachis hypogaea* L.). Exogenous SA and NO improved growth, chlorophyll content, photosynthesis and mineral nutrition, while decreased oxidative stress by increasing the antioxidant system activity. In addition, roots and leaves of both cultivars accumulated Cd in the cell wall and decreased the distribution of Cd in the organelles. Thus, exogenous SA and NO increased the tolerance of peanut plants.

6.5.3.2 Biofertilizers

Biofertilizers contain living microbes that stimulate growth and plant health by increasing the availability of nutrients to the plant. Thus, the most important agronomic practice that can be used in legume crops to overcome abiotic stresses is related with their ability to form symbiotic associations with diazotrophic bacteria, but legume association with other plant growth promoting bacteria (PGPBs) can be also beneficial. The mechanisms triggered by bacteria are direct (nitrogen fixation, phosphorus solubilization, production of siderophores or hormones) or indirect (through secretion of lytic enzymes, volatile compounds or inducing systemic resistance) (Glick 2012). Then, examples of successful cases of stress tolerance evoked by plant inoculation have been widely described (Vurukonda et al. 2016; Backer et al. 2018). Furthermore, arbuscular mycorrhizal fungi (AMF) can provide uptake of water and nutrients under drought stress (Habibzadeh 2015) and thus, have been strongly recommended as inoculants, and even better in combination with PGPBs (Stancheva et al. 2017).

Several legumes have been grown in areas contaminated with metal(loid)s, and were capable to establish associations with metal-tolerant rhizobia. This has increased the interest in deciphering the genetic determinants of metal(oid)s resistance in rhizobia, as well as its effectiveness to increase metal(loid)s tolerance of legume hosts (Fagorzi et al. 2018). Interestingly, the symbiotic interaction that legumes establish with rhizobia allows to improve phytostabilization processes, serving as a promising biotechnological tool to reduce the metal(loid) content in plants (Lafuente et al. 2015 ; Peralta et al. 2019). As a first step in the search for effective biofertilizers for legumes, their isolation from unfavorable environments is usually an adequate preliminary study. Thus, Rangel et al. (2017) isolated 24 strains from nodules of *Crotalaria spectabilis* and *Stizolobium landingimum*, two trap species that grew in a gold mine site contaminated with As. The 16S rRNA sequences analysis allowed to classify the isolated strains to the genera *Bradyrhizobium, Rhizobium, Inquilinus, Labrys, Bosea, Starkeya and Methylobacterium* of the α-proteobacteria group. Strains from *Bradyrhizobium, Rhizobium, Inquilinus* and *Labrys* tolerate As, Zn and Cd, were efficient in the association with their original hosts, and showed plant growth promoting traits *in vitro* (siderophores, indole-3-acetic acid and organic acids production, $Ca_3(PO_4)_2$ solubilization and 1-aminocyclopropane-1-carboxylate deaminase activity). The combined intrinsic abilities (plant growth promotion and metal(loid)s resistance) suggested that these strains could be used as legume inoculants in soils contaminated with toxic elements. On the other hand, studies carried out by Kang et al. (2015) demonstrated the beneficial effects of *Enterobacter asburiae* KE17 inoculation on soybean exposed to 100 μm Cu and 100 μm Zn. These heavy metals induced oxidative stress and a significant accumulation of free amino acids (aspartic acid, threonine, serine, glycine, alanine, leucine, isoleucine, tyrosine, proline and gamma-aminobutyric acid) in uninoculated plants. However, the inoculation with *E. asburiae* KE17 mitigated the stress effects by modulating the response of metabolic processes. Regarding the effect of aluminum (Al), Jaiswal et al. (2018) published a revision that revealed Al tolerance mechanisms in plants, rhizobia and soils. The authors explained that the over expression of the bacterial gene that codified for a citrate synthase in alfalfa, led to Al tolerance. In addition, the synthesis and release of organic acids and metabolic regulation by plant-secreted microRNAs also induced tolerance. In rhizobia, the Al tolerance is related to exopolysaccharides (EPS) and siderophores (limiting Al uptake) production, induction of efflux systems, and the expression of metal-inducible (*dmeRF*) gene clusters. Finally, in soils, the presence of organic matter, lime and the use of Al tolerant species reduced the metal toxicity.

Alfalfa plants exposed to different metal(loid)s (Cd, Cu, Co, Zn, Pb, Ni or Hg) were co-inoculated with equal amounts of *E. meliloti* 1021 and *Ochrobactum cytisi* Azn6.2, metal-sensitive and metal resistant strains, respectively. Interestingly, nodules were mainly occupied by *O. cytisi* Azn6.2, as metals concentration increased, suggesting that metal-resistant bacterial strains can take the place of the natural symbiont. This study demonstrated the importance of select tolerant rhizobia to be inoculated on plants, due to their ability to be competitive in contaminated sites (Navarro-Torre et al. 2019). In a genetic study conducted by Hao et al. (2015) in *Robinia pseudoacacia*, the mutagenesis of the microbial symbiont *Mesorhizobium amorphae* in *copA* genes (encoding a P-type ATPase) did not affect the ability of rhizobia to infect plants under Cu exposure, reducing Cu accumulation on root tissues. Bianucci et al. (2018) evaluated the effect of As on soybean plants inoculated with native and reference strains. They observed that the accumulation of the metalloid in the plant followed the same distribution pattern among strains. However, at a low metalloid concentration (6 μM As), legume inoculation with the native strain *Bradyrhizobium* sp. PER 3.61 revealed a significantly lower TF compared to other inoculated strains, promoting phytostabilization in soybean. Nevertheless, when the plant was subjected to middle As concentration, only the reference strain *Bradyrhizobium diazoefficiens* USDA110 induced nodulation on soybean plants. These findings demonstrated that the selection of specific tolerant strains is advisable for different environmental conditions. In an experiment conducted in soybean plants exposed to a natural dose of As, as found in the groundwater of crop areas of Argentina, Bianucci et al. (2019) reported that As reduced growth and induced metalloid translocation to the aerial part. Besides, As triggered an oxidative burst which further activated the antioxidant system. The inoculation with a rhizobia that promote phytostabilization of As did not ameliorate the deleterious effects of As. These findings contribute to explain the possible adverse effect of realistic As concentrations on crop yields and human health. In addition, these authors reported a strong induction of GST activity in soybean plants grown under natural As doses. Thus, they proposed the use of GST activity as a biomarker of As contamination at field conditions. Therefore, it is necessary to select the best PGPB-plant interaction to achieve promising results and avoid metal toxicity. On the other hand, in *Medicago* plants, the inoculation of As-tolerant rhizobia reversed the toxicity imposed by the metalloid and improved the activity of an enzyme involved in the first step of symbiosis (chalcone synthase). In addition, mycorrhiza have been used as a strategy to counteract metal toxicity due to their ability to accumulate high metal(loid)s concentrations. In this sense, Alam et al. (2019) evaluated the effects of the combined addition of selenium (Se), biochar (BC) and arbuscular mycorrhizal fungi (AMF) in soils. They revealed that the inoculation of AMF was the main factor that contributed to minimize the oxidative stress symptoms in mung bean crops.

In a study conducted by Zhang et al. (2017), red clover plants were inoculated with wild-type or recombinant *R. leguminosarum* expressing an arsenite [As(III)] S-adenosylmethionine methyltransferase gene from the algae *Chlamydomonas reinhardtii (CrarsM)* and exposed to 5 or 10 μM As. Based on their results, it was proven that the recombinant rhizobia got the ability to methylate As forming methylarsenate [MA(V)], dimethylarsenate [DMA(V)] and trimethylarsine oxide [TMA(V)O]. Thus, methylated and volatile inorganic As forms can contribute to As bioremediation mediated by engineered rhizobia. In another study, Deepika et al. (2016) explored the tolerance mechanisms of a native strain, tolerant to As and able to produce EPS, isolated from root nodules of *Vigna radiata*. The strain identified as *Rhizobium radiobacter* confirmed the role of EPS in arsenate sequestration. In a recent study, Das and Sarkar (2018) showed that inoculation of mung beans (*Vigna radiata*), exposed to As with a tolerant strain *Acinetobacter lwoffii* (RJB-2), generated a limited adsorption of the metalloid from the soil and induced plant growth. The tolerant capacity of RJB-2 was associated with the production of plant growth

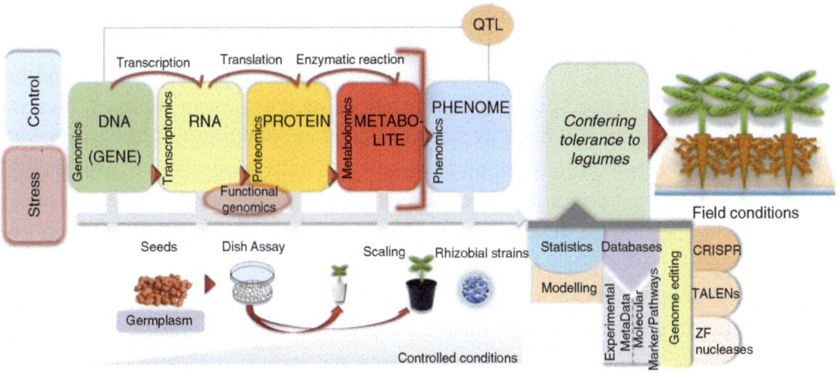

Fig. 6.2 Approaches to confer legume tolerance. The analysis of genome sequences is a very important starting point, however, it is not enough in order to decipher and describe gene functions. Assays at the transcriptomic, proteomic, metabolomic and phenotypic level are required. Data-set analysis using statistical methods and database information (QTL identification) allow identifying and modeling key factors involved in stress responses. In recent years, the genome editing (GEd) (CRISPR, TALENs and ZF nucleases) acquired special importance. Results obtained from legumes grown under controlled conditions can be extrapolated (scaling) to commercially important legume species growing in field conditions and generate agronomic benefits

promoting substances: indole acetic acid (IAA), siderophores, EPS and phosphate solubilization, both in the absence and in the presence of As. In addition, the inoculation reduced ROS production in mung bean plants, reducing oxidative damage caused by the metalloid. The authors proposed the potential use of this strain in large-scale crops for metalloid phytostabilization.

The Fig. 6.2 represents the current knowledge of the different approaches used to confer legume tolerance in response to abiotic challenges such as salinity, drought and metal(loid)s. In this sense, biotechnology, breeding or agronomic tools used are summarized.

6.6 Conclusion

Several abiotic stresses, including salinity, drought and metal(loid)s exposure, adversely affect plant growth and crop development. The increasing amount of experimental data on multiple signalling components, in agronomically important legumes, will further broaden insight into the relevance and significance of tolerance mechanisms mediating plant responses to stress conditions. Moreover, it could lead to improving yield potential, particularly in areas with adverse environmental conditions. In these conditions, legumes exhibit various adaptations to survive, for example, osmotic balance by compatible solutes accumulation, detoxification of harmful oxygen species by antioxidant defense system and orchestrating of coordinated responses by hormonal regulation. In the past decades, an increasing interest to develop legume varieties with improved yield upon unfavourable environmental conditions challenges has been exploited. Exploration, analysis and understanding of undiscovered adaptive traits and their incorporation into advanced perform genotypes are required, being breeding approaches of utmost importance. Also, there is an uprising interest in emerging 'omics' sciences such as transcriptomics, metabolomics and proteomics. These tools improve in an outstanding manner the current knowledge of the abiotic tolerance mechanisms and help researchers to decipher the taking place gene networks and signalling events involved in legume responses. Recent studies identified molecular markers, quantitative trait loci (QTLs) and genes associated with abiotic stress tolerance which can help to improve growth and yield. However, other opportunities for research about new emerging frontiers such as epigenetics, genome editing (GEd) and plant-soil microorganisms interactions are necessary.

References

Abbas G, Murtaza B, Bibi I, Shahid M, Niazi NK, Khan MI, Hussain M (2018) Arsenic uptake, toxicity, detoxification, and speciation in plants: physiological, biochemical, and molecular aspects. Int J Environ Res and Public Health 15(1):59. https://doi.org/10.3390/ijerph15010059

Abid G, Mingeot D, Muhovski Y, Mergeai G, Aouida M, Abdelkarim S (2017) Analysis of DNA methylation patterns associated with stress response in faba bean (*Vicia faba* L.) using methylation-sensitive amplification polymorphism (MSAP). Environ Exp Bot 142:34–44. https://doi.org/10.1016/j.envexpbot.2017.08.004

Acharjee S, Sarmah BK (2013) Biotechnologically generating 'super chickpea' for food and nutritional security. Plant Sci 207:108–116. https://doi.org/10.1016/j.plantsci.2013.02.003

Ahanger MA, Tomar NS, Tittal M, Argal S, Agarwal RM (2017) Plant growth under water/salt stress: ROS production; antioxidants and significance of added potassium under such conditions. Physiol Mol Biol Plants 23(4):731–744. https://doi.org/10.1007/s12298-017-0462-7

Al-Ghzawi AA, Zaitoun S, Gosheh HZ, Alqudah AM (2009) The impacts of drought stress on bee attractively and flower pollination of *Trigonella moabitica* (fabaceae). Arch Agron Soil Sci 55:683–692. https://doi.org/10.1080/03650340902821666

Alam MZ, Hoque MA, McGee R, Boggs LC (2019) Arbuscular Mycorrhizal Fungus (AMF) and reduction of arsenic uptake in lentil crops. BioRxiv 522714. https://doi.org/10.1371/journal.pone.0211441

Allahmoradi P, Mansourifar C, Saiedi M, JalaliHonarmand S (2013) Effect of different water deficiency levels on some antioxidants at different growth stages of lentil (*Lens culinaris* L.). Adv Environ Biol 7:535–543

Alvarez-Gerding X, Cortés-Bullemore R, Medina C, Romero-Romero JL, Inostroza-Blancheteau C, Aquea F (2015) Improved salinity tolerance in carrizocitrange rootstock through overexpression of glyoxalase system genes. Biomed Res Int 2015:1–7. https://doi.org/10.1155/2015/827951

Alyemeni MN, Hayat Q, Hayat S, Faizan M, Faraz A (2016) Exogenous proline application enhances the efficiency of nitrogen fixation and assimilation in chickpea plants exposed to cadmium. Legume Res 39(2). https://doi.org/10.18805/lr.v0iOF.9291

An J, Hu Z, Che B, Chen H, Yu B, Cai W (2017) Heterologous expression of *Panax ginseng* PgTIP1 confers enhanced salt tolerance of soybean cotyledon hairy roots, composite, and whole plants. Front Plant Sci 8:12–32. https://doi.org/10.3389/fpls.2017.01232

Andersen MN, Asch F, Wu Y, Jensen CR, Næsted H, Mogensen VO, Koch KE (2002) Soluble invertase expression is an early target of drought stress during the critical, abortion–sensi-

tive phase of young ovary development in maize. Plant Physiol 130:591–604. https://doi.org/10.1104/pp.005637
Andrésa Z, Pérez-Hormaechea J, Leidia EO, Schlücking K, Leonie S, McLachlanc DH, Schumacher K, Hetherington AM, Kudlab J, Cuberoa B, Pardo JM (2014) Control of vacuolar dynamics and regulation of stomatal aperture by tonoplast potassium uptake. Proc Nat Acad Sci 111:1806–1814. https://doi.org/10.1073/pnas.1320421111
Anjum F, Yaseen M, Rasul E, Wahid A, Anjum S (2003) Water stress in barley (*Hordeum vulgare* L.). I. Effect on chemical composition and chlorophyll contents. Pak J Agric Sci 40:45–49
Annor GA, Ma Z, Boye JI (2014) Crops – legumes. In: Jung SCS, Lamsal B (eds) Food processing: principles and applications, vol 14, 2nd edn. Wiley, New Delhi, pp 305–337. https://doi.org/10.1002/9781118846315.ch14
Apel K, Hirt H (2004) Reactive oxygen species: metabolism, oxidative stress, and signal transduction. Annu Rev Plant Biol 55:373–399. https://doi.org/10.1146/annurev.arplant.55.031903.141701
Araújo SS, Duque AS, Silva JM, Santos D, Silva AB, Fevereiro P (2013) Water deficit and recovery response of *Medicago truncatula* plants expressing the ELIP-like DSP22. Biol Plant 57:159. https://doi.org/10.1007/s10535-012-0235-7
Araújo SS, Beebe S, Crespi M, Delbreil B, González EM, Gruber V, Lejeune-Henaut I, Link W, Monteros MJ, Prats E, Rao I, VadezV VPMC (2015) Abiotic stress responses in legumes: strategies used to cope with environmental challenges. Crit Rev Plant Sci 34(1–3):237–280. https://doi.org/10.1080/07352689.2014.898450
Arbona V, Hossain Z, López-Climent MF, Pérez-Clemente RM, Gómez-Cadenas A (2008) Antioxidant enzymatic activity is linked to waterlogging stress tolerance in citrus. Physiol Plant 132:452–466. https://doi.org/10.1111/j.1399-3054.2007.01029.x
Arbona V, Manzi M, Ollas CD, Gómez-Cadenas A (2013) Metabolomics as a tool to investigate abiotic stress tolerance in plants. Int J Mol Sci 14:4885–4911. https://doi.org/10.3390/ijms14034885
Arzani A, Ashraf M (2016) Smart engineering of genetic resources for enhanced salinity tolerance in crop plants. Crit Rev Plant Sci 35(3):146–189. https://doi.org/10.1080/07352689.2016.1245056
Ashraf M, Iram A (2005) Drought stress induced changes in some organic substances in nodules and other plant parts of two potential legumes differing in salt tolerance. Flora 200:535–546. https://doi.org/10.1016/j.flora.2005.06.005
Babar MM, Zaidi NSS, Azooz MM (2015) Biotechnology approaches to overcome biotic and abiotic stress constraints in legumes. In: Azooz MM, Ahmad P (eds) Legumes under environmental stress: yield, improvement and adaptations, 1st edn. Wiley, Chichester, pp 247–226. https://doi.org/10.1007/s10681-006-6156-9
Backer R, Rokem JS, Ilangumaran G, Lamont J, Praslickova D, Ricci E, Subramanian S, Smith DL (2018) Plant growth-promoting rhizobacteria: context, mechanisms of action, and roadmap to commercialization of biostimulants for sustainable agriculture. Front Plant Sci 9:1473. https://doi.org/10.3389/fpls.2018.01473
Balestrasse KB, Gardey L, Gallego SM, Tomaro ML (2001) Response of antioxidant defence system in soybean nodules and roots subjected to cadmium stress. Aust J Plant Physiol 28:497–504. https://doi.org/10.1071/PP00158
Balestrasse KB, Benavides MP, Gallego SM, Tomaro ML (2003) Effect of cadmium stress on nitrogen metabolism in nodule and roots of soybean plants. Funct Plant Biol 30:57–64. https://doi.org/10.1071/FP02074
Balestrasse KB, Gallego SM, Tomaro ML (2006) Oxidation of the enzymes involved in nitrogen assimilation plays an important role in the cadmium-induced toxicity in soybean plants. Plant Soil 284:187–194. https://doi.org/10.1007/s11104-006-0050-z
Banerjee A, Roychoudhury A (2018) Strigolactones: multi-level regulation of biosynthesis and diverse responses in plant abiotic stresses. Acta Physiol Plant 40(5):86. https://doi.org/10.1007/s11738-018-2660-5

Bao AK, Du BQ, Touil L, Kang P, Wang QL, Wang SM (2016) Co-expression of tonoplast Cation/H^+ antiporter and H^+-pyrophosphatase from xerophyte *Zygophyllum xanthoxylum* improves alfalfa plant growth under salinity, drought and field conditions. Plant Biotechnol J 14(3):964–975. https://doi.org/10.1111/pbi.12451

Baroowa B, Gogoi N (2013) Biochemical changes in two *Vigna* spp. during drought and subsequent recovery. Indian J Plant Physiol 18:319–325. https://doi.org/10.1007/s40502-013-0048-5

Baroowa B, Gogoi N (2014) Biochemical changes in black gram and green gram genotypes after imposition of drought stress. J Food Legumes 27:350–353. Online ISSN: 0976-2434

Barrera-Figueroa B, Gao L, Diop NN, Wu Z, Ehlers JD, Roberts PA, Close TJ, Zhu JK, Liu E (2011) Identification and comparative analysis of drought-associated microRNAs in two cowpea genotypes. BMC Plant Biol 11:127. https://doi.org/10.1186/1471-2229-11-127

Basu PS, Berger JD, Turner NC, Chaturvedi SK, Ali M, Siddique KHM (2007) Osmotic adjustment of chickpea (*Cicer arietinum* L.) is not associated with changes in carbohydrate composition or leaf gas exchange under drought. Ann Appl Biol 150:217–225. https://doi.org/10.1111/j.1744-7348.2007.00119.x

Basu S, Ramegowda V, Kumar A, Pereira A (2016) Plant adaptation to drought stress. F1000 Res 5:1554. https://doi.org/10.12688/f1000research.7678.1

Basu U, Bajaj D, Sharma A, Malik N, Daware A, Narnoliy L, Parida SK (2018) Genetic dissection of photosynthetic efficiency traits for enhancing seed yield in chickpea. Plant Cell Environ 42:158–173. https://doi.org/10.1111/pce.13319

Batley J, Edwards D (2009) Genome sequence data: management, storage, and visualization. Biotechniques 46:333–334. https://doi.org/10.2144/000113134

Belachew KY, Nagel KA, Fiorani F, Stoddard FL (2018) Diversity in root growth responses to moisture deficit in young faba bean (*Vicia faba* L.) plants. Peer J 6:e4401. https://doi.org/10.7717/peerj.4401

Benjamin JG, Nielsen DC (2006) Water deficit effects on root distribution of soybean, field pea and chickpea. Field Crops Res 97:248–253. https://doi.org/10.1016/j.fcr.2005.10.005

Benson DA, Karsch-Mizrachi I, Clark K, Lipman DJ, Ostell J, Sayers EW (2012) GenBank. Nucleic Acids Res 40:48–53. https://doi.org/10.1093/nar/gkr1202

Berger JD, Shrestha D, Ludwig C (2017) Reproductive strategies in mediterranean legumes: trade-offs between phenology, seed size and vigor within and between wild and domesticated lupinus species collected along aridity gradients. Front Plant Sci 8:548. https://doi.org/10.3389/fpls.2017.00548

Bertioli DJ, Cannon SB, Froenicke L, Huang G, Farmer AD, Cannon EK, Ozias-Akins P (2016) The genome sequences of *Arachis duranensis* and *Arachis ipaensis*, the diploid ancestors of cultivated peanut. Nat Genet 48:438–446. https://doi.org/10.1038/ng.3517

Bhatia H, Khemka N, Jain M, Garg R (2018) Genome-wide bisulphite-sequencing reveals organ-specific methylation patterns in chickpea. Sci Rep 8:9704. https://doi.org/10.1038/s41598-018-27979-w

Bianucci E, Fabra A, Castro S (2012) Involvement of glutathione and enzymatic defense system against cadmium toxicity in *Bradyrhizobium* sp. strains (peanut symbionts). Biometals 25(1):23–32. https://doi.org/10.1007/s10534-011-9480-z

Bianucci E, Furlan A, Castro S (2017a) Importance of glutathione in the legume-rhizobia symbiosis. In: Glutathione in plant growth, development, and stress tolerance. Springer, Cham, pp 373–396. https://doi.org/10.1007/978-3-319-66682-2_17

Bianucci E, Furlan A, Tordable MC, Hernández L, Carpena-Ruiz RO, Castro S (2017b) Antioxidant responses of peanut roots exposed to realistic groundwater doses of arsenate: identification of glutathione S- transferase as a suitable biomarker for metalloid toxicity. Chemosphere 181:551–561. https://doi.org/10.1016/j.chemosphere.2017.04.104

Bianucci E, Godoy A, Furlan A, Peralta JM, Hernández LE, Carpena-Ruiz RO, Castro S (2018) Arsenic toxicity in soybean alleviated by a symbiotic species of *Bradyrhizobium*. Symbiosis 74(3):167–176. https://doi.org/10.1007/s13199-017-0499-y

Bianucci E, Furlan A, Hernandez LE, Castro S (2019) Insights into the effect of a natural arsenate dose on growth, nodulation and redox metabolism of soybean plants. Pedosphere 29(4):527–533. https://doi.org/10.1016/S1002-0160(19)60816-7

Blokhina O, Virolainen E, Fagerstedt KV (2003) Antioxidants, oxidative damage and oxygen deprivation stress: a review. Ann Bot 91:179–194. https://doi.org/10.1093/aob/mcf118

Blum A (2017) Osmotic adjustment is a prime drought stress adaptive engine in support of plant production. Plant Cell Environ 40(1):4–10. https://doi.org/10.1111/pce.12800

Bota J, Flexas J, Medrano H (2004) Is photosynthesis limited by decreased Rubisco activity and RuBP content under progressive water stress? New Phytol 162:671–681. https://doi.org/10.1093/aob/mcn244

Briñez B, Perseguini JMKC, Rosa JS, Bassi D, Gonçalves JGR, Almeida C, de Carvalho Paulino JF, Blair MW, Chioratto AF, Carbonell SAM, Valdisser PAMR, Vianello RP, Benchimol-Reis LL (2017) Mapping QTLs for drought tolerance in a SEA 5 x AND 277 common bean cross with SSRs and SNP markers. Gen Mol Biol 40(4):813–823. https://doi.org/10.1590/1678-4685-GMB-2016-0222

Bustingorri C, Noriega G, Lavado RS, Balestrasse K (2017) Protective effect exerted by soil phosphorus on soybean subjected to arsenic and fluoride. Redox Report 22(6):353–360. https://doi.org/10.1080/13510002.2016.1276253

Cabello JV, Giacomelli JI, Gómez MC, Chan RL (2017) The sunflower transcription factor HaHB11 confers tolerance to water deficit and salinity to transgenic Arabidopsis and alfalfa plants. J Biotechnol 257:35–46. https://doi.org/10.1016/j.jbiotec.2016.11.017

Campos T, Oblessuc PR, Sforça DA, Cardoso JMK, Baroni RM, de Sousa ACB, Carbonel SAM, Chioratto AF, Garcia AAF, Rubiano LB (2011) Inheritance of growth habit detected by genetic linkage analysis using microsatellites in the common bean (*Phaseolus vulgaris* L.). Mol Breed 27:549–560. https://doi.org/10.1007/s11032-010-9453-x

Carpena R, Esteban E, Lucena JJ, Peñalosa S, Vázquez P, Zornoza P, Gárate A (2006) Simbiosis y fitorrecuperación de suelos. En: EJ Bedmar, J González, C Lluch, MB Rodelas, Fijación de Nitrógeno: Fundamentos y Aplicaciones. Soc Esp Fijac Nit, Granada, 255–268

Chandrakar V, Dubey A, Keshavkant S (2018) Modulation of arsenic-induced oxidative stress and protein metabolism by diphenyleneiodonium, 24-epibrassinolide and proline in *Glycine max* L. Acta Botanica Croatica 77(1):51–61. https://doi.org/10.2478/botcro-2018-0004

Chang YC, Zouari Y, Lucena GJJ, Abadía J (2003) Effect of cadmium and lead on ferric chelate reductase activities in sugar beet roots. Plant Physiol Biochem 41:999–1005. https://doi.org/10.1016/j.ecoenv.2014.01.036

Charlson DV, Bhatnagar S, King CA, Ray JD, Sneller CH, Carter TE, Purcell LC (2009) Polygenic inheritance of canopy wilting in soybean [*Glycine max* (L.) Merr.]. Theoretic Appl Gen 119:587–594. https://doi.org/10.1007/s00122-009-1068-4

Chatterjee P, Samaddar S, Niinemets Ü, Sa TM (2018) Brevibacterium linens RS16 confers salt tolerance to *Oryza sativa* genotypes by regulating antioxidant defense and H^+ ATPase activity. Microbiol Res 215:89–101. https://doi.org/10.1016/j.micres.2018.06.007

Chen Y, Xu W, Shen H, Yan H, Xu W, He Z, Ma M (2013) Engineering arsenic tolerance and hyperaccumulation in plants for phytoremediation by a PvACR3 transgenic approach. Environl Sci Tech 47(16):9355–9362. https://doi.org/10.1021/es4012096

Chhabra R (2017) Soil salinity and water quality. Routledge. https://doi.org/10.1201/9780203739242

Cho SH, von Schwartzenberg K, Quatrano R (2018) The role of abscisic acid in stress tolerance. Annu Plant Rev Online 282–297 https://doi.org/10.1002/9781119312994.apr0394

Choi WG, Miller G, Wallace I, Harper J, Mittler R, Gilroy S (2017) Orchestrating rapid long-distance signaling in plants with Ca2+, ROS and electrical signals. Plant J 90(4):698–707. https://doi.org/10.1111/tpj.13492

Choudhury FK, Rivero RM, Blumwald E, Mittler R (2017) Reactive oxygen species, abiotic stress and stress combination. Plant J 90(5):856–867. https://doi.org/10.1111/tpj.13299

Clemens S, Ma JF (2016) Toxic heavy metal and metalloid accumulation in crop plants and foods. Ann Rev Plant Biol 67:489–512. https://doi.org/10.1146/annurev-arplant-043015-112301

Conesa A, Madrigal P, Tarazona S, Gomez-Cabrero D, Cervera A, McPherson A, Mortazavi A (2016) A survey of best practices for RNA-seq data analysis. Genome Biol 17(1):13. https://doi.org/10.1186/s13059-016-0881-8

Croser J, Ahmad F, Clarke H, Siddique KHM (2003) Utilisation of wild *Cicer* in chickpea improvement – progress, constraints and prospects. Aust J Agric Res 54:429–444. https://doi.org/10.1071/AR02157

Cruz de Carvalho MH (2008) Drought stress and reactive oxygen species. Production, scavenging and signaling. Plant Signal Behav 3:156–165. https://doi.org/10.4161/psb.3.3.5536

Das K, Roychoudhury A (2014) Reactive oxygen species (ROS) and response of antioxidants as ROS-scavengers during environmental stress in plants. Front Environ Sci 2:53. https://doi.org/10.3389/fenvs.2014.00053

Das J, Sarkar P (2018) Remediation of arsenic in mung bean (*Vigna radiata*) with growth enhancement by unique arsenic-resistant bacterium *Acinetobacter lwoffii*. Sci Total Environ 624:1106–1118. https://doi.org/10.1016/j.scitotenv.2017.12.157

Deepika KV, Raghuram M, Kariali E, Bramhachari PV (2016) Biological responses of symbiotic *Rhizobium radiobacter* strain VBCK1062 to the arsenic contaminated rhizosphere soils of mung bean. Ecotoxicol Environ Saf 134:1–10. https://doi.org/10.1016/j.ecoenv.2016.08.008

Degola F, Fattorini L, Bona E, Sprimuto CT, Argese E, Berta G, di Toppi LS (2015) The symbiosis between *Nicotiana tabacum* and the endomycorrhizal fungus *Funneliformis mosseae* increases the plant glutathione level and decreases leaf cadmium and root arsenic contents. Plant Physiol Biochem 92:11–18. https://doi.org/10.1016/j.plaphy.2015.04.001

Delmer DP (2005) Agriculture in the developing world: Connecting innovations in plant research to downstream applications. Proc Natl Acad Sci USA 102:15739–15746. https://doi.org/10.1073/pnas.0505895102

Diaz LM, Ricaurte J, Tovar E, Cajiao C, Terań H, Grajales M (2018) QTL analyses for tolerance to abiotic stresses in a common bean (Phaseolus vulgaris L.) population. PLoS ONE 13(8):e0202342. https://doi.org/10.1371/journal.pone.0202342

Dong J, Wu FB, Zhang GP (2006) Influence of cadmium on antioxidant capacity and four microelement concentrations in tomato seedlings (*Lycopersicon esculentum*). Chemosphere 64:1659–1666. https://doi.org/10.1016/j.chemosphere.2006.01.030

Du WJ, Fu SX, Yu DY (2009) Genetic analysis for the leaf pubescence density and water status traits in soybean [*Glycine max* (L.) Merr.]. Plant Breeding 128:259–265. https://doi.org/10.1111/j.1439-0523.2008.01588.x

Duan GL, Zhou Y, Tong YP, Mukhopadhyay R, Rosen BP, Zhu YG (2007) A CDC25 homologue from rice functions as an arsenate reductase. New Phytol 174(2):311–321. https://doi.org/10.1111/j.1469-8137.2007.02009.x

Dubey S, Shri M, Gupta A, Rani V, Chakrabarty D (2018) Toxicity and detoxification of heavy metals during plant growth and metabolism. Environm Chem Let 16(4):1169–1192. https://doi.org/10.1007/s10311-018-0741-8

Dwivedi SL, Salvatore C, Blair MW, Upadhyaya HD, Are AK, Ortiz R (2016) Landrace germplasm for improving yield and abiotic stress adaptation. Trends Plant Sci 21(1):31–42. https://doi.org/10.1016/j.tplants.2015.10.012

Dwivedi SL, Siddique KHM, Farooq M, Thornton PK, Ortiz R (2018) Using biotechnology-led approaches to uplift cereal and food legume yields in dryland environments. Front Plant Sci 9:1249. https://doi.org/10.3389/fpls.2018.01249

Fagorzi C, Checcucci A, DiCenzo GC, Debiec-Andrzejewska K, Dziewit L, Pini F, Mengoni A (2018) Harnessing rhizobia to improve heavy-metal phytoremediation by legumes. Genes 9(11):542. https://doi.org/10.3390/genes9110542

Fang X, Turner NC, Yan G, Li F, Siddique KHM (2010) Flower numbers, pod production, pollen viability, and pistil function are reduced and flower and pod abortion increased in chickpea

(*Cicer arietinum* L.) under terminal drought. J Exp Bot 61(2):335–345. https://doi.org/10.1093/jxb/erp307

FAO (2015) FAO land and plant nutrition management service. Food and Agriculture Organization of the United Nations, Rome

Farooq M, Wahid A, Kobayashi N, Fujita D, Basra SMA (2009) Plant drought stress: effects, mechanisms and management. Agron Sustain Dev 29:185–212. https://doi.org/10.1051/agro:2008021

Farooq M, Hussain M, Wakeel A, Siddique KH (2015) Salt stress in maize: effects, resistance mechanisms, and management. A review. Agron Sustain Dev 35(2):461–481. https://doi.org/10.1007/s13593-015-0287-0

Farooq M, Gogoi N, Barthakur S, Baroowa B, Bharadwaj N, Alghamd SS (2016) Stress in grain legumes during reproduction and grain filling. J Agron Crop Sci 203:81–102. https://doi.org/10.1111/jac.12169

Farooq M, Gogoi N, Barthakur S, Baroowa B, Bharadwaj N, Alghamdi SS, Siddique KHM (2017) Drought stress in grain legumes during reproduction and grain filling. J Agro Crop Sci 203:81–102. https://doi.org/10.1111/jac.12169

Fernandez MGS, Bao Y, Tang L, Schnable PS (2017) A high-throughput, field-based phenotyping technology for tall biomass crops. Plant Physiol 174(4):2008–2022. https://doi.org/10.1104/pp.17.00707

Filová A, Krivosudská E (2017) Significance of the osmotic adjustment of soybean (*Glycine max* L.) Selected genotypes during increasing drought conditions. Res J Agric Sci 49(1):188–195

Finnegan P, Chen W (2012) Arsenic toxicity: the effects on plant metabolism. Front Physiol 3:182. https://doi.org/10.3389/fphys.2012.00182

Flora SJ (2011) Arsenic-induced oxidative stress and its reversibility. Free Rad Biol Med 51(2):257–281. https://doi.org/10.1016/j.freeradbiomed.2011.04.008

Fodor F, Sárvári E, Láng F, Szigeti Z, Cseh E (1996) Effects of Pb and Cd on Cucumber Depending on the Fe-Complex in the Culture Solution. J Plant Physiol 148:434–439. https://doi.org/10.1016/S0176-1617(96)80276-8

Foyer CH, Noctor G (2005) Redox homeostasis and antioxidant signaling: A metabolic interface between stress perception and physiological responses. Plant Cell 17:1866–1875. https://doi.org/10.1105/tpc.105.033589

Foyer C, Lam HM, Nguyen H, Siddique K, Varshney R, Colmer TD, Cowling W, Bramley H, Mori T, Hodgson J, Cooper J, Miller A, Kunert K, Vorster B, Cullis C, Ozga J, Wahlqvist M, Liang Y, Shou H, Considine M (2016) Neglecting legumes has compromised human health and sustainable food production. Nature Plants:2. https://doi.org/10.1038/nplants.2016.112

Fraser CM, Chapple C (2011) The phenylpropanoid pathway in *Arabidopsis*. Arab Book 9:e0152. https://doi.org/10.1199/tab.0152

Fu J, Huang B (2001) Involvement of antioxidants and lipid peroxidation in the adaptation of two cool-season grasses to localized drought stress. Environ Exp Bot 45:105–114. https://doi.org/10.1016/S0098-8472(00)00084-8

Fukudome M, Watanabe E, Osuki KO, Imaizumi R, Aoki T, Becana M, Uchiumi T (2018) Stably transformed *Lotus japonicus* plants overexpressing phytoglobin *LjGlb1-1* show decreased nitric oxide levels in roots and nodules as well as delayed nodule senescence. Plant Cell Physiol 4(60):816–825. https://doi.org/10.1093/pcp/pcy245

Furlan A, Bianucci E, Castro S, Dietz KJ (2017) Metabolic features involved in drought stress tolerance mechanisms in peanut nodules and their contribution to biological nitrogen fixation. Plant Sci 263:12–22. https://doi.org/10.1016/j.plantsci.2017.06.009

Garg N, Singla P (2011) Arsenic toxicity in crop plants: Physiological effects and tolerance mechanisms. Environ Chem Let 9(3):303–321. https://doi.org/10.1007/s10311-011-0313-7

Garg R, Shankar R, Thakkar B, Kudapa H, Krishnamurthy L, Mantri N, Varshney RK, Bhatia S, Jain M (2016) Transcriptome analyses reveal genotype- and developmental stage- specific molecular responses to drought and salinity stresses in chickpea. Sci Rep 6:19228. https://doi.org/10.1038/srep19228

Ghanbari AA, Mousavi SH, Mousapour-Gorgi A, Rao IM (2013) Effects of water stress on leaves and seeds of bean (*Phaseolus vulgaris* L.). Turk J Field Crops 18:73–77

Ghanti SKK, Sujata KG, Kumar BV, Karba NN, Janardhan Reddy K, Rao MS, Kishor PK (2011) Heterologous expression of *P5CS* gene in chickpea enhances salt tolerance without affecting yield. Biol Plant 55(4):634. https://doi.org/10.1007/s10535-011-0161-0

Ghosh A (2017) Genome-wide identification of glyoxalase genes in *Medicago truncatula* and their expression profiling in response to various developmental and environmental stimuli. Front Plant Sci 8:836. https://doi.org/10.3389/fpls.2017.00836

Ghosh A, Islam T (2016) Genome-wide analysis and expression profiling of glyoxalase gene families in soybean (*Glycine max*) indicate their development and abiotic stress specific response. BMC Plant Biol 16:87. https://doi.org/10.1186/s12870-016-0773-9

Giarola V, Hou Q, Bartels D (2017) Angiosperm plant desiccation tolerance: hints from transcriptomics and genome sequencing. Trends Plant Sci 22:705–717. https://doi.org/10.1016/j.tplants.2017.05.007

Gill SS, Tuteja N (2010) Reactive oxygen species and antioxidant machinery in abiotic stress tolerance in crop plants. Plant Physiol Biochem 48(12):909–930. https://doi.org/10.1016/j.plaphy.2010.08.016

Gill SS, Khan NA, Tuteja N (2011) Differential cadmium stress tolerance in five Indian mustard (*Brassica juncea* L.) cultivars: an evaluation of the role of antioxidant machinery. Plant Sig Beh 6(2):293–300. https://doi.org/10.4161/psb.6.2.15049

Glick BR (2012) Plant growth-promoting bacteria: mechanisms and applications. Scientifica 2012:Article ID: 963401, 15 pages. https://doi.org/10.6064/2012/963401

Goufo P, Moutinho-Pereira JM, Jorge TF, Correia CM, Oliveira MR, Rosa EAS (2017) Cowpea (*Vigna unguiculata* L.Walp.) metabolomics: osmoprotection as a physiological strategy for drought stress resistance and improved yield. Front Plant Sci 8:586. https://doi.org/10.3389/fpls.2017.00586

Graham PH, Vance CP (2003) Update on legume utilization legumes: importance and constraints to greater use. Plant Physiol 131:872–877. https://doi.org/10.1104/pp.017004

Groppa MD, Tomaro ML, Benavides MP (2001) Polyamines as protectors against cadmium or copper-induced oxidative damage in sunflower leaf discs. Plant Sci 161(3):481–488. https://doi.org/10.1016/S0168-9452(01)00432-0

Guo B (2019) Role of salicylic acid in mitigating cadmium toxicity in plants. In: Cadmium toxicity and tolerance in plants. Academic, pp 349–374. https://doi.org/10.1016/C2017-0-02050-5

Gupta DK, Tohoyama H, Joho M, Inouhe M (2004) Changes in the levels of phytochelatins and related metal-binding peptides in chickpea seedlings exposed to arsenic and different heavy metal ions. J Plant Res 117(3):253–256. https://doi.org/10.1007/s10265-004-0152-8

Gupta K, Sengupta A, Chakraborty M, Gupta B (2016) Hydrogen peroxide and polyamines act as double edged swords in plant abiotic stress responses. Front Plant Sci 7:1343. https://doi.org/10.3389/fpls.2016.01343

Habibzadeh Y (2015) Arbuscular mycorrhizal fungi in alleviation of drought stress on grain yield and yield components of mungbean (*Vigna radiata* L.) plants. Int J Sci 4:34–40. https://doi.org/10.18483/ijSci.651

Hancock J, Desikan R, Harrison J, Bright J, Hooley R, Neill S (2006) Doing the unexpected: proteins involved in hydrogen peroxide perception. J Exp Bot 57(8):1711–1718. https://doi.org/10.1093/jxb/erj180

Hane JK, Ming Y, Kamphuis LG, Nelson MN, Garg G, Atkins CA, Cannon S (2017) A comprehensive draft genome sequence for lupin (*Lupinus angustifolius*), an emerging health food: Insights into plant-microbe interactions and legume evolution. Plant Biotechnol J 15:318–330. https://doi.org/10.1111/pbi.12615

Hanin M, Ebel C, Ngom M, Laplaze L, Masmoudi K (2016) New insights on plant salt tolerance mechanisms and their potential use for breeding. Front Plant Sci 7:1787. https://doi.org/10.3389/fpls.2016.01787

Hao X, Xie P, Zhu YG, Taghavi S, Wei G, Rensing C (2015) Copper tolerance mechanisms of *Mesorhizobium amorphae* and its role in aiding phytostabilization by *Robinia pseudoacacia* in copper contaminated soil. Environ Sci Tech 49(4):2328–2340. https://doi.org/10.1021/es504956a

Hasanuzzaman M, Nahar K, Hossain MS, Mahmud JA, Rahman A, Inafuku M (2017) Coordinated actions of glyoxalase and antioxidant defense systems in conferring abiotic stress tolerance in plants. Int J Mol Sci 18:200. https://doi.org/10.3390/ijms18010200

Havlin JL, Beaton JD, Tisdale SL, Nelson WL (1999) Soil fertility and fertilizers: an introduction to nutrient management, 6th edn. Macmillan, London, p 409

Hayat S, Hayat Q, Alyemeni MN, Wani AS, Pichtel J, Ahmad A (2012) Role of proline under changing environments. Plant Signal Behav 7(11):1456–1466. https://doi.org/10.4161/psb.21949

He Q, Altieri AH, Cui B (2015) Herbivory drives zonation of stress-tolerant marsh plants. Ecology 96(5):1318–1328. https://doi.org/10.1890/14-0937.1

He GH, Xu YJ, Wang XY, Liu MJ, Li SP, Chen M (2016) Drought-responsive WRKY transcription factor genes *TaWRKY1* and *TaWRKY33* from wheat confer drought and/or heat resistance in *Arabidopsis*. BMC Plant Biol 16:116. https://doi.org/10.1186/s12870-016-0806-4

Hemantaranjan A, Katiyar D, Vyas J, Bhanu AN (2018) Alleviating effect of foliar sprayed paclobutrazol and salicylic acid on biochemical changes in mungbean under cadmium stress. J Plant Sci Res 34(1). https://doi.org/10.32381/JPSR.2018.34.01.5

Hernández LE, Sobrino-Plata J, Montero-Palmero MB, Carrasco-Gil S, Flores Cáceres ML, Ortega-Villasante C, Escobar C (2015) Contribution of glutathione to the control of cellular redox homeostasis under toxic metal and metalloid stress. J Exp Bot 66:2901–2911. https://doi.org/10.1093/jxb/erv063

Hirakawa H, Kaur P, Shirasawa K, Nichols P, Nagano S, Appels R, Isobe SN (2016) Draft genome sequence of subterranean clover, a reference for genus *Trifolium*. Sci Reports 6. https://doi.org/10.1038/srep30358

Hirayama T, Shinozaki K (2010) Research on plant abiotic stress responses in the post-genome era: past, present and future. Plant J 61:1041–1052. https://doi.org/10.1111/j.1365-313X.2010.04124.x

Hiz MC, Canher B, Niron H, Turet M (2014) Transcriptome analysis of salt tolerant common bean (*Phaseolus vulgaris* L.) under saline conditions. PLoS ONE 9:e92598. https://doi.org/10.1371/journal.pone.0092598

Hu Y, Schmidhalter U (2005) Drought and salinity: a comparison of their effects on mineral nutrition of plants. J Plant Nutr Soil Sci 168:541–549. https://doi.org/10.1002/jpln.200420516

Huang D, Feurtado JA, Smith MA, Flatman LK, Koh C, Cutler AJ (2017) Long noncoding miRNA gene represses wheat β-diketone waxes. Proc Natl Acad Sci 114(15):E3149–E3158. https://doi.org/10.1073/pnas.1617483114

Huber AE, Bauerle TL (2016) Long-distance plant signaling pathways in response multiple stressors: the gap in knowledge. J Exp Bot 67:2063–2079. https://doi.org/10.1093/jxb/erw099

Jabeen N, Abbas Z, Iqbal M, Rizwan M, Jabbar A, Fari M, Abbas F (2016) Glycinebetaine mediates chromium tolerance in mung bean through lowering of Cr uptake and improved antioxidant system. Arch Agric Soil Sci 62(5):648–662. https://doi.org/10.1080/03650340.2015.1082032

Jaiswal SK, Naamala J, Dakora FD (2018) Nature and mechanisms of aluminium toxicity, tolerance and amelioration in symbiotic legumes and rhizobia. Biol Fertil Soils 54(3):309–318. https://doi.org/10.1007/s00374-018-1262-0

Janadeleh H, Hosseini Alhashemi A, Nabavi SMB (2016) Investigation on concentration of elements in wetland sediments and aquatic plants. Glob J Environ Sci Manag 2(1):87–93. https://doi.org/10.7508/GJESM.2016.01.010

Jensen ES, Peoples MB, Boddey RM, Gresshoff PM, Hauggaard-Nielsen H, Alves BJR, Morrison MJ (2012) Legumes for mitigation of climate change and the provision of feedstock for biofuels and biorefineries. A review. Agron Sustain Dev 32(2012):329–364. https://doi.org/10.1007/s13593-011-0056-7

Jha BK, Mali SS, Naik SK, Kumar A, Singh AK (2015) Optimal planting geometry and growth stage based fertigation in vegetable crops. Technical bulletin NoR-56/Ranchi-25.Research Centre Ranchi. ICAR-Research Complex for Eastern Region, Patna, India. https://doi.org/10.13140/RG.2.1.1717.6564

Jha UC, Bohra A, Jha R, Parida SK (2019) Salinity stress response and 'omics' approaches for improving salinity stress tolerance in major grain legumes. Plant Cell Rep 38(3):255–277. https://doi.org/10.1007/s00299-019-02374-5

Jia B, Sun M, DuanMu H, Ding X, Liu B, Zhu Y, Sun X (2017) GsCHX19. 3, a member of cation/H^+ exchanger superfamily from wild soybean contributes to high salinity and carbonate alkaline tolerance. Sci Rep 7(1):9423. https://doi.org/10.1038/s41598-017-09772-3

Johnsona YK, Fredrick OA, Josiah MK, Sijali IV (2018) Effects of tillage practices on water use efficiency and yield of different effects of tillage practices on water use efficiency and yield of different drought tolerant common bean varieties in machakos county, Eastern Kenya. Am Sci Res J Eng Technol Sci 40:217–234. ISSN (Online) 2313-4402

Jovanovic Z, Stanisavljevic N, Mikic A, Radovic S, Maksimovic V (2013) The expression of drought responsive element binding protein ('DREB2A') related gene from pea (*Pisum sativum* L.) as affected by water stress. Aust J Crop Sci 7:1590–1596. ISSN: 1835-2693

Kaashyap M, Ford R, Bohra A, Kuvalekar A, Mantri N (2017) Improving salt tolerance of chickpea using modern genomics tools and molecular breeding. Curr Genom 18:557–567. https://doi.org/10.2174/1389202918666170705155252

Kang SM, Radhakrishnan R, You YH, Khan AL, Lee KE, Lee JD, Lee IJ (2015) *Enterobacter asburiae* KE 17 association regulates physiological changes and mitigates the toxic effects of heavy metals in soybean. Plant Biol 17(5):1013–1022. https://doi.org/10.1111/plb.12341

Kang Y, Torres-Jerez I, An Z, Greve V, Huhman D, Krom N, Udvardi M (2018) Genome-wide association analysis of salinity responsive traits in *Medicago truncatula*. Plant Cell Environ 42(5):1513–1531. https://doi.org/10.1111/pce.13508

Kanz C, Aldebert P, Althorpe N, Baker W, Baldwin A, Bates K et al (2005) The EMBL nucleotide sequence database. Nucleic Acids Res 33(Data issue):D29–D33. https://doi.org/10.1093/nar/gki098

Karthik S, Tuteja N, Ganapathi A, Manickavasagam M (2019) Pea p68, a DEAD-box helicase, enhances salt tolerance in marker-free transgenic plants of soybean [*Glycine max* (L.) Merrill]. Biotech 9(1):10. https://doi.org/10.1007/s13205-018-1553-z

Karuppanapandian T, Kim W (2013) Cobalt-induced oxidative stress causes growth inhibition associated with enhanced lipid peroxidation and activates antioxidant responses in Indian mustard (*Brassica juncea* L.) leaves. Acta Physiol Plant 35:2429–2443. https://doi.org/10.1007/s11738-013-1277-y

Kashiwagi J, Krishnamurthy L, Upadhyaya HD, Krishna H, Chandra S, Vadez V, Serraj R (2005) Genetic variability of drought-avoidance root traits in the mini-core germplasm collection of chickpea (*Cicer arietinum* L.). Euphytica 146:213–222. https://doi.org/10.1007/s10681-005-9007-1

Kaur G, Asthir B (2015) Proline: a key player in plant abiotic stress tolerance. Biologia plantarum 59(4):609–619. https://doi.org/10.1007/s10535-015-0549-3

Khalid A, Ahmad Z, Mahmood S, Mahmood T, Imran M (2017) Role of Ethylene and Bacterial ACC-Deaminase in Nodulation of Legumes. In: Microbes for Legume Improvement. Springer, Cham, pp 95–118. https://doi.org/10.1007/978-3-319-59174-2_4

Khan HA, Siddique KHM, Munir R, Colmer TD (2015a) Salt sensitivity in chickpea: growth, photosynthesis, seed yield components and tissue ion regulation in contrasting genotypes. J Plant Physiol 182:1–12. https://doi.org/10.1016/j.jplph.2015.05.002

Khan MS, Kanwal B, Nazir S (2015b) Metabolic engineering of the chloroplast genome reveals that the yeast ArDH gene confers enhanced tolerance to salinity and drought in plants. Front Plant Sci 6:725. https://doi.org/10.3389/fpls.2015.00725

Khan HA, Siddique KHM, Colmer TD (2017) Vegetative and reproductive growth of salt-stressed chickpea are carbon-limited: sucrose infusion at the reproductive stage improves salt tolerance. J Expt Bot 68:2001–2011. https://doi.org/10.1093/jxb/erw177

Kim T-H (2014) Mechanism of ABA signal transduction: Agricultural highlights for improving drought tolerance. J Plant Biol 57(1):1–8. https://doi.org/10.1007/s12374-014-0901-8

Kim KD, El Baidouri M, Abernathy B, Iwata-Otsubo A, Chavarro C, Gonzales M (2015) A comparative epigenomic analysis of polyploidy-derived genes in soybean and common bean. Plant Physiol 168:1433–1447. https://doi.org/10.1104/pp.15.00408

Krämer U (2005) Phytoremediation: novel approaches to cleaning up polluted soils. Curr Opin Biotech 16(2):133–141. https://doi.org/10.1016/j.copbio.2005.02.006

Kudla J, Batistic O, Hashimoto K (2010) Calcium signals: the lead currency of plant information processing. Plant Cell 22:541–563. https://doi.org/10.1105/tpc.109.072686

Kumar J, Choudhary AK, Gupta DS, Kumar S (2019) Towards exploitation of adaptive traits for climate-resilient smart pulses. Int J Mol 20(12):2971. https://doi.org/10.3390/ijms20122971

Lafuente A, Pérez-Palacios P, Doukkali B, Molina-Sánchez MD, Jiménez-Zurdo JI, Caviedes MA, Pajuelo E (2015) Unraveling the effect of arsenic on the model *Medicago–Ensifer* interaction: a transcriptomic meta-analysis. New Phyt 205(1):255–272. https://doi.org/10.1111/nph.1300

Lane TW, Morel FM (2000) A biological function for cadmium in marine diatoms. Proc Nat Acad Sci 97:4627–4631. https://doi.org/10.1073/pnas.090091397

Latef AAHA, Ahmad P (2015) Legumes and breeding under abiotic stress: An overview. In: Azooz MM, Ahmad P (eds) Legumes under environmental stress: yield, improvement and adaptations, 1st edn, pp 1–20

Laxa M, Liebthal M, Telman W, Chibani K, Dietz KJ (2019) The role of the plant antioxidant system in drought tolerance. Antioxidants 8:94. https://doi.org/10.3390/antiox8040094

Lee HC, Lai KT, Lorenc MT, Imelfort M, Duran C, Edwards D (2012) Bioinformatics tools and databases for analysis of next generation sequence data. Brief Funct Genomics 11:12–24. https://doi.org/10.1093/bfgp/elr037

Legay S, Guerriero G, André C, Guignard C, Cocco E, Charton S (2016) *MdMyb93* is a regulator of suberin deposition in russeted apple fruit skins. New Phytol 212:977–991. https://doi.org/10.1111/nph.14170

Li Y, Ding X, Wang X, He T, Zhang H, Yang L, Wang T, Chen L, Gai J, Yang S (2017) Genome-wide comparative analysis of DNA methylation between soybean cytoplasmic male-sterile line NJCMS5A and its maintainer NJCMS5B. BMC Genomics 18(1):596. https://doi.org/10.1186/s12864-017-3962-5

Lin F, Xu J, Shi J, Li H, Li B (2010) Molecular cloning and characterization of a novel glyoxalase I gene *TaGlyI* in wheat (*Triticum aestivum* L.). Mol Biol Rep 37:729–735. https://doi.org/10.1007/s11033-009-9578-3

Lipiec J, Doussan C, Nosalewicz A, Kondracka K (2013) Effect of drought and heat stresses on plant growth and yield: a review. Int Agrophys 27:463–477. https://doi.org/10.2478/intag-2013-0017

Liu F, Jensen CR, Andersen MN (2004) Drought stress effect on carbohydrate concentration in soybean leaves and pods during early reproductive development: its implication in altering pod set. Field Crops Res 86(1):1–13. https://doi.org/10.1016/S0378-4290(03)00165-5

Llanes A, Masciarelli O, Ordoñez R, Isla MI, Luna V (2014) Differential growth responses to sodium salts involve different abscisic acid metabolism and transport in *Prosopis strombulifera*. Biol Plant 58(1):80–88. https://doi.org/10.1007/s10535-013-0365-6

Llanes A, Andrade A, Alemano S, Luna V (2016a) Alterations of endogenous hormonal levels in plants under drought and salinity. Am J Plant Sci 7:1357–1371. https://doi.org/10.4236/ajps.2016.79129

Llanes A, Arbona V, Gómez-Cadenas A, Luna V (2016b) Metabolomic profiling of the halophyte *Prosopis strombulifera* shows sodium salt-specific response. Plant Physiol Bioch 108:145–157. https://doi.org/10.1016/j.plaphy.2016.07.010

Lobell DB, Burke MB, Tebaldi C, Mastrandrea MD, Falcon WP, Naylor RL (2008) Priorizing climate change adaptation needs for food security in 2030. Nature 319:607–610. https://doi.org/10.1126/science.1152339

Loscos J, Naya L, Ramos J, Clemente MR, Matamoros MA, Becana M (2006) A reassessment of substrate specificity and activation of phytochelatin synthases from model plants by physiologically relevant metals. Plant Physiol 140(4):1213–1221. https://doi.org/10.1104/pp.105.073635

Maleki A, Naderi A, Naseri R, Fathi A, Bahamin S, Maleki R (2013) Physiological performance of soybean cultivars under drought stress. Bull Env Pharmacol Life Sci 2:38–44. Online ISSN 2277-1808

Manivannan P, Jaleel CA, Sankar B, Kishorekumar A, Somasundaram R, AlaguLakshmanan GM (2007) Growth, biochemical modifications and proline metabolism in *Helianthus annuus* L. as induced by drought stress. Colloids Surf B Biointerfaces 59:141–149. https://doi.org/10.1016/j.colsurfb.2007.05.002

Mano J (2002) Early events in environmental stresses in plants induction mechanisms of oxidative stress. In: Inze D, Montganu MV (eds) Oxidative Stress in Plants. Taylor & Francis, Abingdon, pp 217–246

Mantri N, Basker N, Ford R, Pang E, Pardeshi V (2013) The role of micro-ribonucleic acids in legumes with a focus on abiotic stress response. Plant Genome 6(3). https://doi.org/10.3835/plantgenome2013.05.0013

Maphosa Y, Jideani V (2017) The role of legumes in human nutrition Functional food – improve health through adequate food. Maria Chavarri Hueda. Intechopen 6:103-121. https://doi.org/10.5772/intechopen.69127

Martínez-Alcántara B, Martínez-Cuenca MR, Quiñones A, Iglesias DJ, Primo-Millo E, Forner-Giner MA (2015) Comparative expression of candidate genes involved in sodium transport and compartmentation in Citrus. Environ Exp Bot 111:52–56. https://doi.org/10.1016/j.envexpbot.2014.11.002

Mashima J, Kodama Y, Kosuge T, Fujisawa T, Katayama T, Nagasaki H et al (2016) DNA data bank of Japan (DDBJ) progress report. Nucleic Acids Res 44:D51–D57. https://doi.org/10.1093/nar/gkv1105

Masood A, Per TS, Asgher M, Fatma M, Khan MIR, Rasheed F, Khan NA (2016) Glycine betaine: role in shifting plants toward adaptation under extreme environments. In: Osmolytes and plants acclimation to changing environment: emerging omics technologies. Springer, New Delhi, pp 69–82

Matamoros MA, Loscos J, Coronado MJ, Ramos J, Sato S, Testillano PS, Becana M (2006) Biosynthesis of ascorbic acid in legume root nodules. Plant Physiol 141(3):1068–1077. https://doi.org/10.1104/pp.106.081463

Matsui T, Singh BB (2003) Root characteristics in cowpea related to drought tolerance at the seedling stage. Exp Agric 39:29–38. https://doi.org/10.1017/S0014479703001108

Micheletto S, Rodriguez-Uribe L, Hernandez R, Richins RD, Curry J, Connel MAO (2007) Comparative transcript profiling in roots of *Phaseolus acutifolius* and *P. vulgaris* under water deficit stress. Plant Sci 173:510–520. https://doi.org/10.1016/j.plantsci.2007.08.003

Mir RR, Zaman-Allah M, Sreenivasulu N, Trethowan R, Varshney RK (2012) Integrated genomics, physiology and breeding approaches for improving drought tolerance in crops. Theor Appl Genet 125:625–645. https://doi.org/10.1007/s00122-012-1904-9

Mirouze M, Paszkowski J (2011) Epigenetic contribution to stress adaptation in plants. Curr Opin Plant Biol 14:267–274. https://doi.org/10.1016/j.pbi.2011.03.004

Mishra S, Dubey RS (2006) Inhibition of ribonuclease and protease activities in arsenic exposed rice seedlings: role of proline as enzyme protectant. J Plant Physiol 163(9):927–936. https://doi.org/10.1016/j.jplph.2005.08.003

Mitra G (2018) Molecular approaches to nutrient uptake and cellular homeostasis in plants under abiotic stress. In: Plant nutrients and abiotic stress tolerance. Springer, Singapore, pp 525–590. https://doi.org/10.1007/978-981-10-9044-8_22

Mittal N, Mishra A, Singh R, Kumar P (2014) Assessing future changes in seasonal climatic extremes in the Ganges river basin using an ensemble of regional climate models. Clim Change 123:273–286. https://doi.org/10.1007/s10584-014-1056-9

Mittler R (2002) Oxidative stress, antioxidants and stress tolerance. Trends Plant Sci 7:405–410. https://doi.org/10.1016/S1360-1385(02)02312-9

Mittler R (2017) ROS are good. Trends Plant Sci 22(1):11–19. https://doi.org/10.1016/J.TPLANTS.2016.08.002

Mittler R, Blumwald E (2010) Genetic engineering for modern agriculture: challenges and perspectives. Annu Rev Plant Biol 61:443–462. https://doi.org/10.1146/annurev-arplant-042809-112116

Mittler R, Blumwald E (2015) The roles of ROS and ABA in systemic acquired acclimation. Plant Cell 27:64–70. https://doi.org/10.1105/tpc.114.133090

Mittler R, Vanderauwera S, Gollery M, Van Breusegem F (2004) Reactive oxygen gene network of plants. Trends Plant Sci 9(10):490–498. https://doi.org/10.1016/j.tplants.2004.08.009

Mizoi J, Shinozaki K, Yamaguchi-Shinozaki K (2012) AP2/ERF family transcription factors in plant abiotic stress responses. Biochim Biophys Acta (Special Issue) Gene Regul Mech 1819:86–96. https://doi.org/10.1016/j.bbagrm.2011.08.004

Mokgalaka-Matlala NS, Flores-Tavizon E, Castillo-Michel H, Peralta-Videa JR, Gardea-Torresdey JL (2009) Arsenic tolerance in mesquite (*Prosopis* sp.): low molecular weight thiols synthesis and glutathione activity in response to arsenic. Plant Physiol Biochem 47(9):822–826. https://doi.org/10.1016/j.plaphy.2009.05.007

Møller IM, Jensen PE, Hansson A (2007) Oxidative modifications to cellular components in plants. Annu Rev Plant Biol 58:459–481. https://doi.org/10.1146/annurev.arplant.58.032806.103946

Mondal MMA, Fakir MSA, Juraimi AS, Hakim MA, Islamand MM, Shamsuddoha ATM (2011) Effects of flowering behavior and pod maturity synchrony on yield of mungbean [*Vigna radiata* (L.) Wilczek]. Aust J Crop Sci 5:945–953. ISSN:1835-2707

Montero-Palmero MB, Martín-Barranco A, Escobar C, Hernández LE (2014) Early transcriptional responses to mercury: a role for ethylene in mercury-induced stress. New Phytol 201:116–130. https://doi.org/10.1111/nph.12486

Morales CG, Pino MT, del Pozo A (2013) Phenological and physiological responses to drought stress and subsequent rehydration cycles in two raspberry cultivars. Sci Hortic 162:234–241. https://doi.org/10.1016/j.scienta.2013.07.025

Munns R, Gilliham M (2015) Salinity tolerance of crops–what is the cost? New Phytol 208(3):668–673. https://doi.org/10.1111/nph.13519

Mustafa G, Komatsu S (2016) Toxicity of heavy metals and metal-containing nanoparticles on plants. Biochim Biophys Acta 1864:932–944. https://doi.org/10.1016/j.bbapap.2016.02.020

Nabi RBS, Tayade R, Hussain A, Kulkarni KP, Imran QM, Mun BG, Yun BW (2019) Nitric oxide regulates plant responses to drought, salinity, and heavy metal stress. Environ Exp Bot 161:120–133. https://doi.org/10.1016/j.envexpbot.2019.02.003

Nadeem M, Li J, Yahya M, Sher A, Ma C, Wang X, Qiu L (2019) Research progress and perspective on drought stress in legumes: a review. Int J Mol Sci 20:25–41. https://doi.org/10.3390/ijms20102541

Nageshbabu R, Jyothi MN (2013) Profile of small interfering RNAs from French bean *Phaseolus vulgaris* under abiotic stress conditions. Int J Pharma Bio Sci 4:176–185

Navarro-Torre S, Rodríguez-Llorente ID, Doukkali B, Caviedes MA, Pajuelo E (2019) Competition for alfalfa nodulation under metal stress by the metal-tolerant strain *Ochrobactrum cytisi* Azn6.2. Ann Appl Biol 175(2):184–192. https://doi.org/10.1111/aab.12528

Nayyar H, Gupta D (2006) Differential sensitivity of C3 and C4 plants to water deficit stress: association with oxidative stress and antioxidants. Environ Exp Bot 58:106–113. https://doi.org/10.1016/j.envexpbot.2005.06.021

Niazi NK, Bibi I, Fatimah A, Shahid M, Javed MT, Wang H, Shakoor MB (2017) Phosphate-assisted phytoremediation of arsenic by *Brassica napus* and *Brassica juncea*: morphological and physiological response. Int J Phytorem 19(7):670–678. https://doi.org/10.1080/15226514.2016.1278427

Nie WX, Xu L, Yu BJ (2015) A putative soybean *GmsSOS1* confers enhanced salt tolerance to transgenic *Arabidopsis sos1-1* mutant. Protoplasma 252:127–134. https://doi.org/10.1007/s00709-014-0663-7

Odjegba VJ (2012) Exogenous salicylic acid alleviates arsenic toxicity in *Arabidopsis thaliana*. Indian J Innov Dev 1(7):515–522. ISSN 2277 – 5390

Oracz K, Bouteau HEM, Farrant JM, Cooper K, Belghazi M, Job C, Bailly C (2007) ROS production and protein oxidation as a novel mechanism for seed dormancy alleviation. Plant J 50(3):452–465. https://doi.org/10.1111/j.1365-313X.2007.03063.x

Ortega-Villasante C, Rellán-Álvarez R, Del Campo FF, Carpena-Ruiz RO, Hernández LE (2005) Cellular damage induced by cadmium and mercury in *Medicago sativa*. J Exp Bot 56:2239–2225. https://doi.org/10.1093/jxb/eri223

Ortega-Villasante C, Hernández LE, Rellán-Álvarez R, el Campo FF, Carpena-Ruiz RO (2007) Rapid alteration of cellular redox homeostasis upon exposure to cadmium and mercury in alfalfa seedlings. New Phytol 176:96–107. https://doi.org/10.1111/j.1469-8137.2007.02162.x

Osmon HS, Salim BB (2016) Influence of exogenous application of some phytoprotectants on growth, yield and pod quality of snap bean under NaCl salinity. Ann Agric Sci 61(1):1–13. https://doi.org/10.1016/j.aoas.2016.05.001

Oven M, Raith K, Neubert RHH, Kutchan TM, Zenk MH (2001) Homo-phytochelatins are synthesized in response to cadmium in Azuki beans. Plant Physiol 126:1275–1280. https://doi.org/10.1104/pp.126.3.1275

Panigrahi DP, Sagar A, Dalal S, Randhawa GS (2013) Arsenic resistence and symbiotic efficiencies of alfalfa and cowpea rhizobial strain isolated from arsenic agricultural fields. Pel Res Lib 3:322–333. ISSN: 2248 –9215

Parkhey S, Naithani SC, Keshavkant S (2014) Protein metabolism during natural ageing in desiccating recalcitrant seeds of *Shorea robusta*. Acta Phys Plant 36:1649–1659. https://doi.org/10.1007/s11738-014-1540-x

Parmar N, Singh KH, Sharma D, Singh L, Kumar P, Nanjundan J (2017) Genetic engineering strategies for biotic and abiotic stress tolerance and quality enhancement in horticultural crops: a comprehensive review. 3 Biotech 7:239. https://doi.org/10.1007/s13205-017-0870-y

Paunov M, Koleva L, Vassilev A, Vangronsveld J, Goltsev V (2018) Effects of different metals on photosynthesis: cadmium and zinc affect chlorophyll fluorescence in durum wheat. Int J Mol Sci 19(3):787. https://doi.org/10.3390/ijms19030787

Per TS, Khan NA, Reddy PS, Masood A, Hasanuzzaman M, Khan MIR, Anjum NA (2017) Approaches in modulating proline metabolism in plants for salt and drought stress tolerance: phytohormones, mineral nutrients and transgenics. Plant Physiol Biochem 115:126–140. https://doi.org/10.1016/j.plaphy.2017.03.018

Peralta JM, Travaglia C, Gil R, Furlan A, Castro S, Bianucci E (2019) An effective rhizoinoculation restraints arsenic translocation in peanut and maize plants exposed to a realistic groundwater metalloid dose. In: Zhu Y, Guo H, Bhattacharya P, Ahmad A, Bundschuh J, Naidu R (eds) Environmental arsenic in a changing world, pp 283–286. 714 pp, ISBN 9781351046633. https://doi.org/10.1201/9781351046633

Peralta JM, Travaglia C, Romero-Pertas MC, Furlan A, Castro S, Bianucci E (2020) Unraveling the impact of arsenic on the redox response of peanut plants inoculated with two different *Bradyrhizobium* sp. Strains. Chemosphere 259:127410. https://doi.org/10.1016/j.chemosphere.2020.127410

Pérez-Chaca MV, Rodríguez-Serrano M, Molina AS, Pedranzani HE, Zirulnik F, Sandalio LM, Romero-Puertas MC (2014) Cadmium induces two waves of reactive oxygen species in *Glycine max* (L.) roots. Plant Cell Environ 37(7):1672–1687. https://doi.org/10.1111/pce.12280

Pérez-Palacios P, Romero-Aguilar A, Delgadillo J, Doukkali B, Caviedes MA, Rodríguez-Llorente ID, Pajuelo E (2017) Double genetically modified symbiotic system for improved Cu phytostabilization in legume roots. Envirom Sci Pol Res 24(17):14910–14923. https://doi.org/10.1007/s11356-017-9092-4

Petrović G, Jovičić D, Nikolić Z, Tamindžić G, Ignjatov M, Milošević D, Milošević B (2016) Comparative study of drought and salt stress effects on germination and seedling growth of pea. Genetika 48(1):373–381(9). https://doi.org/10.2298/GENSR1601373P

Piechalak A, Tomaszewska B, Baralkiewicz D, Malecka A (2002) Accumulation and detoxification of lead ions in legumes. Phytochemistry 60(2):153–162. https://doi.org/10.1016/S0031-9422(02)00067-5

Piedra-Aguilera Á, Jiao C, Luna AP, Villanueva F, Dabad M, Esteve-Codina A, Díaz-Pendón JA, Fei Z, Bejarano ER, Castillo AG (2019) Integrated single-base resolution maps of transcriptome, sRNAome and methylome of tomato yellow leaf curl virus (TYLCV) in tomato. Sci Rep 9:2863. https://doi.org/10.1038/s41598-019-39239-6

Pornsiriwong W, Estavillo GM, Chan KX, Tee EE, Ganguly D, Crisp PA (2017) A chloroplast retrograde signal, 3"-phosphoadenosine 5"-phosphate, acts as a secondary messenger in abscisic acid signaling in stomatal closure and germination. eLife 6:e23361 https://doi.org/10.7554/eLife.23361

Pospíšilová J, Synková H, Haisel D, Batková P (2009) Effect of abscisic acid on photosynthetic parameters during ex vitro transfer of micropropagated tobacco plantlets. Biol Plant 53:11–20. https://doi.org/10.1007/s10535-009-0003-5

Pushpavalli R, Zaman-allah M, Turner NC, Baddam R, Rao MV, Vadez V (2015a) Higher flower and seed number leads to higher yield under water stress conditions imposed during reproduction in chickpea. Funct Plant Biol 42(2):162–174. https://doi.org/10.1071/FP14135

Pushpavalli R, Krishnamurthy L, Thudi M, Gaur PM, Rao MV, Siddique KH, Vadez V (2015b) Two key genomic regions harbour QTLs for salinity tolerance in ICCV 2× JG 11 derived chickpea (*Cicer arietinum* L.) recombinant inbred lines. BMC Plant Biol 15(1):124. https://doi.org/10.1186/s12870-015-0491-8

Qadir S, Qureshi MI, Javed S, Abdin MZ (2004) Genotypic variation in phytoremediation potential of *Brassica juncea* cultivars exposed to Cd stress. Plant Sci 167:1171–1181. https://doi.org/10.1016/j.plantsci.2004.06.018

Qi X, Li MW, Xie M, Liu X, Ni M, Shao G, Isobe S (2014) Identification of a novel salt tolerance gene in wild soybean by whole-genome sequencing. Nat Commun 5:4340. https://doi.org/10.1038/ncomms5340

Qurashi AW, Sabri AN (2013) Osmolyte accumulation in moderately halophilic bacteria improves salt tolerance of chickpea. Pak J Bot 45(3):1011–1016

Rahman MA, Lee SH, Kim KY, Park HS, Hwang TY, Choi GJ, Lee KW (2016) Arsenic-induced differentially expressed genes identified in *Medicago sativa* L. roots. J Korean Soc Grassland Forage Sci 36(3):243–247. https://doi.org/10.5333/KGFS.2016.36.3.243

Rajjou L, Lovigny Y, Groot SP, Belghazi M, Job C, Job D (2008) Proteome-wide characterization of seed aging in *Arabidopsis*: a comparison between artificial and natural aging protocols. Plant Physiol 148(1):620–641. https://doi.org/10.1104/pp.108.123141

Raju SK, Shao M, Sanchez R, Xu Y, Sandhu A, Graef G, Mackenzie S (2018) An epigenetic breeding system in soybean for increased yield and stability. Plant Biotechnol J 16:1836–1847. https://doi.org/10.1111/pbi.12919

Ramakrishna B, Rao SSR (2013) 24-Epibrassinolide maintains elevated redox state of AsA and GSH in radish (*Raphanus sativus* L.) seedlings under zinc stress. Acta Physiol Plant 35:1291–1302. https://doi.org/10.1007/s11738-012-1168-7

Ramos J, Clemente MR, Naya L, Loscos J, Pérez-Rontomé C, Sato S, Tabata S, Becana M (2007) Phytochelatin synthases of the model legume *Lotus japonicus*. A small multigene family with differential response to cadmium and alternatively spliced variants. Plant Physiol 143:1110–1118. https://doi.org/10.1104/pp.106.090894

Rangel WDM, de Oliveira Longatti SM, Ferreira PA, Bonaldi DS, Guimaraes AA, Thijs S, Moreira FM (2017) *Leguminosae* native nodulating bacteria from a gold mine As-contaminated soil: Multi-resistance to trace elements, and possible role in plant growth and mineral nutrition. Int J Phytorem 19(10):925–936. https://doi.org/10.1080/15226514.2017.1303812

Ray S, Dansana PK, Giri J (2011) Modulation of transcription factor and metabolic pathway genes in response to water-deficit stress in rice. Funct Integrat Genom 11:157–178. https://doi.org/10.1007/s10142-010-0187-y

Reddy AR, Chaitanya KV, Vivekanandan M (2004) Drought-induced responses of photosynthesis and antioxidant metabolism in higher plants. J Plant Physiol 161:1189–1202. https://doi.org/10.1016/j.jplph.2004.01.013

Reddy AS, Ali GS, Celesnik H, Day IS (2011) Coping with stresses: roles of calcium- and calcium/calmodulin-regulated gene expression. Plant Cell 23:2010–2032. https://doi.org/10.1105/tpc.111.084988

Reddy S, Bhatnagar-Mathur P, Vadez V, Sharma KK (2012) Grain legumes (Soybean, Chickpea, and Peanut): omics approaches to enhance abiotic stress tolerance. In: Tuteja N, Gill SS, Tiburcio AF, Tuteja R (eds) Improving crop resistance to abiotic stress, 1st edn. Wiley-VCH Verlag GmbH & Co, pp 993–1030

Redillas MC, Jeong JS, Kim YS, Jung H, Bang SW, Choi YD (2012) The overexpression of *OsNAC9* alters the root architecture of rice plants enhancing drought resistance and grain yield under field conditions. Plant Biotechnol J 10:792–805. https://doi.org/10.1111/j.1467-7652.2012.00697.x

Rijsberman FR (2006) Water scarcity: fact or fiction? Agric Water Manage 80:5–22. https://doi.org/10.1146/annurev-arplant-042809-112116

Rodríguez-Ruiz M, Aparicio-Chacón MV, Palma JM, Corpas FJ (2019) Arsenate disrupts ion balance, sulfur and nitric oxide metabolisms in roots and leaves of pea (*Pisum sativum* L.) plants. Environ Exp Bot 161:143–156. https://doi.org/10.1016/j.envexpbot.2018.06.028

Rothwell SA, Elphinstone ED, Dodd IC (2015) Liming can decrease legume crop yield and leaf gas exchange by enhancing root to shoot ABA signalling. J Exp Bot 66(8):2335–2345. https://doi.org/10.1093/jxb/erv042

Roychoudhury A, Banerjee A, Lahiri V (2015) Metabolic and molecular-genetic regulation of proline signaling and itscross-talk with major effectors mediates abiotic stress tolerance in plants. Turk J Bot 39(6):887–910. https://doi.org/10.3906/bot-1503-27

Rubiales D, Mikic A (2015) Introduction: Legumes in sustainable agriculture. CRC J Crit Rev Plant Sci 34:2–3. https://doi.org/10.1080/07352689.2014.897896

Sabagh A, Hossain A, Islam MS, Barutçular C, Fahad S, Ratnasekera D, Saneoka H (2018) Role of osmoprotectants and soil amendments for sustainable soybean (*Glycine max* L.) production under drought condition: a review. J Exp Biol Agric Sci 6(1):32–41. https://doi.org/10.18006/2018.6(1).32.41

Sadeghipour O (2017) Amelioration of salinity tolerance in cowpea plants by seed treatment with methyl jasmonate. Legume Res 40(6):1100–1106. https://doi.org/10.18805/lr.v0i0.8394

Sahoo DP, Kumar S, Mishra S, Kobayashi Y, Panda SK, Sahoo L (2016) Enhanced salinity tolerance in transgenic mungbean overexpressing *Arabidopsis* antiporter (NHX1) gene. Mol Breeding 36(10):144. https://doi.org/10.1007/s11032-016-0564-x

Saidi I, Yousfi N, Borgi MA (2017) Salicylic acid improves the antioxidant ability against arsenic-induced oxidative stress in sunflower (*Helianthus annuus*) seedling. J Plant Nut 40(16):2326–2335. https://doi.org/10.1080/01904167.2017.1310888

Samarah NH, Alqudah AM, Amayreh JA, McAndrews GM (2009) The effect of late- terminal drought stress on yield components of four barley cultivars. J Agron Crop Sci 195:427–441. https://doi.org/10.1111/j.1439-037X.2009.00387.x

Sandalio LM, Dalurzo HC, Gomez M, Romero-Puertas MC, del Río LA (2001) Cadmium-induced changes in the growth and oxidative metabolism of pea plants. J Exp Bot 52:2115–2126. https://doi.org/10.1093/jexbot/52.364.2115

Sanitá di Toppi L, Gabbrielli R (1999) Response to cadmium in higher plants. Environ Exp Bot 41:105–130. https://doi.org/10.1016/S0098-8472(98)00058-6

Schmutz J, McClean PE, Mamidi S, Wu GA, Cannon SB, Grimwood J, Jenkins J (2014) A reference genome for common bean and genome-wide analysis of dual domestications. Nat Genet 46:707–713. https://doi.org/10.1038/ng.3008

Schröder P, Lyubenova L, Huber C (2009) Do heavy metals and metalloids influence the detoxification of organic xenobiotics in plants? Environ Sci Pollut Res 16:795–804. https://doi.org/10.1007/s11356-009-0168-7

Shabala S, Munns R (2017) Salinity stress: physiological constraints and adaptive mechanisms. In: Shabala S (ed) Plant Stress Physiology, 2nd edn. CABI; 2017, pp 24–63. http://hdl.handle.net/102.100.100/89125?index=1

Shamsi IH, Wei K, Zhang GP, Jilani GH, Hassan MJ (2008) Interactive effects of cadmium and aluminium on growth and antioxidative enzymes in soybean. Biol Plant 52:165–169. https://doi.org/10.1007/s10535-008-0036-1

Sharma SS, Dietz KJ (2009) The relationship between metal toxicity and cellular redox imbalance. Trends Plant Sci 14:43–50. https://doi.org/10.1016/j.tplants.2008.10.007

Shen Y, Zhang J, Liu Y, Liu S, Liu Z, Duan Z, Wang Z, Zhu B, Guo YL, Tian Z (2018) DNA methylation footprints during soybean domestication and improvement. Genome Biol 19(1):128. https://doi.org/10.1186/s13059-018-1516-z

Shi H, Quintero FJ, Pardo JM, Zhu JK (2002) The putative plasma membrane $Na^{(+)}/H^{(+)}$ antiporter SOS1 controls long-distance $Na^{(+)}$ transport in plants. Plant Cell 14:465–477. https://doi.org/10.1105/tpc.010371

Shrivastava P, Kumar R (2015) Soil salinity: a serious environmental issue and plant growth promoting bacteria as one of the tools for its alleviation. Saudi J Biol Scie 22(2):123–131. https://doi.org/10.1016/j.sjbs.2014.12.001

Sicher RC, Timlin D, Bailey B (2012) Responses of growth and primary metabolism of water-stressed barley roots to rehydration. J Plant Physiol 169:686–695. https://doi.org/10.1016/j.jplph.2012.01.002

Siddique KHM, Loss SP, Regan KL, Jettner RL (1999) Adaptation and seed yield of cool season grain legumes in Mediterranean environments of South-Western Australia. Aust J Agric Res 50:375–388. https://doi.org/10.1071/A98096

Signorelli S, Corpas FJ, Rodríguez-Ruiz M, Valderrama R, Barroso JB, Borsani O, Monza J (2019) Drought stress triggers the accumulation of NO and SNOs in cortical cells of *Lotus japonicus* L. roots and the nitration of proteins with relevant metabolic function. Environ Exp Bot 161:228–241. https://doi.org/10.1016/j.envexpbot.2018.08.007

Sinay H, Karuwal RS (2014) Proline and total soluble sugar content at the vegetative phase of six corn cultivars from Kisar island Maluku, grown under drought stress conditions. Int J Adv Agri Res 2:77–82. ISSN 2053-1265

Singh VP, Srivastava PK, Prasad SM (2013) Nitric oxide alleviates arsenic-induced toxic effects in ridged Luffa seedlings. Plant Physiol Biochem 71:155–163. https://doi.org/10.1016/j.plaphy.2013.07.003

Singh S, Parihar P, Singh R, Singh VP, Prasad SM (2016) Heavy metal tolerance in plants: role of transcriptomics, proteomics, metabolomics, and ionomics. Front Plant Sci 6:1143. https://doi.org/10.3389/fpls.2015.01143

Singh PK, Indoliya Y, Chauhan AS, Singh SP, Singh AP, Dwivedi S, Chakrabarty D (2017) Nitric oxide mediated transcriptional modulation enhances plant adaptive responses to arsenic stress. Sci Reports 7(1):3592. https://doi.org/10.1038/s41598-017-03923-2

Singh M, Rani S, Malhotra N, Katna G, Sarker A (2018) Transgressive segregations for agronomic improvement using interspecific crosses between *C. arietinum* L. x *C. reticulatum* Ladiz. and *C. arietinum* L. x *C. echinospermum* Davis species. PLoS ONE 13(9). https://doi.org/10.1371/journal.pone.0203082

Skórzyńska-Polit E, Drążkiewicz M, Krupa Z (2010) Lipid peroxidation and antioxidative response in *Arabidopsis thaliana* exposed to cadmium and copper. Acta Physiol Plant 32:169. https://doi.org/10.1007/s11738-009-0393-1

Sobrino-Plata J, Ortega-Villasante C, Flores-Cáceres ML, Escobar C, Del Campo FF, Hernández LE (2009) Differential alterations of antioxidant defenses as bioindicators of mercury and cadmium toxicity in alfalfa. Chemosphere 77(7):946–954. https://doi.org/10.1016/j.chemosphere.2009.08.007

Songsri P, Jogloy S, Holbrook CC, Kesmala T, Vorasoot N, Akkasaeng C, Patanothai A (2009) Association of root, specific leaf area and SPAD chlorophyll meter reading to water use efficiency of peanut under different available soil water. Agric Wat Manag 96:790–798. https://doi.org/10.1016/j.agwat.2008.10.009

Srivastava S, Tripathi RD, Dwivedi UN (2004) Synthesis of phytochelatins and modulation of antioxidants in response to cadmium stress in *Cuscuta reflexa* an angiospermic parasite. J Plant Physiol 161:665–674. https://doi.org/10.1078/0176-1617-01274

Srivastava S, Srivastava AK, Suprasanna P, D'Souza SF (2010) Comparative antioxidant profiling of tolerant and sensitive varieties of *Brassica juncea* L. to arsenate and arsenite exposure. Bull Environ Cont Toxicol 84(3):342–346. https://doi.org/10.1007/s00128-010-9937-8

Srivastava R, Hemantaranjan A, Lalotra S (2018) Effect of methyl jasmonate on morpho-physiological and biochemical attributes of chickpea (*Cicer arietinum* L.) under salinity stress. J Plant Sci Res 34(2). https://doi.org/10.32381/JPSR.2018.34.02.12

Stancheva I, Geneva M, Hristozkova M, Sichanova M, Donkova R, Petkova G, Djonova E (2017) Response of *Vigna unguiculata* grown under different soil moisture regimes to the dual inoculation with nitrogen-fixing Bacteria and Arbuscular Mycorrhizal Fungi. Commun Soil Sci Plant Anal 48:1378–1386. https://doi.org/10.1080/00103624.2017.1358740

Stoeva N, Bineva T (2003) Oxidative changes and photosynthesis in Oat plants grown in as contaminated soil. Bulg J Plant Physiol 29:87–95

Stritzler M, Elba P, Berini C, Gomez C, Ayub N, Soto G (2018) High-quality forage production under salinity by using a salt-tolerant *AtNXH1*-expressing transgenic alfalfa combined with a natural stress-resistant nitrogen-fixing bacterium. J Biotechnol 276:42–45. https://doi.org/10.1016/j.jbiotec.2018.04.013

Subbarao GV, Nam NH, Chauhan YS, Johansen C (2000) Osmotic adjustment, water relations and carbohydrate remobilization in pigeon pea under water deficits. J Plant Physiol 157:651–659. https://doi.org/10.1016/S0176-1617(00)80008-5

Surekha C, Kumari KN, Aruna LV, Suneetha G, Arundhati A, Kavi Kisho PB (2014) Expression of the *Vigna aconitifolia P5CSF129A* gene in transgenic pigeonpea enhances proline accumulation and salt tolerance. Plant Cell Tissue Organ Cult 116:27–36. https://doi.org/10.1007/s11240-013-0378-z

Švábová L, Lebeda A (2005) In vitro selection for improved plant resistance to toxin-producing pathogens. J Phytopathol 153:52–64. https://doi.org/10.1111/j.1439-0434.2004.00928.x

Szabados L, Savouré A (2010) Proline: a multifunctional amino acid. Trends Plant Sci 15(2):89–97. https://doi.org/10.1016/j.tplants.2009.11.009

Talukdar D (2015) Functional interplay between glutathione and hydrogen sulfide in regulation of thiol cascade during arsenate tolerance of common bean (*Phaseolus vulgaris* L.) genotypes. 3. Biotech 5(5):819–829. https://doi.org/10.1007/s13205-015-0285-6

Talukdar D (2016) Exogenous thiourea modulates antioxidant defence and glyoxalase systems in lentil genotypes under arsenic stress. J Plant Stress Physiol:9–21. https://doi.org/10.19071/jpsp.2016.v2.3041

Tiwari S, Lata C (2018) Heavy metal stress, signaling, and tolerance due to plant-associated microbes: an Overview. Front Plant Sci 9:452. https://doi.org/10.3389/fpls.2018.00452

Tiwari S, Sarangi BK (2017) Comparative analysis of antioxidant response by *Pteris vittata* and *Vetiveria zizanioides* towards arsenic stress. Ecol Engine 100:211–218. https://doi.org/10.1016/j.ecoleng.2016.12.007

Toker C, Mutlu N (2011) Breeding for abiotic stress. In: Pratap A, Kumar J (eds) Biology and breeding of food legumes. CAB International, Wallingford, pp 241–261

Torres MA, Dangl JL (2005) Functions of the respiratory burst oxidase in biotic interactions, abiotic stress and development. Curr Opin Plant Biol 8(4):397–403. https://doi.org/10.1016/j.pbi.2005.05.014

Tripathi DK, Singh S, Singh S, Srivastava PK, Singh VP, Singh S, Chauhan DK (2017) Nitric oxide alleviates silver nanoparticles (AgNps)-induced phytotoxicity in *Pisum sativum* seedlings. Plant Physiol Biochem 110:167–177. https://doi.org/10.1016/j.plaphy.2016.06.015

Tuomainen M, Ahonen V, Kärenlampi SO, Schat H, Paasela T, Svanys A (2011) Characterization of the glyoxalase 1 gene TcGLX1 in the metal hyperaccumulator plant *Thlaspicae rulescens*. Planta 233:1173–1184. https://doi.org/10.1007/s00425-011-1370-7

Vadez V, Soltani A, Krishnamurthy L, Sinclair TR (2012) Modelling possible benefit of root related traits to enhance terminal drought adaption of chickpea. Field Crops Res 137:108–115. https://doi.org/10.1016/j.fcr.2012.07.022

Vandecasteele C, Teulat-Merah B, Morére-Le Paven MC, Leprince O, Ly Vu B, Viau L, Buitink J (2011) Quantitative trait loci analysis reveals a correlation between the ratio of sucrose/raffinose family oligosaccharides and seed vigour in *Medicago truncatula*. Plant Cell Environ 34:1473–1487. https://doi.org/10.1111/j.1365-3040.2011.02346.x

Varshney RK, Hiremath PJ, Lekha P, Kashiwagi J, Balaji J, Deokar AA, Vadez V, Xio Y, Srinivna R, Gau PM, KHM S, Town CD, Hoisington DA (2009) A comprehensive resource of drought- and salinity- responsive ESTs for gene discovery and marker development in chickpea (*Cicer arietinum* L.). BMC Genomics 10:523. https://doi.org/10.1186/1471-2164-10-523

Vazquez S, Goldsbrough P, Carpena RO (2009) Comparative analysis of the contribution of phytochelatins to cadmium and arsenic tolerance in soybean and white lupin. Plant Physiol Biochem 47:63–67. https://doi.org/10.1016/j.plaphy.2008.09.010

Vazquez MN, Guerrero YR, de la Noval WT, Gonzalez LM, Zullo MAT (2019) Advances on exogenous applications of brassinosteroids and their analogs to enhance plant tolerance to salinity: a review. Aust J Crop Sci 13(1):115. https://doi.org/10.21475/ajcs.19.13.01.p1404

Vazquez-Reina S, Esteban E, Goldsbrough P (2005) Arsenate induced phytochelatins in white lupin: influence of phosphate status. Physiol Plant 124:41–49. https://doi.org/10.1111/j.1399-3054.2005.00484.x

Verbruggen N, Hermans C, Schat H (2009) Mechanisms to cope with arsenic or cadmium excess in plants. Curr Opin Plant Biol 12:364–372. https://doi.org/10.1016/j.pbi.2009.05.001

Verslues PE, Zhu JK (2005) Before and beyond ABA, upstream sensing and internal signals that determine ABA accumulation and response under abiotic stress. Bioch Soc Trans 33:375–379. https://doi.org/10.1042/BST0330375

Vinocur B, Altman A (2005) Recent advances in engineering plant tolerance to abiotic stress: achievements and limitations. Curr Op Biotechnol 16(2):123–132. https://doi.org/10.1016/j.copbio.2005.02.001

Vlasova A, Capella-Gutiérrez S, Rendon-Anaya M, Hernández-Oñate M, Minoche AE, Erb I, Sanseverino W (2016) Genome and transcriptome analysis of the Mesoamerican common bean and the role of gene duplications in establishing tissue and temporal specialization of genes. Genome Biol 17:32. https://doi.org/10.1186/s13059-016-0883-6

Voytas DF (2013) Plant genome engineering with sequence- specific nucleases. Annu Rev Plant Biol 64:327–350. https://doi.org/10.1146/annurev-arplant-042811-105552

Vu WT, Chang PL, Moriuchi KS, Friesen ML (2015) Genetic variation of transgenerational plasticity of offspring germination in response to salinity stress and the seed transcriptome of *Medicago truncatula*. BMC Evol Biol 15(1):59. https://doi.org/10.1186/s12862-015-0322-4

Vurukonda SS, Vardharajula S, Shrivastava M, Skz A (2016) Enhancement of drought stress tolerance in crops by plant growth promoting rhizobacteria. Microbiol Res 184:13–24. https://doi.org/10.1016/j.micres.2015.12.003

Wahid A, Rasul E (2005) Photosynthesis in leaf, stem, flower and fruit. In: Pessarakli M (ed) Photosynthesis, 2nd edn. CRC Press, Boca Raton, pp 479–497

Wang T, Chen L, Zhao M, Tian Q, Zhang WH (2011) Identification of drought-responsive microRNAs in *Medicago truncatula* by genome-wide high-throughput sequencing. BMC Genomics 12:367. https://doi.org/10.1186/1471-2164-12-367

Wu C, Ma C, Pan Y, Gong S, Zhao C, Chen S (2013) Sugar beet M14 glyoxalase I gene can enhance plant tolerance to abiotic stresses. J Plant Res 126:415–425. https://doi.org/10.1007/s10265-012-0532-4

Xu LL, Fan ZY, Dong YJ, Kong J, Bai XY (2015) Effects of exogenous salicylic acid and nitric oxide on physiological characteristics of two peanut cultivars under cadmium stress. Biol Plant 59(1):171–182. https://doi.org/10.1007/s10535-014-0475-9

Yadav SK (2010) Heavy metals toxicity in plants: an overview on the role of glutathione and phytochelatins in heavy metal stress tolerance of plants. S Afr J Bot 76(2):167–179. https://doi.org/10.1016/j.sajb.2009.10.007

Yaish MW, Lawati AA, Al-Harrasi I, Patankar HV (2018) Genome-wide DNA Methylation analysis in response to salinity in the model plant caliph medic (*Medicago truncatula*). BMC Genomics 19:78. https://doi.org/10.1186/s12864-018-4484-5

Yang DS, Zhang J, Li MX, Shi LX (2017) Metabolomics analysis reveals the salt-tolerant mechanism in *Glycine soja*. J Plant Growth Reg 36(2):460–471. https://doi.org/10.1007/s00344-016-9654-6

You J, Chan Z (2015) ROS regulation during abiotic stress responses in crop plants. Front Plant Sci 6:1092. https://doi.org/10.3389/fpls.2015.01092

Zandalinas SI, Sales C, Beltrán J, Gómez-Cadenas A, Arbona V (2017) Activation of secondary metabolism in citrus plants is associated to sensitivity to combined drought and high temperatures. Front Plant Sci 7:1954. https://doi.org/10.3389/fpls.2016.01954

Zhang GC, Zhu WL, Gai JY, Zhu YL, Yang LF (2015) Enhanced salt tolerance of transgenic vegetable soybeans resulting from overexpression of a novel D^{1}-pyrroline-5-carboxylate synthetase gene from *Solanum torvum* Swartz. Hortic. Environ Biotechnol 56:94–104. https://doi.org/10.1007/s13580-015-0084-3

Zhang J, Xu Y, Cao T, Chen J, Rosen BP, Zhao FJ (2017) Arsenic methylation by a genetically engineered Rhizobium-legume symbiont. Plant Soil 416(1-2):259–269. https://doi.org/10.1007/s11104-017-3207-z

Zhang H, Zuelsdorf C, Penneys D, Fan S, Kofsky J, Song BH (2018) Transcriptome profiling of a beach-adapted wild legume for dissecting novel mechanisms of salinity tolerance. Sci Data 5:180290. https://doi.org/10.1038/sdata.2018.290

Zheng M, Tao Y, Hussain S, Jiang Q, Peng S, Huang J (2016) Seed priming in dry direct-seeded rice: consequences for emergence, seedling growth and associated metabolic events under drought stress. Plant Growth Regul 78:167–178. https://doi.org/10.1007/s10725-015-0083-5

Zhou H, Lin H, Chen S, Becker K, Yang Y, Zhao J, Kudla J, Schumaker KS, Guo Y (2015) Inhibition of the *Arabidopsis* salt overly sensitive pathway by 14-3-3 proteins. Plant Cell 26:1166–1182. https://doi.org/10.1105/tpc.113.117069

Zlatev Z, Lidon FC (2012) An overview on drought induced changes in plant growth, water relations and photosynthesis. Emir J Food Agric 24:57–72. https://doi.org/10.9755/ejfa.v24i1.10599

Chapter 7
Biotic Stress to Legumes: Fungal Diseases as Major Biotic Stress Factor

Ghada Abd-Elmonsef Mahmoud

Abstract Legumes represent one of the most valuable crops all over the world because of their high content of protein, carbohydrates, fibers and minerals. Legumes significantly affect animal and human health along with soil fertility enhancement. Despite large areas under legume cultivation, the per year legume produces are insufficient due to crop losses induced by several biotic and abiotic stress factors. Abiotic stress factors include numerous environmental parameters like temperature, soil compaction, soil contamination, chilling, frost, drought, humidity, and soil water content. Biotic stress factors involve parasites like nematodes, weeds, insects, viruses, bacteria, and fungi. Fungal plant pathogens are considered one of the major reasons of legume crop losses because of their high spreading instances and aggressive infection via seeds or entire plant body. Fungal diseases can even lead to 100% legume yield losses. Various natural, chemical and biological control methods of regulating fungal diseases have been reported. The most common fungal diseases include rust, powdery mildew, wilt, ascochyta blight, chocolate spot and root rot disease. Present article focuses on various biotic stress inducers of legumes with attention to fungal diseases and its management strategies.

Keywords Legumes · Stress · Biotic · Phytopathogens · Viruses · Bacteria · Nematodes

G. A.-E. Mahmoud (✉)
Botany and Microbiology Department, Faculty of Science, Assiut University, Assiut, Egypt
e-mail: ghadamoukabel@aun.edu.eg

P. Guleria et al. (eds.), *Sustainable Agriculture Reviews 51*, Sustainable Agriculture Reviews 51, https://doi.org/10.1007/978-3-030-68828-8_7

7.1 Introduction

A legume belongs to family Fabaceae (Leguminosae). It is among the most important crops in the world, highly utilized for human nutrition, silage, livestock, and as a green manure for soil productivity (Farooq et al. 2017). Legumes are significant reservoirs of proteins, carbohydrates, fibers, and minerals. The protein content of legume ranges from 17 to 20% of dry weight in pea and beans, 38–40% of soybean and lupin dry weight (Bouchenak and Lamri-Senhadji 2013). Carbohydrates represented by amylose and starch servings constitute 30–40% of legume dry weight (Thorne et al. 1983). Legumes also contain significant oligosaccharides, stachyose, raffinose, and verbascose levels from 1–12% of legume dry weight (Rebello et al. 2014). Total legume fiber content ranges from 8 to 27.5%, with soluble fiber 3.3–13.8% of legume dry weight (Guillon and Champ 2002). Legumes are also considered as a good vitamin source for niacin, riboflavin, thiamin, pyridoxine, and folic acid that play significant roles in the energy metabolism and metabolic pathways of fatty acids. Beans only provide 400 μg of folate that represents the daily vitamin requirement (Rebello et al. 2014).

Legumes cover 12–15% of agriculture lands all over the world, giving approximately 27% of the major crop production and serving 33% of our protein dietary demands (Mishra et al. 2014; Farooq et al. 2017). Association between *Rhizobium* sp. and legumes give a renewable nitrogen source for agriculture (Damiani et al. 2016). Legume crops provide nitrogen supplementation to the different biological life forms present in the soil by fixing the atmospheric nitrogen ($N \equiv N$) and converting it from non-available physiologically form to available form (Qureshi et al. 2010). The productivity of legumes is affected by numerous abiotic stresses like drought, frost, chilling, heat, salinity waterlogging, and mineral toxicities and major biotic stresses produced by weeds, nematodes, insects, viruses, bacteria and fungal diseases (Reddy et al. 2004). These stresses cause huge crop loses. However, fungal diseases cause aggressive quantities of drop in crop productivity. Fungal diseases affect the yield of crops qualitatively and quantitatively by inducing 10–100% yield losses of legumes (Karimi et al. 2012).

Plant diseases are generally referred as the occurrence of morphological and physiological disorders appeared by biotic (microbial infection) or abiotic agents (environmental factors). Fungi represent the major biotic stress factors affecting legumes production. However parasites, viruses, bacteria and weeds can also decrease the production of legumes (Dita et al. 2006). Fungal spores can enormously spread in large areas all over the world leading to significant yield losses and considerable reduction in both seed quality and quantity than other biotic stresses (Dita et al. 2006). Hence, fungal agents have been identified as the major biotic stress factors affecting production of legumes. The foliar fungal diseases usually affecting legumes include biotrophic pathogens like downy mildews, rusts, and powdery mildews (Dita et al. 2006). Among these, *Fusarium* root rot is one of the most common and deadly fungal disease affecting legumes. *Fusarium solani* can infect the plant seedling and seeds even in the growing plant (Nene 1979). The initial symptoms are reddish-brown streaks on the plant root with leaves yellowing or necrosis at the late disease stage. *Fusarium* root rot is different from *Fusarium* wilt in symptoms and mode of action. It can cause yield losses up to 30–57% (Kraft et al. 1981).

The legume defense strategies include anatomical, physiological, biochemical, and genetic defense responses (Pande et al. 2005). The resistant genotypes show anatomical defense responses including cuticle thickening, epithelium, palisade cells (that provide the mechanical walls for the pathogen penetration), higher xylem and its parenchyma, and the presence of hypodermis compared to susceptible genotypes. Host physiology includes the respiration rate, total carbohydrate content, reducing sugars and phosphorous that reportedly increases in the resistant types than the susceptible ones. Phenolic compounds as biochanin A and formononetin show rapid accumulation in large quantities in resistant plants representing their role in plant defense. Phytoalexin accumulation is also identified as a part of defense mechanism as its occurrence has enhanced tolerance against phytopathogenic fungi (Angelini et al. 1993; Vogelsang et al. 1994; Pande et al. 2005).

Management of plant disease is achieved mainly by developing and cultivating resistant and tolerant legume varieties instead of the current used ones (Boyd et al. 2013). Crop rotation, draining wet soils, fertilization, soil preparation by tillage, and weed control also represents mechanisms of the legume disease management (Foroud et al. 2014). Intercropping of faba bean with other crops was found to decrease the microbial number in rhizosphere and phyllosphere, which consequently means decreasing the abundance and number of fungal pathogens (Mahmoud et al. 2018). Using of natural oils like garlic oil was found to decrease the count of most common plant pathogen *Fusarium* sp. (Mansour et al. 2018), also essential oils of Thyme, oregano and their extracts have inhibiting activities against various pathogenic fungi and bacteria (Ahmed et al. 2020). New antifungal agents like calcium and copper based compounds have proven efficiency against *Fusarium* species pathogens by inhibited their fungal hydrolytic enzymes (Mahmoud et al. 2019; Ibrahim and Mahmoud 2019). The metal ion generate fungal morphological deformations that could inhibit all the fungal growth and function (Ibrahim et al. 2020), or disturb the cell metabolic activity through affecting on the antioxidant microbial system (Mahmoud et al. 2020). Biological control of fungi by using different new biological agents including rhizobacteria, *Bacillus pumilus, Pseudomonas putida, Aspergillus awamori, A. niger* and *Trichoderma harzianum* represent the most ecofriendly microbes of disease management (Gardiner et al. 2013; Giorgio et al. 2016).

7.2 Biotic Stressors in Legumes

7.2.1 Nematodes

Nematodes are roundworms sharing similarity with animal parasites encountered in the livestock and pets. Many nematodes have the ability to attack legume plants through root hair leading to *in planta* stabilization or migration of endoparasite through the stem and the bulb. Faba bean is attacked by numerous types of nematodes such as, *Ditylenchus dipsaci* (stem-nematode), *Heterodera goettingiana* (cyst-nematode), *Pratylenchus* spp. (root-lesion nematode) and *Meloidogyne* spp. (root-knot nematode) (Sillero et al. 2010). These nematodes specifically target faba beans. The nematodal infection of weeds like *Chenopodium album* and *Lamium album* has also been reported. Seeds infected with nematodes have the potential to completely damage the faba bean crop. However significant prevention of nematode infection in resistant legume varities has also been reported (Di Vito et al. 2001). *Meloidogyne* spp. is a fatal parasite spreading worldwide and constitutes large species of nematode that can infect legumes (Fourie et al. 2015).

The peanut root infected with *Meloidogyne arenaria* was the first record of nematode parasitizing local crop by Van der Linde et al. (1959). Coetzee (1968) listed *M. hapla, M. incognita*, and *M. javanica* as soybean root-knot nematodes. Infected plants generate symptoms including gall formation characterized by elongated or roundish, knot-like protuberances (Bridge and Starr 2007). The size of galls is determined by numerous factors like the species type and the host status (Karssen et al. 2013). These galls can be confused with *Rhizobium* nodules present in nitrogen fixing legumes, but they can be easily distinguished as *Rhizobium* nodules can be easily rubbed off the plant roots while galls establish contact with root integral parts so it is hard to remove them without damaging the roots (Liebenberg 2012). To control nematode plant infection, genetic plant resistance represents cost-effective and eco-friendly management strategy to reduce the yield losses resulting from nematode infection (Bridge and Starr 2007). Use of synthetic nematicides is a common and popular way to kill the pathogenic nematodes (Fourie and Mc Donald 2007).

7.2.2 Viruses

Viruses are the smallest plant obligate parasites. The simple virus particle consists of protein coat surrounding the genomic material of either deoxyribo-nucleic acid (DNA) or ribonucleic acid (RNA) (Schoelza et al. 2011). Pea, faba bean, lentil, and chickpea are significant food crops attacked by numerous viruses (Schmidt and Naumann 1981; Hampton 1984). Viruses can not establish in the annual crops so they can survive only by means of hosts alternations and seed infections. Among perennial legumes, *Trifolium* species is considered as an effective reservoir of the viral inoculums (Hampton 1967; Barnett and Diachun 1984). Lucerne is also considered as a major viral (bean leaf roll virus) and aphid reservoir (*Acyrthosiphon pisum*). The plants infected with bean leaf roll virus do not clearly show viral infection, but it remains uncertain whether the vein yellowing is due to the viral infection or not (Cockbain and Gibbs 1973). Lucerne crops are also a source of alfalfa mosaic virus and pea streak virus that are spread by aphids to pea legumes (Hampton 1983).

Viruses can have wide host ranges like gladiolus and freesia, non-leguminous plants loaded with broad bean wilt virus (Zettler and Abo El-Nil 1977; Derks et al. 1980).Viruses depend on the living cells for its survival and multiplication, so the symptoms vary greatly and confound them mostly with abiotic and biotic stress induced symptoms. Pea mosaic virus causes leaf rolling or narrowing with various degrees, yellowing, necrosis, and finally plant death (Baggett and Hampton 1972; Bos et al. 1988). However, the symptoms could be similar or identical with the symptoms for other viruses infecting pea including pea streak virus, bean yellow mosaic virus, broad bean wilt virus, alfalfa mosaic virus, beet mosaic virus, clover yellow vein virus, pea early-browning virus and tomato spotted wilt virus (Bos et al. 1988).

Faba bean crops are also damaged by many viruses e.g. faba bean necrotic yellow virus, bean yellow mosaic virus, bean leaf roll virus, pea enation mosaic virus, true broad bean mosaic virus, broad bean mottle virus and broad bean stain virus causing leaf yellowing, rolling, plant stunting and necrosis (Bond et al. 1993; van Leur et al. 2006). These viruses have wide host range and needs bridge between seasonal cropping for transmission (van Leur et al. 2006). To control the viral infection of legumes, there is no well-known specific method for curing the plants. The current control strategy includes prevention or decreasing the infection rate by selecting resistance plants for cultivation and removing the infection sources (Hamilton et al. 1983).

7.2.3 Bacteria

Bacteria are the most abundant and simplest prokaryotic microorganisms that do not possess specific nucleus or complex organelles with size less than 10 μm (Htike and Win 2013). Searching for bacteria affecting legumes, plant growth promoting bacteria will take the greatest part of the study, however plant pathogenic bacteria present in the soil can cause huge damage to the legume crops like the gram-negative *Xanthomonas* sp. which cause common bacterial blight as a significant bean borne disease. Genus *Xanthomonas* belongs to family Xanthomonadacea, order Xanthomonadales and includes 27 disease species having host range upto approximately 400 different plants with high degree of plant host specificity (Benson et al. 2009). *Xanthomonas axonopodis* pv. *Phaseoli* and *Xanthomonas fuscans* subsp. *fuscans* are the most dominant bacterial pathogens of legumes (Fourie 2002; Schaad et al. 2006; Belete and Bastas 2017), however *X. phaseoli* has been reported to be highly aggressive (Opio et al. 1996).

These two pathogens spread in numerous regions in the world whenever bean is cultivated (Schaad et al. 2000). In 2003, these bacteria caused severe crop losses in Rift Valley of Ethiopia and Eastern Ethiopia (Belete and Bastas 2017). Warm temperature with humid weather raises the disease severity that causes losses in bean yield and quality (Fininsa 2003). The yield losses from 10% to 45% have been reported to occur depending on the environmental conditions and plant genotype (Gillard et al. 2009). In high disease severity, the pathogen causes yield losses from 5.2 kg ha^{-1} to 9.1 kg ha^{-1} (Fininsa 2003). Pathogenic bacteria enter the plant through leaf stomata and form inoculums. The inoculum multiplicities lead to the appearance of water-soaked spots on leaves, enlarging them. The infected leaf tissue becomes flaccid lemon-yellow developing necrosis until causing defoliation and stem girdle (Belete and Bastas 2017). In warm weather, lesions are rapidly performed; they enlarge and merge with varying sizes depending on the disease and plant stage. Highly infected leaves become tattered with winds and later drops off the plant (Buruchara et al. 2010).

Infected seeds appeared with butter yellow to brown spots in the seed coat or were restricted only to the seed hilum area. In severe infection, seeds shriveled with poor germination or vigor (Agrios 2005; Buruchara et al. 2010). *Xanthomonas* infection occurs in natural openings and/or wounds; rain facilitates the bacterial entry into the openings, stomata, wounds which represent favorable infection sites especially at 28 °C to 32 °C as optimal infection temperature. Once it enters the plant, bacteria multiplies rapidly in the cell spaces taking from 10 to14 day to establish the initial infection (Schwartz et al. 2006). Management of *Xanthomonas* disease include reduction of the initial inoculum by field sanitation, application of a foliar bactericide, weeds elimination, crop rotation, healthy seed plantation, burning of crop residues, seed treatment, and using resistant cultivars (Schwartz et al. 2006; Buruchara et al. 2010).

Using copper hydroxide, copper sulphate, and potassium methyl dithiocarbamate, the foliage infection was regulated by reducing the bacterial population, especially when spraying it every 7–10 days during moist cold weather (Gilbertson and Maxwell 1992; Fininsa 2003). Biological control has been considered environmental friendly way to reduce the plant disease using the microbial activities. *Pseudomonas* sp. reduces *Xanthomonas* disease up to 39% by applying on the seeds (da Silva et al. 2008). *Rahnella aquatilis* showed promising results in the disease suppression by the antagonistic activities (Sallam 2011). Osdaghi et al. (2011) used the ability of *Rhizobium leguminosarum* to control the *Xanthomonas* disease as a seed treatment. Another study in Italy recorded the ability of bean *Rhizobacteria* to inhibit the *Xanthomonas* disease *in vitro* (Giorgio et al. 2016).

7.2.4 *Fungus*

Fungi are eukaryotic organisms with unicellular or filamentous structure with chitin cell wall and ergosterol cell membrane. They survive as free saprophytic organisms or in combination with another organism in parasitic or symbiotic relationship. They are classified according to the sexual or asexual reproduction abilities (McConnaughey 2007). The interaction between plant and fungi causes the disease induced physiological disorders. Infected plants have external and internal symptoms like wilting, cells disintegration or collapse, abscission, increased respiration, abnormal cell division, cells enlargement, and chlorophyll degeneration (Hancock and Huisman 1981). Fungal disease cycle represented by series of events lead to infection symptoms including inoculation, pre-penetration, penetration, infection, colonization, growth, and pathogen reproduction (Dixon et al. 1994).

Inoculation is the contact between the fungal partner and the plant at specific sites to initiate the pathogen infection by sclerotia, spores, and even mycelia fragments. There are two inoculum types primary inoculum that causes original fungal infections, while secondary inoculum is produced from the primary infections. Pre-penetration involves fungal attachment with the external plant surface before the penetration. Fungi like powdery mildew releases cutinase enzyme at adhesion sites making them more hydrophilic which cements the spore with the plant surface (Romantschuk 1992). Spore starts to germinate by utilizing the stored foods like lipids, and carbohydrates to form germ tube, and then the germ tube extends to various places in the plant (Stanley et al. 2002). Appressoria are formed to penetrate the plant cell wall to get nutrition via lysozymal secretions (Emmett and Parbery 1975).

Fungal pathogens penetrate the plant surface usually through natural openings, then start the infection process. Successful infections are symptomized by leaf discoloration, necrosis, stem dwarfism, and other symptoms until the last disease stage of plant death (Nielsen et al. 2000). Plant invasion occurs by the growth of fungal mycelia between the plant cuticle and epidermis. Finally most phytopathogens like fungi invade the plant tissues, grow and spread until plant cells death occurs (Meredith 1973; Perfect and Green 2008).

7.3 Common Fungal Diseases of Legumes

Legumes (Fabaceae) economically represent the second most significant family of crop plants followed by the grasses (Poaceae). Legumes account for about 27% of the world crop production and provide the humans with 33% of the needed dietary protein and also for animal feeding like pasture or forage legumes (Smýkal et al. 2015). The three most significant legume crops all over the world are chickpea (*Cicer arietinum* L.), bean (*Phaseolus vulgaris* L.) and pea (*Pisum sativum*) (Diapari et al. 2014). Fungal diseases have been reported to affect the quality and quantity of legume crops very badly. The foliar fungal diseases generated by biotrophic

Table 7.1 Common fungal diseases of legumes.

Disease	Causal pathogen	Symptoms and the crop losses percentage	References
Chocolate spot Disease	*Botrytis cinerea* *Botryotinia fabae* (faba bean)	Small circular brown spots surrounded by red border on the leaves. In aggressive stages spots become coalescent lesions evolving later in dark brown rot. It caused 61% losses in yield for the susceptible plants and 34% for tolerant plants.	Dereje (1999), Dereje and Yaynu (2001), and Berhanu et al. (2005)
Downy mildew disease	*Peronospora viciae* (faba bean) *Peronospora pisi* (pea)	Lower surface of leaf has a cottony gray layer of conidiophores and conidia. This layer begins as spots then expanded to cover all over the lower leave surface.	Biddle (2001) and Sillero et al. (2006)
Legume blight disease	*Didymella rabiei* (chickpea) *Didymella fabae* (faba bean)	Irregular lesions appeared on leaves, first with dark brown color then turn to gray in the center with a darker border. It causes up to 100% yield loss in some crops.	Singh and Sharma (1998), Akem (1999), Nene and Reddy (1987), and Nasir et al. (2000)
Powdery mildew disease	*Erysiphe* sp. *Erysiphe pisi* (pea)	Small spots appeared on the leaflets and the stipules especially in the lowest part then starts to spread with white forming heavily conidia in pale grey powdery characters. *Erysiphe pisi* cause 25–50% yield losses of legumes, reducing yield biomass, pods number per plant, seeds, nodes and the plant height.	Gritton and Ebert (1975), Warkentin et al. (1996), Falloon and Viljanen-Rollinson (2001), and Fondevilla and Rubiales (2012)
Root rot disease	*Fusarium avenacearum* *Fusarium culmorum* *Fusarium graminearum* *Fusarium oxysporum* *Fusarium solani*	Infection starts via natural openings, then infects the attachment area of cotyledonary. Root rotting generated resulting black roots followed by complete plant death. In other plants, primary roots show reddish lesions with outer root cracks and finally destruction of the main root. It cause 30–57% yield reduction in pea, 84% in faba bean.	Bywater (1959), Nelson (1964), Helsper et al. (1994), Otsyula et al. (1998), and Naseri (2008, 2014)

(continued)

Table 7.1 (continued)

Disease	Causal pathogen	Symptoms and the crop losses percentage	References
Rust disease	*Uromyces viciae-fabae* (faba bean) *Uromyces ciceris-arietini* (chickpea) *Uromyces appendiculatus* (bean) *Uromyces striatus* (alfalfa) *Uromyces pisi* (pea) *Uromyces vignae* (cowpea)	Small leaf pustules, elongated or rounded, white pinkish at first then become brown reddish. Pustules on pods are wider with irregular shapes. Rust severity cause reducing in the faba bean seed yield by 8.39%.	Negussie and Pretorius (2005), Dita et al. (2006), and Negussie and Pretorius (2008)
Wilt disease	*Fusarium oxysporum* *Fusarium oxysporum* f. sp. *ciceris* (chickpea)	Wilt symptoms generated at any plant growth stage by patches spreading, flaccidity of leaves with discoloration, desiccation, and finally the entire plant collapsing. Yield losses by wilt disease 20–100% depending on the disease stage.	Mengistu and Negussie (1994), Nene et al. (1996), Stevenson et al. (1997), and Karimi et al. (2012)

pathogens like downy mildews, rusts, wilt, seedling blight, and powdery mildews are commonly present in the legume cultivated areas (Dita et al. 2006). These fungal diseases cause high yield losses approximately 10–100% every year depending on the environmental conditions (Karimi et al. 2012). The common fungal diseases of legumes, their causative agents and disease symptoms are discussed in Table 7.1.

7.3.1 Rust Disease

Rust diseases are caused by special group of fungi that specifically targets the aerial plant parts especially the leaves, but it can also affect plant stems. Various rust species infecting legumes mostly belonging to genus *Uromyces*, like *U. ciceris-arietini* (chickpea), *U. appendiculatus* (bean), *U. striatus* (alfalfa), *U. pisi* (pea), *U. viciae-fabae* (faba bean and lentil) and *U. vignae* (cowpea). Additionally, *Phakopsora pachyrhizi* and *P. meibomiae* infect soybean and *Puccinia arachidis* infect groundnut (Dita et al. 2006).The disease symptoms include pale leaf spots especially in the lower leaf surface forming pustules that produces huge spore numbers. The pustule color varies between orange, white, yellow, brown, or black, while rust brown color is most common (Carmona et al. 2005). Rust caused by *P. pachyrhizi* has caused huge yield losses of soybean across the globe, especially in Asia (Pivonia and Yang 2004; du Preez et al. 2005).

U. fabae rust disease represents an economically important disease of legumes. Its spread on a wide scale results in complete yield loss. Each one percent increase in the rust severity caused reduction in the seed yield by 8.39% (Negussie and Pretorius 2005). The pathogen lives in the debris of infected plants and contaminant seeds. It attacks leaves, pods and stems of lentil plants causing complete elimination of lentil farms area of 2500 ha in Gimbichu Wereda (Negussie and Pretorius 2008). The optimum infection temperature is 20 °C with leaf wetness, and then the infection starts within three hours (Negussie et al. 2005a). To control the rust disease, complete removal of the infected plant debris as well as infected seeds and use of resistant plant varieties is recommended (Negussie and Pretorius 2005). Also, the use of systemic chemical products like tebuconazole and propiconazole control the rust disease (Negussie et al. 2005b; Tadesse 2008).

7.3.2 Powdery Mildew Disease

Powdery mildew is an air-borne disease spread worldwide damaging various legume plants in their late maturation phase. It is caused by the obligate pathogen of *Erysiphe* species. It affects the plant green parts by causing small spots on the leaflets and stipules especially in the lowest part. Later, it spreads with white forming heavily conidia in pale grey powdery characters (Falloon and Viljanen-Rollinson 2001). Heavily conidia in the fungal debris can cause serious human allergy problems with breathing troubles to the field worker (Fondevilla and Rubiales 2012). Powdery mildew of pea caused by several important pathogens, such as *Erysiphe pisi* (Smith et al. 1996), *E. baeumleri* (Ondřej et al. 2005) and *E. trifolii* (Attanayake et al. 2010). *Erysiphe pisi* is a significant fungal pathogen infecting several legumes. Its wide distribution causes 25–50% yield losses of legumes, reducing yield biomass, pods number per plant, seeds, nodes and plant vegetative parameters (Gritton and Ebert 1975; Warkentin et al. 1996).

For management and control of the powdery mildews infection of legumes, we need to plant the crops in early season or use rapid maturing cultivar so that the fungus do not gets enough time to establish and spread (Fondevilla and Rubiales 2012). Use of *Erysiphe* -free seeds do not have significant effect in the disease management, as the pathogen could be present in the soil and attack the plant easily (Tiwari et al. 1999).The incidence of powdery mildews reportedly, were increased by the nitrogen availability in soil and reduced by increasing phosphorous (Jarvis et al. 2002). Chemical control using sulphur compounds have been used successfully in the prevention of powdery mildews (Warkentin et al. 1996). Triazole compounds were proven effective in controlling the powdery mildews of pea (Ransom et al. 1991). Strobirulins, spiroxamine, quinoxyfen and anilinopyri-midines have also been used as powdery mildews chemical control agents (Hollomon and Wheeler 2002).

Biological control of powdery mildews in legumes represents the most significant environmental friendly regulatory strategy with high safety properties. Fungi, bacteria, yeasts, and arthropods have been used in the biological control of powdery mildews (Paulitz and Bélanger 2001; Kiss 2003). Biological control fungi like *Acremonium alternatum* (Koitabashi 2005), *Ampelomyces quisqualis* (Hofstein et al. 1996), *Paecilomyces fumosoroseus* (Kavková and Curn 2005), *Pseudozyma flocculosa* (Paulitz and Bélanger 2001), *Verticillium lecanii* (Verhaar et al. 1996), and *Trichoderma harzianum* (Hijwegen 1992) have proved its efficiency to control the powdery mildews disease with promising results. Biological control bacteria, such as *Bacillus subtilis* and *B. pumilus* (Marrone 2002), *Pseudomonas fluorescens* and *P. aeruginosaalone,* have also been used successfully in the disease management strategies (Singh et al. 2000).

7.3.3 Wilt Disease

Wilt disease is widely distributed in about 32 countries producing severe yield losses from 20 to 100% depending on the stage of disease (Nene et al. 1996). Wilt disease has caused 30% losses in chickpea fields in the Shewa region (Mengistu and Negussie 1994).Chickpea is an important food and feed source of human and animal foods/ feeds (Millan et al. 2006). It increases soil fertility through nitrogen fortification using the plant association with nitrogen fixing bacteria and breaks the disease life cycle of some pathogens (Pande et al. 2011) that makes it most convenient for crop rotation (Geletu and Yadeta 1994). Chickpea cultivated on about 13.5 million hectares of land produces nearly 13.1 million tons of grain legumes (FAOSTAT 2014). However, chickpea fungal wilt disease caused 10% to 100% losses in the crop yield with 61% seedling damage depending on the environmental conditions (Chand and Khirbat 2009). If the conditions are favorable, *Fusarium* wilt disease causes high yield (10–50%) losses every year, that often reaches to 100% (Trapero-Casas and Jimenez-Diaz 1985; Karimi et al. 2012).

Fusarium oxysporum f. sp. *ciceris* induced wilt disease decreased the chickpea yield with 30% (Meki et al. 2008). Geletu et al. (1996) recorded that *F. oxysporum* decrease chickpea yield with 50–80% in certain areas. *Fusarium oxysporum* f. sp. *ciceris* causes wilt in wide host range of legumes like chickpea, bean, faba bean, lentil, pea and pigeon pea (Trapero-Casas and Jimenez-Díaz 1985). The pathogen could survive in the chickpea debris and soil by its chlamydospores for about 6 years. The abundance of chlamydospores induced severe symptoms on the plant as its condensing infection could limit the growth of other fungal saprophytes decreasing the natural plant protection. The optimal infection conditions include temperature range from 25 to 27.5 °C and pH range from 5.1 to 5.9, depending on the pathogen type and adaptation (Haware 1982).

The disease starts with the direct infection of *F. oxysporum* in germinating seeds and seedlings after sowing. It later invades mainly through the zones of epicotyl or hypocotyl and root elongation zone until it has complete establishment in the plant

(Stevenson et al. 1997). Wilt disease symptoms developed at any stage of the plant growth includes visual plant spread patches, flaccidity of leaves with discoloration, desiccation, and finally the entire plant collapsing (Nene and Reddy 1987). In the late wilted stage of the plants, petioles, leaflets, foliage become yellow or necrotic and rachis fall (Haware 1990). Roots and stem develop dark-brown discoloration in the xylem tissues as observed by longitudinal sections through the xylem vessels (Jimenez-Díaz et al. 1989).

The *Fusarium* wilt can be regulated by using resistant cultivars, adjusting the sowing dates, and seed treatment with various fungicides (Navas-Cortes et al. 1998; Sharma et al. 2007). Delaying the sowing of legume seeds (at 6 cm depth) improved the plant growth significantly and decreased the incidence of wilt disease. Intercropping in same field also reduced the incidence of wilt (Seid et al. 1990; Singh et al. 2006).

Soaking of seeds in salicylic acid and bion are also recommended for wilt disease control (Sarwar et al. 2005). Chitosan treatments reduced the wilt symptoms by 45–59% (Chowdhury and Sinha 2000) and salicylic acid addition reduced the wilt disease by 26–50% (Saikia et al. 2003). Biological control has a promising results in the disease control like using *Bacillus subtilis, Gliocladium virens, Trichoderma viride, T. harzianum* and *Streptomyces griseus* (De et al. 1996; Hervas et al. 1998). *Trichoderma harzianum, T. koningii, Rhizobium* and *Bacillus subtilis* can be use as biological control agents of *F. oxysporum* wilt disease using its ability to mycoparasitism and antibiosis (Poddar et al. 2004; Arfaoui et al. 2007). More over new antifungal agents could be used to protect the plants from *Fusarium* pathogens like calcium and copper based compounds (Mahmoud et al. 2019; Ibrahim and Mahmoud 2019).

7.3.4 Ascochyta Blight Disease

Ascochyta rabiei, is the causal agent of ascochyta blight of the chickpea, present in the form of anamorph and teleomorph. The anamorphic form is characterized by pear-shaped black fruiting bodies known as pycnidia containing unicellular hyaline pycnidiospores on short conidiophores (Punithalingam and Holliday 1972). Pycnidiospores are oblong in shape with 6–12 μm in diameter and straight to bent at the ends (Nene 1982). The pathogen teleomorphic stage of *Didymella rabiei* is characterized by bipolar heterothallic fungi giving rise to pseudothecia that infects legume crop residues. Pseudothecia are characterized by dark brown color, subglobose shape with 120–270 μm diameter containing binucleate asci with 8 unequally hyaline bicellular ascospores surrounded by paraphyses (Armstrong et al. 2001; Pande et al. 2005).

Ascochyta blight symptoms are developed on all the aerial plant parts. Infected seeds show brown lesions at the seedlings base, and brown girdle at the base of the plant stem with pycnidia necrotic lesions resulting plant breakage and death (Singh and Sharma 1998). Cultivated legumes can be attacked at any growth stage by the pathogen, while most prominent attacking stage is the flowering stage. The air-borne ascospores attack leaves producing water- necrotic spots that rapidly enlarges on the plant. Diseased pods fail to develop seeds and the infected seeds appear discolored with deep cankers (Nene 1982; Akem 1999). *Ascochyta rabiei* survives in the seeds and plant debris using its mycelia and pycnidia (Kaiser 1997), while *D. rabiei* sustains as free pathogen in the soil at 10–35 °C, in legume debris for maximum of 8 months (Nene and Reddy 1987) and for 5 months in the seeds (Singh et al. 1995). Low temperatures with high moisture content are essential for pseudothecia formation (Trapero-Casas and Kaiser 1992; Navas-Cortes et al. 2007).

Ascochyta rabiei cause the ascochyta blight of numerous crops like lentil, vetch, field pea, bean, cowpea (Khan et al. 1999b), *Vigna unguiculata*, *P. vulgaris* (Kaiser 1973; Khan et al. 1999a), *Lamium amplexicaule*, *Lactuca serriola*, *Melilotus alba*,*Medicago sativa*, *Thlapsi arvense*, *Descurainia sophia*, *Brassica nigra*, *Lamium amplexicaule*, *Galium apanine*, and *Triticum aestivum* (Kaiser 1997). It causes up to 100% yield losses in some crops (Nene and Reddy 1987). There are many methods to control the disease especially by cultivating resistant plant varieties, removing the infected plants and plant parts and burning it, and using non-contaminated legume seeds (Nasir et al. 2000; Pande et al. 2005). Foliar application of chlorothalonil and mancozeb as fungicides was highly efficient in the disease control (Bretag et al. 2003), treatment of seeds with calixin (Reddy et al. 1982), methyl benzimidazole (Kaiser and Hannan 1988), captan, and propiconazole (Rauf et al. 1992) also reported promising results in disease management.

7.3.5 Chocolate Spot Disease

Chocolate leaf spot disease is caused by both *Botrytis cinerea, B. fabae* and *Botryotinia fabae* (infected faba bean). *Botrytis cinerea* is a saprophytic parasite having wide range of host plants, while *B. fabae* causes major diseases in faba bean crops (Berhanu et al. 2005). These parasites cause yield losses of upto 61% in the susceptible plants and 34% in tolerant types. It infects the plant leaf, petioles, seeds, stems and lives as sclerotia (remains 2 years viable) in the debris of the infected plants (Dereje 1999). Soil type, humidity and cultivation depth affects the disease establishment. High humidity ≥70% and warm temperature with rainy weather are the most favorable growth conditions for incidences of chocolate spot disease (Dereje and Yaynu 2001). The disease favored temperature ranges from 15 °C to 22 °C with high humidity (Sahile et al. 2008).

The disease symptoms appears as lesions in the upper and lower leaf surfaces with large number of spots called as "chocolate spots" on the leaflets, petals, pods and stems. Chocolate spots appear reddish in color that later turns to chocolate brown color. The disease spread is very fast and aggressive in favorable conditions (Harrison 1988) as shown in Fig. 7.1. The progression of disease is indicated by the irreversible changes in the form of large patches that blight the leaves, defoliation and loss of flowers and pods. The cultivation of resistant plant varieties has been

Fig. 7.1 Chocolate Spot of faba bean caused by *Botrytis fabae*

identified as an easy way for disease management. Biological control agents like *Trichoderma ovalisporum* and *T. longibrachiatum* showed significant inhibiotry effect against *B. fabae* (Boubekeur et al. 2012; Sahile et al. 2012). Chemical control of chocolate spots disease by chemical fungicides like Benomyl, chlorothalonil, mancozeb has also been considered effective (Stoddard et al. 2010).

7.3.6 Fusarium *Root Rot Disease of Legumes*

Fusarium species represent cosmopolitan filamentous fungi that cause several plant diseases like root rots, cortical rots, leaf spots, head blights, fruit rots, and vascular wilt. The species is present in various soil and temperature zones e.g. moderate, temperate, tropical, arctic and deserts areas (Nelson et al. 1983). According to the American Phytopathological Society, from each 101 economically plants, 81 plants were infected with *Fusarium* (Leslie and Summerell 2016). It is the most common root rot pathogen of legumes. Variuos isolated species include *F. avenacearum* (Corda ex Fr.) Sacc., *F. culmorum* (Smith) Sacc, *F. graminearum* Schwabe, *F. oxysporum* Schlecht and *F. solani* (Mart.) Sacc., (Nelson 1964; Helsper et al. 1994; Naseri 2008).

7.3.6.1 *Fusarium* Taxonomy

Fusarium belongs to Hyphomycetes, order Hypocreales and class Ascomycetes. Genus *Fusarium* was introduced by Link in 1809 as a non-septate *(Fusarium roseum)* species on a stroma with fusiform spores (Booth 1971a). After the emergence of new species, the presence of stroma was no longer considered as the basic genus character. Now the diagnostic features were the presence of septate macroconidia with marked foot-shaped basal cell, microconidia and sometimes the presence of chlamydospores (Snyder and Toussoun 1965). From 1809 to1935, scientists described approximately one thousand *Fusarium* species. Wollenweber & Reinking (1935) has classified *Fusarium* genus into 16 sections, 65 species and 55 varieties. Snyder and Hansen (1940) re-arranged the species into nine species. Booth (1971b) arranged *Fusarium* into 48 species and 104 varieties. Leslie and Summerell (2016) classified *Fusarium* into 70 species.

7.3.6.2 Symptoms of *Fusarium* Root Rot Disease

The general symptoms of the root rot disease are reddish discoloration of tap roots of the young plants that later becomes dark leading to ageing. This discoloration not only covers the tap root but also the stem part below the soil surface or it could appear as emerging streaks extending up to the soil line surface. The longitudinal cracks rise above the main root indicating the cell death of lateral roots. The growth rate of plant decreases and the leaves become yellow with higher instances of senescence. In certain cases, infected plants could produce secondary roots below the soil surface to ensure plants growth upto maturation stage. However in some cases, the infected plant dies even without the appearance of the disease symptoms (Agrios 2005).

7.3.6.3 Description of Most Common *Fusarium* Root rots

Several *Fusarium* pathogens causing root rot disease in legumes have been reported. *Fusarium* identification was done on the basis of anamorph morphological characteristics like colony color, presence or absence of micro/macro conidia, size, shape, septa number, conidia, and presence/absence of chlamydospores (Gupta et al. 1986). *Fusarium oxysporum* and *Fusarium solani* are the most significant and common soil borne pathogens causing severe root rot in almost all soil types (Helsper et al. 1994; Akrami et al. 2009).

Fusarium oxysporum Schlechtendahl Emend (Snyder& Hansen)

It is also known as *F. redolens* Wollenweber 1913, *F. oxysporum* Schlechtendahl 1824, *F. oxysporum* Schlecht. emend. Snyd. & Hans. var. *redolens* Wollenweber Gordon 1952. Growth features (macroscopic) on potato dextrose agar medium (PDA) characterized by growth rate ranged from 7 to 8.25 cm (in 9 cm plate) after 10 days at 28 °C ± 1 cultivation. The produced mycelia abundant with initial white color turn to violate degrees when it became old (Van-der Maesen 1987). It also produces pale violet pigmentation in most isolates. The microscopic features represented by thin walled macroconidia, straight or slightly curved, having 3–5 septate as cleared in Fig. 7.2. Microconidia are abundant, elliptical, straight or curved, usually not septate and produced in short monophialidic conidiogenous cells. Chlamydospores formed usually singly in terminal or intercalary position with smooth or rough walled (Booth 1971a and Nelson et al. 1983; Mohamed and Mahmoud 2018).

Fusarium solani (Martius) Appel & Wollenweber Emend

It synonyms as *Fusisporium solani* Martius 1842, *F. javanicum* Koorders 1907, *F. ventricosum* Appel & Wollenweber 1910, *F. tumidum* var. *coeruleum* Bugn. 1939, *F. illudens* Booth 1971. Unlike *Fusarium oxysporum* it has known telemorph *Haemanectria haematococca* (Berkeley & Broome) Samuels & Nirenberg 2001. Growth features (macroscopic); on potato dextrose agar medium (PDA) characterized by growth range from 7.2 to 9.55 cm length after 10 days at 28 °C ± 1 with abundant white aerial mycelia, not produce pigments. Microscopic features are macroconidia present abundant in sporodochia, have from 3 to 7 septa as cleared in Fig. 7.3. Microconidia abundant, reniform and have 0–1 septa produced in long monophialidic conidiogenous cells. Chlamydospores formed with globose to oval shape, with sooth or rough walled and usually singly (Booth 1971a and Nelson et al. 1983). *Fusarium solani* root rot can cause considerable losses of faba bean yield as reported by Akrami et al. (2009).

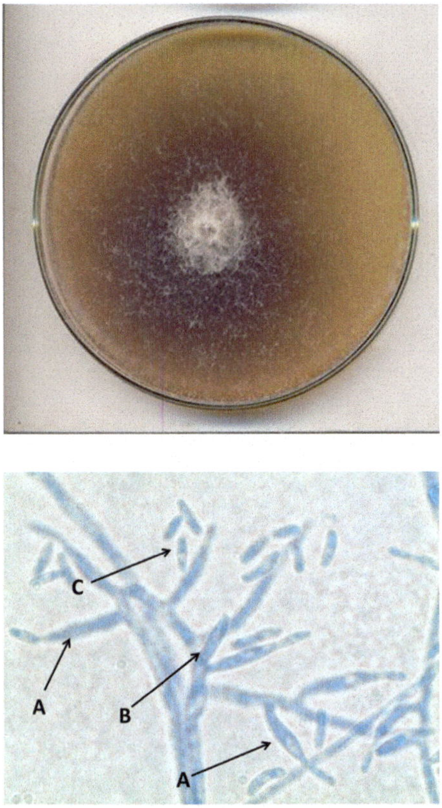

Fig. 7.2 *Fusarium oxysporum* Schlecht. fungal growth on PDA medium for 10 days at 28 °C ± 1with violet reverse (left photo); the microscopic features includes (write photo); (**a**) Short monophilaidic conidiogenous cell; (**b**) Macroconidia and (**c**) Microconidia; Bars, 10 μm

7.3.6.4 *Fusarium* Root Rot of Bean

Fusarium root rot of bean caused by *F. solani* f. sp. *phaseoli* (Naseri 2008). Bean (*Phaseolus vulgaris* L.) has been reported as a major legume crops threats in United States, Mexico, Brazil, Nicaragua, and coastal Peru (Naseri 2014). The fungus infects the bean root by mycelial germination from chlamydospores which increase the fungus survives in the soil for long time (Kraft et al. 1981; Naseri 2008). The disease generated severe root rotting resulting of black color of the roots followed by whole plant death as cleared in Fig. 7.4. This infection depends on the interaction between the pathogen, the host plant, and soil conditions (Otten and Gilligan 2006). Soil pH, texture, organic materials, root-colonizing bacteria, soil microorganisms and pests like bean fly are known to influence on the root rot disease of bean (Ghorbani et al. 2008). Also, presence of saprophytic fungi with antagonistic activities in the soil could protect the beans crop from *Fusarium* root rot pathogens

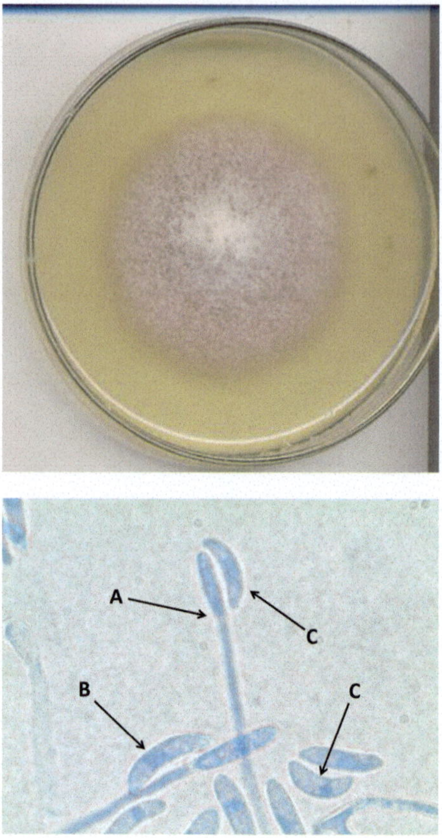

Fig. 7.3 *Fusarium solani (Martius)* Appel & Wollenweber emend; fungal growth on PDA medium for 4 days at 28 °C ± 1 with white to pale pink mycelia and no reverse (left photo); the microscopic features includes (write photo); (**a**) Long monophilaidic conidiogenous cell; (**b**) Macroconidia with two septa and (**c**) Reniform microconidia; Bars, 10 μm

Fig. 7.4 *Fusarium* root rot of bean; in the left the infected plant while in the right the healthy plant

(Buerkert et al. 1990; Tu 1992). Naseri and Marefat (2011) found that the agronomic practices like previous crop planted, irrigation, urea application, date and depth of bean inoculation control root rot disease.

7.3.6.5 *Fusarium* Root Rot of Pea

Fusarium solani f. sp. *pisi* and *F. oxysporum* f. sp. *pisi* responsible for pea root rot diseases (Kerr 1963; Kraft 1978). Root rot disease was first described in 1918, USA by Jones (1923), considered the most damaging disease of pea in both irrigated and dry areas (Kraft 1984) causing 30–57% yield reduction (Kraft et al. 1981). *Fusarium solani* f. sp. *pisi* initiated the infection via natural plant openings like stoma of epicotyl and hypocotyl (Bywater 1959), then infects the cotyledonary attachment area, epicotyl and hypocotyl of pea. The pathogen also can penetrate the cuticular epicotyls surface of pea via enzyme degradation directly (Koller et al. 1982). Pathogen spores start germination within 20 hour of seed planting in soil contains 9% moisture content (Cook and Flentje 1967).The underground symptoms were initiated with reddish-brown streaks on the primary and secondary roots, and then primary root red discolouration occurred in the cotyledonary attachment area. The above ground symptoms were lower leaves yellowing or at the disease end necrosis with shoot growth stunted (Agrios 2005). The optimum conditions of *Fusarium* for the infection and disease development were optimum temperature range from 25 °C to 30 °C, inadequate soil rotation, soil moisture contents, soil compaction, poor soil fertility, and soil pH 5.1–6.2 (Kraft et al. 1981).

7.3.6.6 *Fusarium* Root Rot of Faba Bean

Root rots are spread diseases of faba bean (*Vicia faba* L.) caused mainly by *Fusarium solani* (black root rot). It could destruct complete field by the *F. solani* black root rot pathogen attack (Madkour et al. 1983; Akrami et al. 2009) especially in favourable conditions of the disease. *Fusarium oxysporum* f. sp. *fabae* and *F. solani* f. sp. *fabae* also causing root rot of faba bean in the Sudan that considered the most important faba bean diseases (Hussein 1983). *Fusarium avenaceum* was reported to be the cause of faba bean root rot by Lamari and Bernier (1985). In Ethiopia, the black root rots causes major losses in the faba bean reach to 70% yield loss (Mukankusi et al. 2011).

Also, Otsyula et al. (1998) reported lossing to 84% of legume yield by *Fusarium solani* root rot. The optimum soil temperature to start the infection by *F.solani* is 25 °C (Negussie 1989). In early stage, the disease developed seedling death, in growing plant severe root rotting generated resulting black discoloration of most roots followed by complete plant death. Other symptoms could observe like primary roots reddish lesions, outer root longitudinal cracks and finally destruction of the main root (Agrios 2005) as cleared in Fig. 7.5. Management of faba bean root rot depends on cultivation of resistant genetic plant varities, chemical control using

Fig. 7.5 *Fusarium* root rot of faba bean; in the left the infected plant while in the right the healthy plant

fungicides especially with seeds, good soil drainage and crop rotation (El-Mougy and Abdel-Kader 2009; Goellner et al. 2010). Habtegebriel and Boydom (2016) reported that biological control of faba bean using *Trichoderma harzianum* as seed coating or soil treatment decrease faba bean death with 65.64%.

7.3.6.7 *Fusarium* Root Rot of chickpea and Lentil

Chickpea (*Cicer arietinum* L.) is an important legume cultivated in large areas; *Fusarium solani* cause black root rot of chickpea at favorable conditions the disease causes losses in crops up to 50% (Beniwal et al. 1989; Tadesse et al. 1999). *Fusarium solani* represents significant pathogen developed severe root rot of chickpea (Nene 1979). The disease development is strongly associated with soil temperature, moisture, inoculum density and the crop age. General optimum temperature of *F. oxysporum* is 25 °C while *F. solani* favour 30 °C (Negussie 1989). *Fusarium* inoculums increased in wet soils which consequently will increase the disease spreading and severity (Mengistu and Negussie 1994).

Lentil (*Lens culinaris*) occupies unique position in the human food as high proteins, carbohydrates, minerals, vitamins source for human nutrition and also used in animal feeding. Lentil is influenced by numerous fungal diseases reduce the productivity through damaging the plant leaves, roots, stems, and pods. *Fusarium oxysporum* f. sp. *lentis* represents the most serious *Fusarium* wilt disease of lentil spread in India and in Russia (Kraft et al. 1981). It appeared the same root rot symptoms as the previous legume crops. To manage the disease and control it, cultivation methods of inoculum destruction un-decomposed organic matter like wheat stubble offers effective results of root rot control (Ali-Khan et al. 1974; Ahmed et al. 1990). Cultivating chickpea genotypes resistance to root rots pathogens (Beniwal et al. 1992). Now most of chickpea cultivars possess resistance genes mainly against to *Fusarium* wilt and root rot (Ahmed et al. 1990). Biological control also offers alternative method of controlling the plant disease, *Trichoderma* spp. indicated potential to control *Fusarium* wilt and root rot of chickpea, also some types of *Aspergillus* and *Penicillium* have antagonistic effect against *F. oxysporum* (Beniwal et al. 1992). Chemical treatments of chickpea seeds with fungicides like vitavax and thiram gives good disease reduction for *Fusarium* wilt, root rot and collar rot (Habtu and Zadoks 1995).

7.3.6.8 *Fusarium* Root Rot Management

Effective management of plant root rot achieved by developing resistant and tolerant legume varieties (Goellner et al. 2010; Boyd et al. 2013). Crop rotation, draining wet soils, fertilization, soil preparation by tillage, and weed control represented tools to control the root rot diseases (Naseri and Marefat 2011). Planting seed without overcrowding decrease the disease, crop rotation using Brassicaceae members suppress the disease with its fungicidal effects (El-Mougy and Abdel-Kader 2009).

Reducing the weed plants using herbicides, which help the pathogenic fungi to survive in crops (Foroud et al. 2014). Cultivation methods of the inoculum destruction with un-decomposed organic matter like wheat stubble offers were effective results of root rot control (Ahmed et al., 1990). Biological control using different microorganisms also helps to decrease the disease like *Bacillus pumilus, Pseudomonas putida, Aspergillus awamori, A. niger* and *Trichoderma harzianum* used efficiently to control *Fusarium* root rot (González-Sánchez et al. 2004; Alabouvette et al. 2009; Gardiner et al. 2013; Habtegebriel and Boydom 2016). Using natural oils like garlic oil found to be decrease the counts of *Fusarium* sp. (Mansour et al. 2018). Chemical treatment with methyl bromide, pyraclostrobin, fludioxonil, trifloxystrobin, vitavax and thiram gives good disease reduction for *Fusarium* diseases like wilt; root rot and collar rot (Habtu and Zadoks 1995; Someya et al. 2000; Gullino et al. 2005).

7.4 Conclusion

Legumes are considered as significant source for our dietary protein and the animal nutrition, legume crops also improves the soil fertility by their biological nitrogen fixation. Our needs for these crops increase by the time so the growing areas increased to sufficient our demands. These increasing in the growing areas still not sufficient due to the crop losses by fungal diseases that could reach to complete loss of the total crop. Rust, powdery mildew, wilt, ascochyta blight, chocolate spot and root rot are common fungal diseases infect legumes. The severity of infection depends on many environmental conditions (temperature, soil pH, humidity, wind, rains, and etc.) and the plant defense strategies (anatomical, physiological, biochemical, and genetic defense characters). Many natural disease controlling methods were discussed like intercropping, crop rotation, draining wet soils, fertilization, soil preparation by tillage, and weed. Chemical control with common and new antifungal agents proved efficiency against the fungal pathogens like calcium and copper based compounds. Natural oils also used more recently as safe controlling agent. Biological controls using another microorganism like *Bacillus pumilus, Pseudomonas putida, Trichoderma harzianum* and etc. efficiently control *Fusarium* root rot and other legume diseases. Although all these controlling strategies we mentioned researchers must search for more safe controlling agents to protect the environment and faces the pathogen resistance development.

References

Agrios GN (2005) Plant pathology, 5th edn. San Diego, Copyright © Elsevier Academic Press. https://www.elsevier.com/books/plant-pathology/agrios/978-0-08-047378-9

Ahmed S, Beniwal SPS, Tadesse N (1990) Field screening of chickpea for resistance to wilt/ root rots in Ethiopia. Int Chickpea Newsletter 22:34–36. https://www.cabdirect.org/cabdirect/abstract/19921626762

Ahmed AF, Mahmoud GA-E, Ma C, Kang W (2020) Antioxidant and antimicrobial activities of essential oils and extracts of thyme (*Thymus vulgaris* L.) and oregano *(Origanum vulgare* L.) plants grown in Egypt. J Henan Uni (Medic Sci) 39(3):157–170

Akem C (1999) Ascochyta blight of chickpea: present status and future priorities. Int J Pest Manage 45:131–137. https://doi.org/10.1080/096708799227923

Akrami M, Ibrahimov AS, Zafari DM, Valizadeh E (2009) Control of *Fusarium* Rot of bean by combination of *Trichoderma harzianum* and *Trichoderma asperillum* in green house conditions. Agric J 4:121–123. http://medwelljournals.com/abstract/?doi=aj.2009.121.123

Alabouvette C, Olivain C, Migheli Q, Steinberg C (2009) Microbiological control of soil-borne phytopathogenic fungi with special emphasis on wilt-inducing *Fusarium oxysporum*. New Phytol 184:529–544. https://doi.org/10.1111/j.1469-8137.2009.03014.x

Ali-Khan ST, Zimmer RC, Kenaschuk EO (1974) Reaction of pea introductions to Ascochyta foot rot and powdery mildew. Can Plant Disease Survey 53: 155–156. http://scholar.google.com.eg/scholar_url?url=http://phytopath.ca/wp-content/uploads /2014/10/cpds-archive/vol53/CPDS_Vol_53_No_3_1973.pdf%23page%3D39&hl=ar&sa=X&scisig=AAGBfm27trs_7aLM2t3o9Sc3HISwDtQ0CQ&nossl=1&oi=scholar

Angelini R, Bragaloni M, Federico R, Infantino A, Porta-Puglia A (1993) Involvement of polyamines, diamine oxidase and peroxidase in resistance of chickpea to Ascochyta rabiei. J Plant Physiol 142:704–709. https://doi.org/10.1016/S0176-1617(11)80906-5

Arfaoui A, El Hadrami A, Mabrou Y, Sifi B, Boudabous A, El Hadrami I, Daayf F, Cherif M (2007) Treatment of chickpea with *Rhizobium* isolates enhances the expression of phenyl- propanoid defense-related genes in response to infection by *Fusarium oxysporum f. sp. ciceri*. Euphytica 45:470–479. https://doi.org/10.1016/j.plaphy.2007.04.004

Armstrong CL, Chongo G, Gossen BD, Duczek LJ (2001) Mating type distribution and incidence of the teleomorph of *Ascochyta rabiei (Didymella rabiei)* in Canada. Can J Plant Pathol 23:110–113. https://doi.org/10.1080/07060660109506917

Attanayake RN, Glawe DA, McPhee KE, Dugan FM, Chen W (2010) *Erysiphe trifolii* – a newly recognized powdery mildew pathogen of pea. Plant Pathol 59:712–720. https://doi.org/10.1111/j.1365-3059.2010.02306.x
Baggett JR, Hampton RO (1972) Plant introduction lines of *Pisum sativum* resistance to *Pea* fizzle top disease. Plant Disease Rep 56:171–175
Barnett OW, Diachun S (1984) Clover science and technology, Agronomy monograph no. 25. ASA-CSSA-SSSA, Madison, Wisconsin, pp 235–268
Belete T, Bastas KK (2017) Common bacterial blight (*Xanthomonas axonopodis* pv. *phaseoli*) of beans with special focus on Ethiopian condition. Review article Belete and Bastas. J Plant Pathol Microbiol 8:3. https://doi.org/10.4172/2157-7471.1000403
Beniwal SPS, Ahmed S, Tadesse N (1989) Effect of sun drying of lentil seeds on the control of seed-borne *Ascochyta lentis*. Lens News 16(2):27-8-51. http://agris.fao.org/agris-search/search.do?recordID=QV9100020
Beniwal SPS, Ahmed S, Gorfu D (1992) Wilt root rot diseases of chicklpea in Ethiopia. Trop Pest Manag 38(1):48–51. https://doi.org/10.1080/09670879209371644
Benson DA, Karsch-Mizrachi I, Lipman DJ, Ostell J, Sayers EW (2009) GenBank. Nucleic Acids Res 37:26–31. https://doi.org/10.1093/nar/gkn723
Berhanu B, Kumari SG, Ali K,Yusuf A, Makkouk KM, Aslake M, Ayalew M, Girma G, Hailu D (2005) Survey of viruses affecting legume crops in Amhara and Oromiya regions of Ethiopia. Phytopathol Mediterr 44:235–246. https://doi.org/10.14601/Phytopathol_Mediterr-1803
Biddle AJ (2001) Downy mildew. In: Kraft JM, Pfleger FL (eds) Compendium of pea diseases and pests, 2nd edn. APS Press, St. Paul, pp 29–30
Bond DA, Jellis GJ, Rowland GG, Le Guen J, Robertson, LD, Khalil SA, Li-Juan L, (1993) Present status and future strategy in breeding faba beans (*Vicia faba* L.) for resistance to biotic and abiotic stresses. Euphytica 73:151–166. https://doi.org/10.1007/BF00027191
Booth C (1971a) The genus *Fusarium*. Common Wealth Mycological Institute, Kew, p 114
Booth C (1971b) The genus *Fusarium*. Common Wealth Mycological Institute, Kew, p 137
Bos L, Hampton RO, Makkouk KM (1988) Viruses and virus diseases of pea, lentil, faba bean and chickpea. World crops: Cool season food legumes, pp 591–615. https://doi.org/10.1007/978-94-009-2764-3_49
Boubekeur SB, Mahiout D, Benzohra IE, Benkada MY (2012) Antagonism of three *Trichoderma* species against *Botrytis fabae* and *B. cinerea*. The causal agents of chocolate spot of faba bean (*Vicia faba* L.) in Algeria. World Appl Sci J 17(3):278–283
Bouchenak M, Lamri-Senhadji M (2013) Nutritional quality of legumes, and their role in cardiometabolic risk prevention: a review. J Med Food 16:185–198. https://doi.org/10.1089/jmf.2011.0238
Boyd LA, Ridout C, O'Sullivan DM, Leach JE, Leung H (2013) Plant – pathogen interactions: disease resistance in modern agriculture. Trends Genet 29(4):233–240. https://doi.org/10.1016/j.tig.2012.10.011
Bretag TW, Meredith KA, Knights E, Materne MA (2003) Control of ascochyta blight in chickpeas using disease resistance and foliar fungicides. In Proceedings of the 8th international congress of plant pathology, Christchurch, New Zealand, 2–7 February, pp 291 (International Society for Plant Pathology, UK)
Bridge J, Starr JL (2007) Plant nematodes of agricultural importance. Academic/CRC Press, Boston, p 152. https://doi.org/10.1201/b15142
Buerkert A, Cassman KG, Piedra R, de la Munns DN (1990) Soil acidity and liming effects on stand, nodulation and yield of common bean. Agron J 82:749–754. https://doi.org/10.2134/agronj1990.00021962008200040020x
Buruchara R, Mukankusi C, Ampofo K (2010) Bean disease and pest identification and management, handbooks for small-scale seed producers International Center for Tropical Agriculture. (CIAT), Pan-Africa Bean Research Alliance (PABRA), Uganda, Kampala. http://www.fao.org/sustainable-food-value-chains/training-and-learning-center/details-materials/en/c/276602/

Bywater J (1959) Infection of peas by*Fusarium solani* var.*martii* forma 2 and the spread of the pathogen. Trans Brit Mycol Soc 42:201–212. https://doi.org/10.1016/S0007-1536(59)80029-2
Carmona MA, Gally ME, Lopez SE (2005) Asian soybean rust: incidence, severity, and morphological characterization of *Phakopsora pachyrhizi* (uredinia and telia) in Argentina. Plant Dis 89:109–109. https://doi.org/10.1094/PD-89-0109B
Chand H, Khirbat SK (2009) Chickpea wilt and its management – a review. Agric Rev 30(1):1–12
Chowdhury AK, Sinha AK (2000) Increase resistance in chickpea plants to *Fusarium*-wilt following treatment with chitosan. J Mycopathol Res 38:105–108
Cockbain AJ, Gibbs AJ (1973) Host range and overwintering sources of bean leaf roll and pea enation mosaic viruses in England. Ann appll Biolo 73:177–187. https://doi.org/10.1111/j.1744-7348.1973.tb01323.x
Coetzee V (1968) The distribution of the family *Heteroderidae* (Filipjev, 1934) in South Africa and some host records of *Meloidogyne* species. S Afr J Agric Sci 11:775–788
Cook RJ, Flentje NT (1967) Chlamydospore germination and germling survival of *Fusarium solani* f. *pisi* in soil as affected by soil water and pea seed exudation. Phytopathology 57:178–182
da Silva EG, Moura AB, Deuner CC, Farias DR (2008) Study of biocontrol mechanisms of bacterial bean blight by bacteria. Revista Ceres 55:377–383. https://doi.org/10.1016/j.biocontrol.2014.02.013
Damiani I, Pauly N, Puppo A, Brouquisse R, Boscari A (2016) Reactive oxygen species and nitric oxide control early steps of the legume – rhizobium symbiotic interaction. Front Plant Sci 7:454. https://doi.org/10.3389/fpls.2016.00454
De RK, Choudhary RG, Naimuddin (1996) Comparative efficiency of biocontrol agents and fungicides for controlling chicken pea wilt by *F. oxysporum* f. sp. *ciceri*. Indian J Agri Sci 60(6):370–373
Dereje G (1999) Survival of Botrytis fabae Sard. Between seasons on the crop debris in field soils at Holetta, Ethiopia. Phytopathol Meditter 38:68–75. https://www.jstor.org/stable/42685359?seq=1#page_scan_tab_contents
Dereje G, Yaynu H (2001) Yield losses of crops due to plant diseases in Ethiopia. Pest Manag J Ethiopia 5:55–67. http://agris.fao.org/agris-search/search.do?recordID=ET2003000103
Derks AFLM, Vink-van den Abeele JL, Muller PJ (1980) Bean yellow mosaic virus in some iridaceous plants. Acta Horti 110: 31–37. https://doi.org/10.17660/ActaHortic.1980.110.2
Di Vito M, Abbad Andaloussi F, Caubel G, Greco N (2001) Resistance of faba bean and chickpea to root nematodes. In: European Association for Grain Legume Research-AEP (eds) Legumed grain legumes in Mediterranean agriculture, pp 57–60
Diapari M, Sindhu A, Bett K, Deokari A, Warkentin DT, Tar'an B (2014) Genetic diversity and association mapping of iron and zinc concentration in chickpea (*Cicer arietinum* L). Genome 57:459–468. https://doi.org/10.1139/gen-2014-0108
Dita MA, Rispail N, Prats E, Rubiales D, Singh KB (2006) Biotechnology approaches to overcome biotic and abiotic stress constraints in legumes. Euphytica 147:1–24. https://doi.org/10.1007/s10681-006-6156-9
Dixon RA, Harrison MJ, Lamb CJ (1994) Early events in the activation of plant defense responses. Annu Rev Phytopathol 32:479–501. https://doi.org/10.1146/annurev.py.32.090194.002403
du Preez ED, van Rij NC, Lawrance KF (2005) First report of soybean rust caused by *Phakopsora pachyrhizi* on dry beans in South Africa. Plant Dis 89:206–206. https://doi.org/10.1094/PD-89-0206C
El-Mougy NS, Abdel-Kader MM (2009) Seed and soil treatments as integrated control measure against Faba bean root rot pathogens. Plant Pathol Bull 18:75–87. http://scholar.google.com.eg/scholar_url?url=http://140.112.183.1/cpps/pdf/18-2/75-87.pdf&hl=ar&sa=X&scisig=AAGBfm3PV05DnFeqtujZjbhPzUZijbytig&nossl=1&oi=scholarr
Emmett RW, Parbery DG (1975) Appressoria. Annu Rev Phytopathol 13:147–167. https://doi.org/10.1146/annurev.py.13.090175.001051

Falloon RE, Viljanen-Rollinson SLH (2001) Powdery mildew. In:Kraft JM, Plfleger FL (eds) Compendium of pea diseases and pests. American phytopathol Soc:28–29. https://my.apsnet.org/ItemDetail?iProductCode=42694

Farooq M, Gogoi N, Hussain M, Barthakur S, Paul S, Bharadwaj N, Migdadi HM, Alghamdi SS, Siddique KHM (2017) Effects, tolerance mechanisms and management of salt stress in grain legumes. Plant Physiol Biochem 118:199–217. https://doi.org/10.1016/j.plaphy.2017.06.020

Fininsa C (2003) Relationship between common bacterial blight severity and bean yield loss in pure stand and bean-maize intercropping systems. Int J Pest Manag 49:177–185. https://doi.org/10.1080/0967087021000049269

Fondevilla S, Rubiales D (2012) Powdery mildew control in pea. A review. Agro Sust Dev 32:401–409. https://doi.org/10.1007/s13593-011-0033-1

Food and Agriculture Organization STAT (2014) Available online: http://faostat.fao.org/site/567/DesktopDefault.aspx

Foroud NA, Chatterton S, Reid LM, Turkington TK, Tittlemier SA et al (2014) *Fusarium* diseases of Canadian grain crops: impact and disease management strategies. In: Future challenges in crop protection against fungal pathogens. Springer, New York, pp 267–316. https://doi.org/10.1007/978-1-4939-1188-2

Fourie D (2002) Distribution and severity of bacterial disease of dry beans (*Phaseolus vulgaris*) in South Africa. J Plant Pathol 150:220–226. https://doi.org/10.1046/j.1439-0434.2002.00745.x

Fourie H, Mc Donald AH (2007) Chemical control options for plant-parasitic nematodes associated with soybean in South Africa, Report on project M151/60. Agricultural Research Council–Grain Crops Institute (ARC-GCI), Potchefstroom

Fourie H, De Waele D, Mc Donald AH et al (2015) Nematode pests threatening soybean production in South Africa, with reference to Meloidogyne. S Afr J Sci:111. https://doi.org/10.17159/sajs.2015/20140212

Gardiner DM, Stephens AE, Munn AL, Manners JM (2013) An ABC pleiotropic drug resistance transporter of *Fusarium graminearum* with a role in crown and root diseases of wheat. FEMS Microbiol Lett 348(1):36–45. https://doi.org/10.1111/1574-6968.12240

Geletu B and Yadeta A (1994) Breeding chickpea for resistance to drought. International symposium on pulse research, April 2–6, New Delhi, India, pp 145–146

Geletu B, Million E, Yadeta A (1996) Improved cultivars and production technology of chickpea in Ethiopia. Debre-Zeit Agricultural Research Center Research Bulletin. No. 2: p. 60. http://agris.fao.org/agris-search/search.do?recordID=XF2016018765

Ghorbani R, Wilcockson S, Koocheki A, Leifert C (2008) Soil management for sustainable crop disease control: a review. Environ Chemist Lett 6:149–162. https://doi.org/10.1007/s10311-008-0147-0

Gilbertson RL, Maxwell DP (1992) Common bacterial blight of bean. Prentice Hall Inc, Englewood Cliffs. https://www.cabdirect.org/cabdirect/abstract/19932328355

Gillard CL, Conner RL, Howard RJ, Pauls KP, Shaw L et al (2009) The performance of dry bean cultivars with and without common bacterial blight resistance in field studies across Canada. Can J Plant Sci 89:405–416. https://doi.org/10.4141/CJPS08045

Giorgio A, Lo Cantore P, Shanmugaiah V, Lamorte D, Sante Iacobellis N (2016) Rhizobacteria isolated from common bean in southern Italy as potential biocontrol agents against common bacterial blight. Eur J Plant Pathol 144:297–309. https://doi.org/10.1007/s10658-015-0767-8

Goellner K, Loehrer M, Langenbach C, Conrath U, Koch E et al (2010) *Phakopsora pachyrhizi*, the causal agent of Asian soybean rust. Mol Plant Pathol 11:169–177. https://doi.org/10.1111/j.1364-3703.2009.00589.x

González-Sánchez MA, Cazorla FM, Ramos C, de Vicente A, Pérez-Jiménez RM (2004) Studies of soil and rhizosphera bacteria to improve biocontrol of avocado white root rot caused by *Rosellinia necatrix*. Management of plant diseases an arthropod pests by BCAs and their integration in agricultural. IOBC/WPRS Bulletin 27:169–172. https://doi.org/10.1007/978-1-4020-8571-0_7

Gritton ET, Ebert RD (1975) Interaction of planting date and powdery mildew on pea plant performance. Am Soc Horti Sci 100:137–142. http://agris.fao.org/agris-search/search.do?recordID=US201302734107

Guillon F, Champ MM (2002) Carbohydrate fractions of legumes: uses in human nutrition and potential for health. Br J Nutr 88(3):293–306. https://doi.org/10.1079/BJN2002720
Gullino ML, Clini C, Garibald A (2005) Life without methyl bromide: the Italian experience in replacing the fumigant. Commun Agri Appl Biol Sci 70:13–25
Gupta OM, Khare MN, Kotasthane SR (1986) Variability among six isolates of *F. oxysporum f. sp ciceris* causing vascular wilt of chickpea. Ind Phytopathol 39:279–281. https://doi.org/10.1080/03235400802214836
Habtegebriel B, Boydom A (2016) Integrated Management of Faba Bean Black Root rot (*Fusarium solani*) through varietal resistance, drainage and adjustment of planting time. J Plant Pathol Microbiol 7(363):2157–7471. https://doi.org/10.4172/2157-7471.1000363
Habtu A, Zadoks JC (1995) Components of partial resistance in Phaseolus beans against an Ethiopian isolate of bean rust. Euphytica 83:95–102. https://doi.org/10.1007/BF01678035
Hamilton WDO, Bisaro DM, Coutts RHA, Buck KW (1983) Demonstration of the bipartite nature of the genome of a single-stranded DNA plant virus by infection with the cloned DNA components. Nucleic Acids Res 11(21,11):7387–7396. https://doi.org/10.1093/nar/11.21.7387
Hampton RO (1967) Natural spread of viruses infectious to beans. Phytopathology 5:476–481
Hampton RO (1983). Pea streak and alfalfa mosaic viruses in alfalfa: reservoir of viruses infectious to *Pisum peas*. Phytopathology 67(3):308–310. https://doi.org/10.1094/PD-67-308
Hampton RO (1984) In Hagedorn DJ (ed) Compendium of pea diseases. American Phytopathological Society, Minneapolis, pp 31–37. https://books.google.com.eg/books/about/Compendium_of_Pea_Diseases.html?id=3gVJAQAAIAAJ&redir_esc=y
Hancock JG, Huisman OC (1981) Nutrient movement in host pathogen systems. Annu Rev Phytopathol 19:309–331. https://doi.org/10.1146/annurev.py.19.090181.001521
Harrison JG (1988) The biology of Botrytis spp. on vicia beans and chocolate spot disease: a review. Plant Pathol 37:168–201. https://doi.org/10.1111/j.1365-3059.1988.tb02064.x
Haware MP (1982) *Fusarium* wilt and its managements. Phytopathology 68:1364–1367. https://doi.org/10.1094/Phyto-68-1364
Haware MP (1990) *Fusarium* wilt and other important disease of chickpea in the Mediterranean area. Options Mediterranean Seminar 9:61–64. http://agris.fao.org/agris-search/search.do?recordID=QC9165012
Helsper JPFG, Van Norelydi'x A, Burgermeyer K, Hoogendijk JM (1994) Effect of the absence of condensed tannins in faba beans (*Vicia faba*) on resistance to foot rot, ascochyta blight and chocolate spot. J Agric Sci 123 349–355. https://doi.org/10.1017/S0021859600070350
Hervas A, Landaa B, Datnoff LE, Jiménez-Dı́az RM (1998) Effects of commercial and indigenous microorganisms on *Fusarium* Wilt development in chickpea. Biolog Cont 13 : 166–176. https://doi.org/10.1006/bcon.1998.0659
Hijwegen T (1992) Biological control of cucumber powdery mildew with *Tilletiopsis minorunder* greenhouse conditions. Neth J Plant Pathol 98:221–225. https://doi.org/10.1007/BF02000089
Hofstein R, Daoust RA, Aeschlimann JP (1996) Constraints to the development of biofungicides: the example of'AQ-10', a new product for controlling powdery mildews. Entomophaga 41:455–460. https://doi.org/10.1007/BF02765797
Hollomon DW, Wheeler IE (2002) Controlling powdery mildews withchemistry. In: Bélanger RR, Bushnell WR, Dik AJ, TLW C (eds) The powdery mildews, a comprehensive treatise. APS Press, St. Paul, pp 249–255. https://my.apsnet.org/ItemDetail?iProductCode=42910&WebsiteKey=2661527a-8d44-496c-a730-8cfeb6239be7
Htike ZZ, Win SL (2013) Classification of eukaryotic splice-junction genetic sequences using averaged one-dependence estimators with Subsumption resolution. Procedia Comput Sci 23:36–43. https://doi.org/10.1016/j.procs.2013.10.006
Hussein MM (1983) In Proceedings of the international workshop on Faba Beans, Kabuli Chickpeas, and Lentils in the 1980s: 119–127 (Saxena MC, Varma S, eds). ICARDA, Aleppo. https://www.worldcat.org/title/faba-beans-kabuli-chickpeas-and-lentils-in-the-1980s-an-international-workshop-16-20-may-1983-proceedings/oclc/12704270

Ibrahim ABM, Mahmoud GA-E (2019) Nonstoichiometric Mesoporous Cu1.90S nanoparticles hydrothermally prepared from a copper Anthranilato complex inhibit Cellulases of Phytopathogenic Fungi. J Inorg Organometa Polymers Materials 29(4):1280–1287. https://doi.org/10.1007/s10904-019-01091-6

Ibrahim ABM, Zidan ASA, Aly AAM, Mosbah HK, Mahmoud GA-E (2020) Mesoporous cadmium sulfide nanoparticles derived from a new cadmium anthranilato complex: characterization and induction of morphological abnormalities in pathogenic fungi. Appl Organometal Chem 34(2):e5391. https://doi.org/10.1002/aoc.5391

Jarvis WR, Gubler WD, Grove GG (2002) Epidemiology of powdery mildews in agricultural pathosystems. In: Bélanger RR, BushnellWR, Dik AJ, carver TLW (eds) the powdery mildews, acomprehensive treatise. APS Press, St. Paul, pp. 169–199. https://my.apsnet.org/ItemDetail?iProductCode=42910&WebsiteKey=2661527a-8d44-496c-a730-8cfeb6239be7

Jimenez-Díaz RM, Basallote-Ureba MJ, Rapoport H (1989) Colonization and pathogenesis in chickpea infected by races of *Fusarium oxysporum* f. sp. ciceris. In: Tjamos EC, Beckman C (eds) Vascular wilt diseases of plants, vol H28. Springer, Berlin, pp 113–112. https://doi.org/10.1007/978-3-642-73166-2_7

Jones FR (1923) Phythium as a causal factor in (pea blight) Phytopath. of *Fusarium*. Int Agri Res 26:459–476. https://scholar.google.com/scholar_lookup?author=FR.%20Jones&journal=J.%20Agric.%20Res.&volume=26&pages=459-476&publication_year=1923

Kaiser WJ (1973) Factors affecting growth, sporulation, pathogenicity and survival of Ascochyta rabiei. Mycologia 65:444–457. https://doi.org/10.2307/3758115

Kaiser WJ (1997) Inter- and intranational spread of Ascochyta pathogens of chickpea, faba bean, and lentil. Can J Plant Pathol 19:215–224. https://doi.org/10.1080/07060669709500556

Kaiser WJ, Hannan RM (1988) Seed transmission of Ascochyta rabiei in chickpea and its control by seed-treatment fungicides. Seed Sci Technol 16:625–637. https://www.cabdirect.org/cabdirect/abstract/19891129253

Karimi K, Amini J, Harighi B, Bahramnejad B (2012) Evaluation of biocontrol potential of *Pseudomonas* and *Bacillus* spp. against *Fusarium* wilt of chickpea. Aust J Crop Sci 6(4):695–703

Karssen G, Wesemael W, Moens M (2013) Root-knot nematodes. In: Perry RN, Moens M (eds) Plant nematology. CAB International, Wallingford, pp 73–108

Kavková M, Curn V (2005) *Paecilomyces fumosoroseus* (Deuteromy-cotina: Hyphomycetes) as a potential mycoparasite on *Sphaer-otheca fuliginea* (Ascomycotina: Erysiphales). Mycopathology 159:53–63. https://doi.org/10.1007/s11046-003-0787-3

Kerr A (1963) The root rot-*Fusarium* wilt complex of peas. Aust I Biol Sci 16:55–59. https://scholar.google.com/scholar_lookup?author=A.%20Kerr&journal=Aust.%20J.%20Biol.%20Sci.&volume=16&pages=55-59&publication_year=1963

Khan MSA, Ramsey MD, Scott ES (1999a) Host range studies with Australian isolates of Ascochyta rabiei. Aust Plant Pathol 28:170–173. https://doi.org/10.1071/AP99028

Khan MSA, Ramsey MD, Corbiere R, Infantino A, Porta-Puglia A, Bouznad Z, Scott ES (1999b) Ascochyta blight of chickpeas in Australia: identification, pathogenicity and mating type. Plant Pathol 48:230–234. https://doi.org/10.1046/j.1365-3059

Kiss L (2003) A review of fungal antagonists of powdery mildews and their potential as biocontrol agents. Pest Manag Sci 59:475–483. https://doi.org/10.1002/ps.689

Koitabashi M (2005) New biocontrol method for parsley powdery mildew by the antifungal volatiles-producing fungus Kyu-W63. J Gen Plant Pathol 71:280–284. https://doi.org/10.1007/s10327-005-0198-z

Koller W, Allan CR, Kolattukudy PE (1982) Protection of *Pisum sativum* from *Fusarium solani* f. sp. *pisi* by inhibition of Cutinase with Organophosphorus pesticides. Phytopathology 72:1425–1430. https://doi.org/10.1094/Phyto-72-1425

Kraft JM (1978) Effects of root rot pathogens on Fusarium wilt of peas. Plant Dis Rep 62:216–221

Kraft JM (1984) In Hagedorn DJ (ed) Compendium of pea diseases. American Phytopathological Society, St Paul, pp 30–31. https://scholar.google.com/scholar_lookup?title=Fusarium%20root%20rot&author=J.M..%20Kraft&pages=30-31&publication_year=1984

Kraft JM, Burke DW, Haglund WA (1981) *Fusarium* diseases of beans, peas, and lentils. In: Nelson PE, Toussoun TA, Cook RJ (eds) *Fusarium*: diseases, biology, and taxonomy. Pennsylvania State University, University Park, pp 142–156. https://doi.org/10.1007/s13225-010-0075-8

Lamari L, Bernier CC (1985) Etiology of seedling blight and root rot of faba bean (Vicia faba) in Manitoba. Can J Plant Pathol 7:139–145. https://doi.org/10.1080/07060668509501490

Leslie JF, Summerell BA (2016) The Fusarium laboratory manual. *Blackwell Publishi* Rangaswamy*ng*, 1st ed. https://doi.org/10.1002/9780470278376

Liebenberg AJ (2012) Soybean production manual: your guide to successful soybean production. Agricultural Research Council–Grain Crops Institute (ARC–GCI), Potchefstroom

Madkour MA, Abou-Taleb EM, Oka-sha AM (1983) Aceton inhibition of *Rhizoctonia solani* growth. Phytopathol Z 107:111–116. https://doi.org/10.1111/j.1439-0434.1983.tb00057.x

Mahmoud GA-E, Hamd-Alla WA, Hefzy M (2018): Intercropping Faba bean with some crops and its impact on the production, phyllosphere and rhizosphere microorganisms under modern irrigation systems. Assiut J Agri Sci 52(4):44–63. https://doi.org/10.21608/ajas.2018.26930

Mahmoud GA-E, Zidan ASA, Aly AAM, Mosbah HK, Ibrahim ABM (2019) Calcium and strontium anthranilato complexes as effective *Fusarium moniliforme* controlling agents. Appl Organometal Chem 33(2):e4740. https://doi.org/10.1002/aoc.4740

Mahmoud GA-E, Ibrahim ABM, Mayer P (2020) Zn(II) and Cd(II) thiosemicarbazones for stimulation/inhibition of kojic acid biosynthesis from *Aspergillus flavus* and the fungal defense behavior against the metal complexes' excesses. JBIC J Biol Inorganic Chem 25:797–809. https://doi.org/10.1007/s00775-020-01802-2

Mansour AHA, Mahmoud GA-E, Mohamed AA (2018) Evaluation of natural oils impact on flame seedless grape quality at harvest and storage marketing process. Asian J Biol Res 11(4):228–235. https://doi.org/10.3923/ajbs.2018.228.235

Marrone PG (2002) An effective biofungicide with novel modes of action. Pestic Outlook 13:193–194. https://doi.org/10.1039/b209431m

McConnaughey M (2007). Physical and chemical properties of fungi. The Comprehensive Pharma. Ref. pp 1–3 https://doi.org/10.1016/B978-0-12-801238-3.05231-4

Meki S, Seid A, Sakhuja PK (2008) Pathogenic variability in Ethiopian isolates of *Fusarium oxysporum* f. sp. *ciceris* and reaction of chickpea improved varieties to the isolates. Inter J Pest Manage 54(2)143–149. https://doi.org/10.1080/09670870701780153

Mengistu H, Negussie T (1994) Chickpea and lentil disease research in Ethiopia. In: Asfaw T, Geletu B, Saxena MC, Solh MB, (eds) Cool-season food legumes of Ethiopia. Proceedings of the First National Cool-Season Food Legumes Review Conference, 1993 December 16–20, ICARDA, Addis Ababa, pp 315–366

Meredith DS (1973) Significance of spore release and dispersal mechanisms in plant disease epidemiology. Annu Rev Phytopathol 11:313–342. https://doi.org/10.1146/annurev.py.11.090173.001525

Millan T, Clarke H, Siddique K, Buhariwalla H, Gaur P, Kumar J, Gil J, Kahl G, Winter P (2006) Chickpea molecular breeding: new tools and concepts. Euphytica 147:81–103. https://doi.org/10.1007/s10681-006-4261-4

Mishra S, Panda SK, Sahoo L, (2014) Transgenic Asiatic grain legumes for salt tolerance and functional genomics. Rev Agri Sci 2:21–36. https://doi.org/10.7831/ras.2.21

Mohamed MH, Mahmoud GA-E (2018) Microbial gibberellins impact of on *Zea mays* (L.) plants under different levels of water salinity. Egyptian J Soil Sci 58(3):373–382. https://doi.org/10.21608/ejss.2018.5229.1207

Mukankusi CM, Melis RJ, Derera J, Buruchara RA, Mark D (2011) A screening technique for resistance to *Fusarium* root rot of common bean. Afri J Plant Sci 5:152–161

Naseri B (2008) Root rot of common bean in Zanjan, Iran: major pathogens and yield loss estimates. Australas Plant Pathol 37:546–551. https://doi.org/10.1071/AP08053

Naseri B (2014) Bean production and *Fusarium* root rot in diverse soil environments in Iran. J Soil Sci Plant Nutri 14(1):177–188. https://doi.org/10.4067/S0718-95162014005000014

Naseri B, Marefat A (2011) Large-scale assessment of agricultural practices affecting *Fusarium* root rot and common bean yield. Eur J Plant Pathol 131:179–195. https://doi.org/10.1007/s10658-011-9798-y

Nasir M, Bretag TW, Kaiser WJ, Meredith KA, Brouwer JB (2000) Screening chickpea germplasm for ascochyta blight resistance. Aust Plant Patholo 29:102–107. https://doi.org/10.1071/AP00018

Navas-Cortes JA, Hau B, Jimenez-Diaz RM (1998) Effect of sowing date, host cultivar, and race of *Fusarium oxysporum f. sp. ciceris* on development of *Fusarium* wilt of chickpea. Phytopathology 88:1338–1346. https://doi.org/10.1094/PHYTO.1998.88.12.1338

Navas-Cortes JA, Trapero-Casas A, Jimenez-Diaz RM (2007) Influence of relative humidity and temperature on development of *Didymella rabiei* on chickpea debris. Plant Patholo 47:57–66. https://doi.org/10.1046/j.1365-3059.1998.00208.x

Negussie T (1989) Effect of temperature on the growth of important wilt and root rot fungi of chickpea. Page 24 in Abstracts of the 14th annual meeting of the Ethiopian phytopathological Committee, 30–31 March, Ambo, Ethiopia

Negussie T, Pretorius ZA (2005) A settling tower for quantitative deposition of urediniospores of *Uromyces viciae-fabae*. South Africa J. Plant Soil 22(3):141–144. https://doi.org/10.1080/02571862.2005.10634696

Negussie T, Pretorius ZA (2008) Yield loss of lentil caused by *Uromyces viciae-fabae*. South Afri J Plant Soil 25(1):32–41. https://doi.org/10.1080/02571862.2008.10639892

Negussie T, Pretorius ZA, Bender CM (2005a) Effect of some environmental factors on in vitro germination of urediniospores and infection of lentils by rust. J Phytopathol 153:43–47. https://doi.org/10.1111/j.1439-0434.2004.00926.x

Negussie T, Pretorius ZA, Bender CM (2005b) Components of rust resistance in lentil. Euphytica 142:55–64. https://doi.org/10.1007/s10681-005-0511-0

Nelson PE (1964) Carnation as a symptomless carrier of *Fusarium oxysporum* f. dianthi. Phytopathology 54:323–329

Nelson PE, Toussoun TA, Marasas WFO (1983) Fusarium species: an illustrated manual for identification. Pennsylvania State University Press, University Park, p 193. https://www.psupress.org/books/titles/0-271-00349-9.html

Nene YL (1979) Proceedings of the Consultant's group discussion on the resistance to soil-borne diseases of legumes. ICRISAT, Hyderabad, p 165. http://agris.fao.org/agris-search/search.do?recordID=US201300336111

Nene YL (1982) A review of Ascochyta blight of chickpea. Tropical Pest Manage 28:61–70. https://doi.org/10.1080/09670878209370675

Nene YL and Reddy MV (1987) Chickpea diseases and their control. In: Saxena MC, Singh KB (eds) The Chickpea. CABI Publishing, Wallingford, pp 233–270

Nene YL, Sheila VK, Sharma SB (1996) A world list of chickpea and pigeon pea pathogens, 5th edn. ICRISAT, Patancheru, p 27. http://agris.fao.org/agris-search/search.do?recordID=XF2016011971

Nielsen KA et al (2000) First touch: an immediate response to surface recognition in conidia of Blumeria graminis. Physiol Mol Plant Pathol 56:63–70. https://doi.org/10.1006/pmpp.1999.0241

Ondřej M, Dostálová R, Odstrčilová L (2005) Response of *Pisum sativum* germplasm resistant to *Erysiphe pisito* inoculation with *Erysiphe baeumleri*, a new pathogen of peas. Plant Prot Sci 41:95–103. https://www.agriculturejournals.cz/publicFiles/16843.pdf

Opio AF, Allen DJ, Teri JM (1996) Pathogenic variation in the causal agent of common bacterial blight in phaseolus bean. Plant Pathol 45:1126–1133. https://doi.org/10.1046/j.1365-3059.1996.d01-187.x

Osdaghi E, Shams-Bakhsh M, Alizadeh A, Lak MR, Maleki HH (2011) Induction of resistance in common bean by *Rhizobium leguminosarum* bv. *phaseoli* and decrease of common bacterial blight. Phytopathol Mediterr 50:45–54. https://doi.org/10.14601/Phytopathol_Mediterr-8524

Otsyula RM, Ajanga SI, Buruchara RA, Wortmann CS (1998) Development of bean root rot control strategy for western Kenya. Afri Crop Sci J 6:61–67. https://doi.org/10.4314/acsj.v6i1.27825

Otten W, Gilligan CA (2006) Soil structure and soil borne diseases: using epidemiological concepts to scale from fungal spread to plant epidemics. Eur J Soil Sci 57:26–37. https://doi.org/10.1111/j.1365-2389.2006.00766.x

oxysporum f. sp. *ciceris*). Physiol Molec Plant Pathol 50:167–178. https://doi.org/10.1006/pmpp.1997.0082

Stoddard FL, Nicholas AH, Rubiales D, Thomas J, Villegas AM (2010) Integrated pest management in faba bean. Field Crops Res 115:308–318. https://doi.org/10.1016/j.fcr.2009.07.002

Tadesse A (ed) (2008) Increasing crop production through improved plant protection – volume I. Plant Protection Society of Ethiopia (PPSE), 19–22 December 2006. Addis Ababa, Ethiopia. PPSE and EIAR, Addis Ababa, 598 pp

Tadesse N, Ali K, Gorfu D, Yusuf A, Abraham A, Ayalew M, Lencho A, Makkouk K, Kumari SG (1999) Survey for chickpea and lentil virus diseases in Ethiopia. Phytopathol Mediterr 38:149–158. https://www.jstor.org/stable/42685994?seq=1#page_scan_tab_contents

Thorne MJ, Thompson LU, Jenkins DJ (1983) Factors affecting starch digestibility and the glycemic response with special reference to legumes. Am J Clin Nutr 38:481–488. https://doi.org/10.1093/ajcn/38.3.481

Tiwari KR, Warkentin TD, Penner GA, Menzies JG (1999) Studies on winter survival strategies of *Erysiphe pisiin* Manitoba. Can J Plant Pathol 21:159–164. https://doi.org/10.1080/07060669909501207

Trapero-Casas A, Jiménez-Díaz RM (1985) Fungal wilt and root rot diseases of chickpea in southern Spain. Phytopathology 75:1146–1151

Trapero-Casas A, Kaiser WJ (1992) Development of *Didymella rabiei*, the teleomorph of *Ascochyta rabiei*, on chickpea straw. Phytopathology 82:1261–1266. https://doi.org/10.1094/Phyto-82-1261

Tu JC (1992) Management of root rot diseases of peas, beans and tomatoes. Can J Plant Pathol 14:92–99. https://doi.org/10.1080/07060669209500910

Van der Linde WJ, Clemitson JG, Crous ME (1959) Host-parasite relationships of South African root-knot eelworm (*Meloidogyne* spp.). Dep Agric Technol Serv Repub S Afr Ent Ser 44:3–16

van Leur JAG, Kumari SG, Makkouk KM, Rose IA (2006) Viruses on faba bean in north-east Australia and strategies for virus control. In: Avila CM, Cubero JI, Moreno MT, Suso MJ, Torres AM (eds) International workshop on Faba Bean Breeding and Agronomy, Cordoba, pp 129–131

Van-der Maesen LJG (1987) Cicer L. origin, history and taxonomy of chickpea. In: Saxena MC, Singh KB (eds) The chickpea. C.A.B. International, Cambrian News Ltd, Aberystwyth, pp 11–34

Verhaar MA, Hijwegen T, Zadoks JC (1996) Glasshouse experiments on biocontrol of cucumber powdery mildew (*Sphaerotheca fuliginea*) by the mycoparasites *Verticillium lecanii* and *Sporothrix rugulosa*. Biol Control 6:353–360. https://doi.org/10.1006/bcon.1996.0045

Vogelsang R, Berger E, Hagedorn T, Muhlenbeck U, Tenhaken R, Barz W (1994) Characterization of metabolic changes involved in hypersensitive-like browning reactions of chickpea (Cicer arietinum L.) cell cultures following challenge by Ascochyta rabiei culture filtrate. Physiol Molec Plant Pathol 44:141–155. https://doi.org/10.1016/S0885-5765(05)80107-9

Warkentin TD, Rashid KY, Xue AG (1996) Fungicidal control of powdery mildew in field pea. Can J Plant Sci 76:933–935. https://doi.org/10.4141/cjps96-156

Wollenweber HW, Reinking OA (1935) Die Fusarien, ihr Beschreibung, Schadwirkung und Bekampfung. Verlag Paul Parey, Berlin

Zettler FW, Abo El-Nil MM (1977) Bean yellow mosaic virus infections of gladiolus in Florida. Plant Dis Rep 61:243–247. http://agris.fao.org/agris-search/search.do?recordID=US201302966924

Chapter 8
Molecular Mechanism Underlying Chickpea – *Fusarium oxysporum* f. sp. *ciceri* Interaction

K. P. Suthar, B. K. Rajkumar, Preeti R. Parmar, and Diwakar Singh

Abstract Chickpea is pivotal source of protein for vegetarian diet, however its lower productivity is the major hurdle to achieve nutritional security. The major reason for the low productivity is fungal disease, *Fusarium* wilt caused by *Fusaruim oxysporum* f. sp. *ciceris* (Padwick) Matuo & K. Sato (*Foc*). Different disease management modules have been implemented but the development of resistant genotypes is environment friendly strategy to combat yield loss and also offer sustainability. This can be achieved thorough understanding of chickpea – *Foc* interaction.

This chapter reviews current know how of *Foc* infection process in the chickpea such as spore germination, root infection and colonization. The resistance mechanism adopted by plant against *Foc* is a step insight into resistance genes and its systematic exploitation to breed disease resistance has been discussed. Further, available molecular marker techniques and its achievements in fastening breeding approach through assisting in selection are reviewed. The saturated generic maps are developed using different populations. It has made it possible to find closely linked markers to be exploited in breeding. The current developments in "Omics" technology further enlightened our understanding on wilt-chickpea interaction; hence the recent understanding on transcript modulation is also reviewed in this chapter. Overall, the systematic work on chickpea – *Foc* interaction, and its exploitation has significantly contributed in production of grain in chickpea. The current review focuses on *Foc* variability, mode of action and symptomatology, the genetics of disease resistance and linkage studies, and at last the transctiptome research status underlying chickpea – *Foc* interaction.

K. P. Suthar (✉) · D. Singh
Department of Plant Molecular Biology and Biotechnology, ASPEE College of Horticulture and Forestry, Navsari Agricultural University, Navsari, Gujarat, India
e-mail: kiransuthar@nau.in; diwakar@nau.in

B. K. Rajkumar · P. R. Parmar
Main Cotton Research Station, Navsari Agricultural University, Surat, Gujarat, India
e-mail: rajkumar@nau.in; prparmar@nau.in

P. Guleria et al. (eds.), *Sustainable Agriculture Reviews 51*, Sustainable Agriculture Reviews 51, https://doi.org/10.1007/978-3-030-68828-8_8

Keywords Chickpea · *Fusarium oxysporum* f. sp. *ciceri* interaction · *Foc* diversity · Genetics of wilt resistance · QTLs for wilt · Marker assisted backcrossing · Transcriptome

8.1 Introduction

Chickpea (*Cicer arietinum* L.; 2n = 2x = 16; genome size of 750 Mbp) is an important legume crop in most of the developing countries of the world and ranks third among food legumes in production. South west Asia and the Mediterranean were identified as two primary centres of origin for chickpea, while Ethiopia as secondary centre of origin (Ladizinsky and Adler 1976). Chickpea is pivotal source of protein for vegetarian diet, however its lower productivity is major hurdle to achieve nutritional security. The major reasons for the low productivity of cultivated chickpea is its narrow genetic bases and its sexual incompatibility with other wild species of *Cicer* in natural inter specific cross. Furthermore, various biotic (*Fusarium* wilt, *Aschochyta* blight, nematodes and pests) and abiotic (heat, salinity, drought and cold) stresses severely reduce the yield (Winter et al. 2000). Among the biotic stresses, the *Fusarium* wilt and *Ascochyta* blight are the most devastating fungal diseases causing serious yield losses. *Fusarium* wilt caused by *Fusaruim oxysporum* f. sp. *ciceris* (Padwick) Matuo & K. Sato (*Foc*) is most destructive vascular disease of chickpea leads to 10–15% annual yield losses but its epidemics can be devastating to individual crops and even sometime it may cause up to 100% loss under favorable conditions (Chaube and Pundhir 2005).

The significant pathogenic variability observed for *Foc* as the pathogen can survive in soil for long period during which it may have to go through different environmental stresses; further biological competition makes it difficult to develop resistant genotypes. The understanding regarding prevailing pathogen variability and their mode of action with symptomatology can facilitate breeding strategies to develop durable resistant genotypes. The resistance being genetic phenomena the understanding regarding the genes involved in disease resistance and its inheritance is essential for breeding. Further, the resistance character is governed by expression and regulations of different disease resistance related genes, their understanding have additional benefit for developing sustainable resistance. The objectives of this review, therefore, are to discuss the current understanding of *Foc* variability, mode of action and symptomology, to outline the genetics of disease resistance and recent molecular development for developing resistance genotype and to highlight the status of transctiptome research underlying chickpea – *Foc* interaction. It is anticipated that this work will provide helpful information to come out with sustainable wilt resistance material.

8.2 Molecular Insight into *Fusarium oxysporum* f sp. *ciceri*

8.2.1 Formae Species, Races and Genetic Diversity of Foc

In India, *Fusarium* wilt of chickpea was first reported by Butler (1918) and corrected etiology determined by Padwick (1940). Synder and Hansen in 1940 introduced the concept of *formae* species. There are some high levels of host specificity within the species of *Fusarium* causing wilt in agricultural crops. Based on virulence to its specific host, fungal strains have been divided into *formae* species. *Fusarium oxysporum f. sp. ciceris* (*Foc*) *is* one of the *formae* species of monophyletic origin in *F. oxysporum* complex of the *Gibberella* clade. Races are the further subdivision of *formae* species. For the classification of organisms, proper identification and differentiation of species or races is an essential step. Diversity of *Foc* races can be accessed based on morphology, pathogenic variability, geographic distribution and molecular techniques. The visual examination of pathogenic plant and microscopic tools tentatively helps to identify the type of fungus. It is based on types of fungal hyphae and growth on microbiological media, spores, size and shape of spores and pigmentation of the fungus. Fungal hyphae are septate and branched when grown on potato dextrose agar with yellow, brown or crimson, white to violet, reddish violet, greenish violet, yellowish pink and dark green pigmentation (Saxena and Singh 1987). Pigmentations are based upon type of races, type of media, temperature and age of fungi (Honnareddy and Dubey 2006). Singh et al. (2010) noted quantitative difference in colony characters in *Foc*-1, *Foc* -2 and *Foc*-3. *Foc*-1 has dull white, thin and flat hairy growth or thread like growth; a *Foc*-2 possesses white, fluffy growth with irregular margin; while *Foc*-3 was pinkish white, fluffy with regular margin. *F. oxysporum* produces microconidia, macroconidia and chlamydospores. Macroconidia are slightly curved or straight, slender, thin walled with three or four septa, foot-shaped basal celled and curved apical cell (Couteaudier and Alabouvette 1990). Microconidia are ellipsoidal with single septa or without septa, forming phialides in false head by basipetal division and play significant role in secondary infection. Chlamydospores are globose and thick walled and they are formed from hyphae or alternatively by modification of hyphal cells and important as endurance organ in soils (Leslie and Summerell 2006).

Pathogenic variability can be tested based on the virulence of pathogen to a particular set of differential host cultivars that vary in disease resistance (isolate x cultivar interaction) (Correll 1991). For that, ten different chickpea lines such as 12-071/10054, JG-62, C-104, JG-74, CPS-1, BG-212, WR-315, ICCV-2, ICCV-4 and P-2245 were used to test pathogenic races of *Foc* (Haware and Nene 1982; Jiménez-Díaz et al. 1989, 1993). On the basis of pathogenicity, *Foc* is divided into eight pathogenic races (biotypes of a pathogen which able to cause disease on host cultivars carrying specific resistance genes) *viz*., 0, 1A, 1B/C, 2, 3, 4, 5, and 6. Pathogenic races induce two types of syndromes on chickpea cultivar: first is wilting syndrome and second is yellowing syndrome. Races 0 and 1B/C induce yellowing syndrome while 1A, 2, 3, 4, 5 and 6 causes wilting syndrome in the chickpea.

Recently, *Fusarium redolens* was reported for the first time to cause yellowing syndrome in Tunisian in chickpea (Bouhadida et al. 2017). Figure 8.1 and Table 8.1 summaries the geographical distribution and pathogenicity diversity of eight races of *Foc*.

Besides pathogenic variability and geographic distribution, genetic diversity of *Foc* can be studied using different molecular techniques e.g. Polymerase chain reaction (PCR) assays, Restriction fragment length polymorphism (RFLP) pattern for mitochondrial DNA and Inter genic spacer (IGS) region, development of Sequence characterized amplified region (SCAR) PCR markers, different microsatellites, study of genes encoding β-tubulin, TEF, histone 3, actin, calmodulin, xylanase gene and its transcriptional factor (*xlnR*), genes for endopolygalacturonase *pg1* and exopolygalacturonase *pgx41*. Identification of races 0, 5 and 6 can made by using specific PCR assay from fungal mycelia and primers *Foc*R0-M15f/*Foc*R0-M15r, *Foc*R5-L10f/*Foc*R5-L10r, or *Foc*R6-02f/*Foc*R6-02r which amplifies 900, 938 and 1000-bp race markers respectively (Jiménez-Gasco and Jiménez-Díaz 2003). Further, it was found that all races, pathotypes or geographic origin shared identical RFLP pattern for mitochondrial DNA and IGS region, gene β-tubulin, TEF, histone 3, actin and calmodulin (Jiménez-Gasco et al. 2002). Gurjar et al. (2009) performed ITS-RFLP to differentiate the pathogenic races of *Foc*. Sharma et al. (2014) developed DArT (Diversity Array Technology) markers and characterized set of 110 *Foc* isolates representing diverse agro-ecological zones of India. Demers et al. (2014) studied with large samples of *Foc* and found some minor differences between races for the gene β-tubulin, endopolygalacturonase *pg1*, exopolygalacturonase *pgx41* and IGS region of rDNA which might be a critical tool to differentiate *Foc* in future.

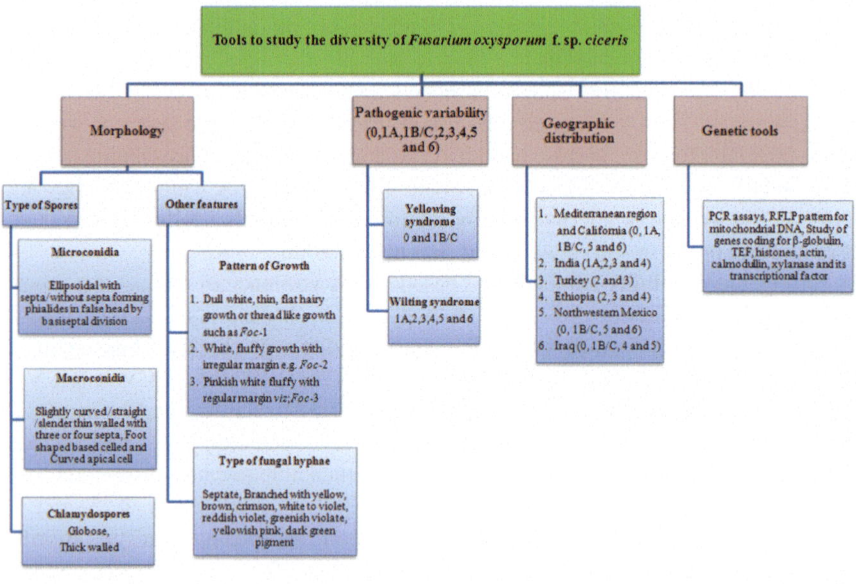

Fig. 8.1 Tools to study the diversity of *Fusarium oxysporum* f. sp. *ciceris*

Table 8.1 Geographic distribution of *Foc* races of chickpea

Sr. no.	Races	Geographical distribution	References
1.	0, 1A, 1B/C, 5 and 6	Mediterranean region and California	Halila and Strange (1996), Jiménez-Díaz et al. (1993), Jiménez-Gasco and Jiménez-Díaz (2003), Jiménez-Gasco et al. (2001)
2.	1A, 2, 3 and 4	India	Haware and Nene (1982)
3.	2 and 3	Turkey	Bayraktar and Dolar (2012), Dolar (1997)
4.	2, 3 and 4	Ethiopia	Shehabu et al. (2008)
5.	0, 1B/C, 5 and 6	Northwestern Mexico	Arvayo-Ortiz et al. (2011)
6.	0, 1B/C, 4 and 5	Iraq	Al-Taae et al. (2013)

Poornima et al. (2017) differentiated the race 4 and race 2 *Foc* using gene specific ITS regions; and race 2 and 5 using a SCAR marker; and also found unique band of 500 bp for race 4 by restriction analysis of ITS region. Additionally, they have used CAPS (Cleaved Amplified Polymorphic Sequences) and multiplex PCR profiles to differentiate the different races of *Foc*. Thus, molecular marker identified for *Foc* races will serve as an additional resource for disease resistance breeding.

8.2.2 Controversies Regarding Evolution of Races

Inferred phylogeny on the basis of mapping of the pathogenic races to different cultivars of the chickpea indicates that races evolve in a stepwise manner. Each races evolving from another by gaining the ability to cause disease in resistance chickpea cultivars. It is based to gain or loss the virulence power, mostly to gain virulence rather than loss.

Jiménez-Gasco et al. (2004a, b) proposed two different theories regarding common ancestor of all races. One that race 0 is being ancestral of all races because of following reasons;

1. It appears consistent with yellowing and wilting type disease.
2. It is pathogenic on fewest race differentials among all races.
3. Most wide spread race in Mediterranean region but not found in India.
4. Showing highest molecular diversity among all races.

Second proposal is that race 1A is the common ancestor of all races because it is widespread geographically, being consistent and subsequent development of pathogenic races within the lineage. In 2014, Demers used two specific microsatellite markers to compare wilting races 1A, 5, 6 and yellowing races 0 and 1B/C. It was found that wilting races (1A, 5 and 6) having very little intra-race diversity but these divert was high when compared to yellowing races (0 and 1 B/C) which hypothesized that race 0 and 1 B/C might be ancestral for wilting races. Further, it seems that some of the races appeared in those geographical area in which resistance cultivars of chickpea are not being cultivated, which refuge the theory of races evolving from another by gaining the ability to cause disease in resistance chickpea cultivars (Halila and Strange 1996; Jiménez-Díaz et al. 1993; Jiménez-Gasco and Jiménez-Díaz 2003; Jiménez-Gaco et al. 2001). In India, numerous molecular tools used to identify and differentiate races of *Foc* using large samples of *Fusarium* and wide diversity reported among all isolates (Jiménez-Díaz et al. 2015). Thus, specific knowledge regarding methods like environmental controlled conditions, molecular tools and virulence test on chickpea cultivar, types of cultivars and statistical analysis are critical pre-requisites to understand diversity of *Foc* races in depth.

8.2.3 Foc *Mode of Action*

The life cycle of the pathogen possesses parasitic as well as saprophytic modes (Backman and Turmer 1989). According to hypothesis of Beckman (1987), parasitism and saprophytism modes of fungi possess different functions (Table 8.2 and Fig. 8.2).

Infection due to *F. oxysporum* races to its host plant is directly from the soil through spores/fragments of mycelia or infected seeds (Haware et al. 1990). The asexual spores of *Foc* can survive for longer period of time and disperse through various processes (Agrios 2005). The Fig. 8.3 represents the stepwise process of

Table 8.2 Different phases during life cycle of *F. oxysproum* causing wilt in the plants

Sr. no.	Mode of fungi during life cycle	Different phases	Functions	References
1.	Parasitism	Determinative phase I	Success or failure of pathogen to penetrate in the outer tissues and entering inside the vascular system of plant	Talboys (1957)
		Determinative phase II	Determination of success or failure of pathogen to colonize in the vascular system	
		Expressive phase	Successfully development of pathogen into the plant to cause disease	
2.	Saprophytism	Sub-cycles	Senescing dead tissues of host by pathogen	Beckman (1987)
			Production of resting spores (Chlamydospores) and release of spores into the soil	
			Renewal of resting structures by saprophytic growth on organic debris and root exudates or by infection to roots of host or non-host plants	

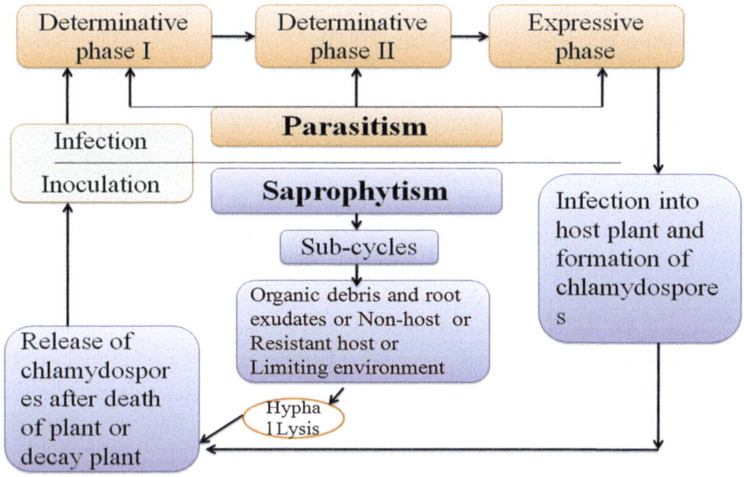

Fig. 8.2 Schematic representation of phases during life cycle of *Fusarium oxysporum*

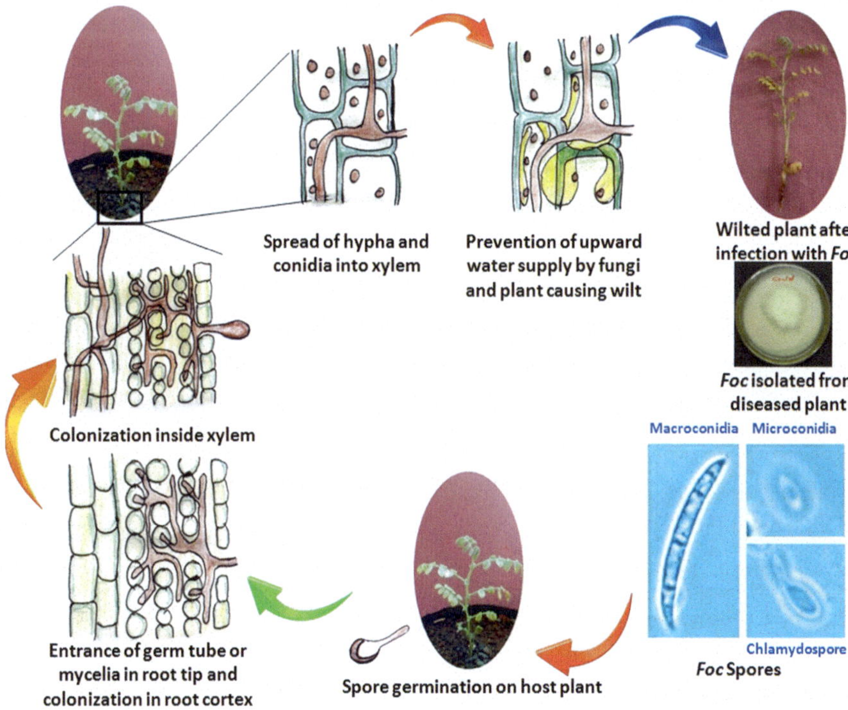

Fig. 8.3 Mode of *Fusarium oxysporum* f. sp. ciceris action in chickpea

wilt infection by *Foc* into the chickpea such as spore germination, root infection and colonization (Sharma et al. 2016; Agrios 2005).

8.2.3.1 Spore Germination

Chlamydospore is the dormant stage of the *Foc* and conversion of dormant stage into the vegetative form is the crucial step for the development of wilt in the host plant. External signals, pre-germination phase and formation of germ tube are the fundamental steps to break the dormancy of spore. Root of the chickpea/host plant releases broad types of nutrients like sugars and amino acids that can initiate pre-germination and germ tube growth (Lockwood 1977; Schroth and Hildebrand 1964; Leroch et al. 2013). Once the germ tube germinated, growth of germ tube towards the root tip of chickpea or identification of specific host plant root are the crucial step and need to be studied in detail. According to Sharma et al. (2016), each step for wilt infection governed by intercellular pathway and signals generated due to different gene expression.

8.2.3.2 Root Infection

Identification of specific host root by the pathogen is a strategic step. Available information indicates that the type of specific plant exudates, types of *Foc*, physio-chemical and microbial factors of soils, time required for spore germination, extension of germ tube and its direction towards the root tips, rate of root growth and different environmental factors are the critical factors that exactly determines mechanism behind successful entry of *Foc* into the host plant (Huisman 1982; Watt et al. 2006; Gordon 2017). Germ tube or mycelia of *Foc* can directly enter the host root tip through wounds or at the point of lateral roots. Once it comes in contact with the plant roots, fungus produce hyphal branches to form mycelial network. *Foc* generally produces hyphal swelling at the site of infection (Bishop and Cooper 1983; Fang et al. 2012; Liu et al. 2011). Degradative enzymes secretion by fungal pathogen makes successful penetration inside the host cell wall and epidermal cell (Michielse and Rep 2009). Pectate lyase expected as a *Foc* pathogenicity factor in chickpea wilt (Jorge et al. 2006).

8.2.3.3 Colonization of *Foc*

After adhesion to the host surface, series of nuclear events take places including mitosis in the fusing hypha and nuclear migration into the receptor hypha followed by degradation of the resident nucleus, establishment of hyphal network through vegetative hyphal fusion and differentiation of infection hyphae. Mycelium of the fungus grows in the root cortex intercellularly and proliferated within xylem vessels. Inside the vessels, mycelium forms branches and produces microconidia that germinate and penetrates the vessel wall. Mycelium also penetrates laterally into the adjacent vessels, penetrating them through pits. It is presumed that due to mycelium growth or colonization in the vessels, it prevents the water supply and nutrient to the upper part of the plants which results into wilt disease (Agrios 2005).

8.3 Molecular Understanding of Chickpea – *Foc* Interaction

8.3.1 Genetics of Chickpea Wilt Resistance in Chickpea

The development of high yielding *Fusarium* wilt (Fw) resistance is an effective method to manage the disease. This aim can be fulfilled with a thorough understanding of interactions of cultivars with the pathogen at molecular level. Chickpea resistance to Fw is race specific and controlled by major resistance genes (reviewed by Jendoubi et al. 2017) depending on the number of genes, the race or source of resistance. Intensive race specific inheritance studies of Fw resistance in chickpea started only after 1982 when Haware and Nene characterized four races of *Foc*. An overview of genetics of chickpea *Fusarium* wilts resistance has been presented in Table 8.3.

A good number of reports on the inheritance of resistance to Fw race 1 in chickpea are available. Earlier studies on genetics of wilt resistance were restricted to race 1 where it was shown to be inherited by a recessive gene (Ayyar and Iyer 1936; Kumar and Haware 1982; Sindhu et al. 1983). The identification of phenomenon of late wilting in some genotypes susceptible to race 1, (Upadhyaya et al. 1983a, b) had shift the focus to genetics of late wilting. The late wilting was found to be a monogenic trait and was controlled by three independent genes named as *h1*, *h2* and *H3*, each of which delayed onset of disease symptoms (Singh et al. 1987a, b; Smithson et al. 1983). Late wilting resistance is conferred by any one of these three genes, but a combination of any two of the late wilting genes (*h1h2* or *h1H3* or *h2H3*) confers complete resistance (Upadhyaya et al. 1983a, b; Singh et al. 1987a, b). However, other studies indicated resistance to race 1 is governed by two recessive genes under homozygous condition (Anupama 2001), trigenic (Girase and Deshmukh 2002) and duplicate recessive genes (Bayraktar and Dolar 2012).

Very little information is available regarding race 2 and other races of *Fusarium* wilt. Similar to race1A (Race 1 from India described as Race 1A), resistance to race 2 was initially found to be conferred by a single recessive gene (Pathak et al. 1975; Sharma et al. 2005b), however, later studies revealed involvement of two (Gumber et al. 1995) or three independent genes (a, b, and c) (Kumar 1998). Others studies showed that resistance to race 3 (Sharma et al. 2004, 2005a) and race 5 governed by a single gene (Tekeoglu et al. 2000; Sharma et al. 2005a), while resistance to race 4 was reported as monogenic recessive by Tullu et al. (1998) and Sharma et al. (2005a); and digenic recessive by Tullu et al. (1999). Similar to races 1 and 2, the phenomenon of late wilting was also detected for race 4. Tekeoglu et al. (2000) and

Table 8.3 Overview of genetics of *Fusarium* wilt resistance to different races in chickpea

Fusarium race	Nature of wilt resistance gene	Gene symbol	Effect of resistance gene on wilting	References
Race 1A	Trigenic	*h1* (syn *Foc*-1), *h2*, *H3*	Recessive at first two loci and dominant at the third locus	Singh et al. (1987a, b)
	Digenic	*h1* and *h2*	Homozygous recessive at both loci (*h1h1h2h2*)	Upadhyaya et al. (1983a, b)
Race 1	Trigenic	*H1*/*h1*, *H2*/*h2*, *H3*/*h3*	3 independent genes	Smithson et al. (1983)
	Trigenic	–	Two recessive genes under homozygous	Anupama (2001), Girase and Deshmukh (2002)
	Duplicate recessive genes	*Foc*-1	Two genes (*h1h1h2h2*)	Bayraktar and Dolar (2012)
	Monogenic	*Foc*-2	Complete resistance	Pathak et al. (1975), Sharma et al. (2005a)
Race 2	Digenic	–	–	Gumber et al. (1995)
	Trigenic	a, b, C	Individually confers late wilting, first two genes impart complete resistance	Kumar (1998)
	Polygenic	–	Influencing the major WR genes	Sharma and Muehlbauer (2007)
	Monogenic	*Foc*-3/ *Foc*-3	Complete resistance	Tekeoglu et al. (2000), Sharma et al. (2004, 2005a, b)
Race 3	Monogenic recessive	*Foc*-4	Complete resistance	Tullu et al. (1998), Sharma et al. (2005a)
Race 4	Digenic recessive	–	Complete resistance	Tullu et al. (1999)
	Monogenic	*Foc*-5/ *Foc*-5	Complete resistance	Tekeoglu et al. (2000), Sharma et al. (2005a, b)
Race 5	Monogenic or digenic	*Foc*-01/ *Foc*-01	Complete resistance	Tekeoglu et al. (2000)
Race 0	Monogenic or digenic	*Foc*-02/ *Foc*-02	–	Rubio et al. (2003)
	Polygenic	–	Act along with vertical resistance genes	Halila et al. (2009)

Rubio et al. (2003) showed that resistance to race-0 is monogenic, while, Halila et al. (2009) reported as polygenic. However, genetic control of resistance to three races (race 1B, race 1C and race 6) is not reported so far. The thorough understanding of the genetics of resistance to each race will augment successful disease resistance breeding and pyramiding of resistance genes.

8.3.2 Genetic Mapping of Wilt Resistance Genes and QTLs

Conventional breeding approach has produced chickpea cultivars with improved resistance but progress has been considerably slow. Initially, mass or pure-line selection methods were employed. Later, crossing programs and various modifications of pedigree and bulk methods were used. Effective field, laboratory and greenhouse techniques to screen for resistance to *Fusarium oxysporum* have been developed and it has been found that significant proportion of the resistant chickpea accessions were Desi types rather than Kabuli (Haware et al. 1992).

The information available on genetics of resistance to Fw in chickpea suggests that resistance is controlled by one or a few major genes. Thus, identification of molecular markers (genes/QTLs) associated with resistance to Fw, is a prerequisite for molecular breeding. Further, development of genetic linkage maps are the fundamental for the localization of genes or genomic regions, quantitative traits loci (QTLs) conferring resistance to Fw. Linkage mapping is the process of systematically scanning the genome of both intraspecific (within a species) and interspecific (across different species) populations segregating for the Fw using polymorphic DNA markers. Due to the limited genetic diversity in the cultivated gene pool coupled with inadequate genomic resources earlier studies used interspecific crosses to generate linkage maps. However, identification of sources of resistance to all races in the cultigen *C. arietinum* and advent of genetic markers with high-throughput genotyping systems efforts were also made to map Fw resistance gene using intraspecific populations, which not only represents the true recombination distance and map order, also useful in marker-assisted selection for disease resistance (Winter et al. 2000; Benko-Iseppon et al. 2003).

8.3.3 Molecular Markers Linked to Fusarium *Wilt Resistance Genes*

Markers linked to *Fusarium* wilt resistance genes in chickpea have been presented in Table 8.4. Mayer et al. (1997) first reported linkage of gene for *Foc* race 1 to RAPD markers UBC-170_{550} and CS-27_{700}; one of them (CS27) was converted to allele-specific associated primer (ASAP). Analysis of the RILs (Recombinant Inbred Lines) of an intraspecific cross (C-104 x WR315) with these primers showed that gene *H1* was located 7.0 cM from RAPD markers CS-27_{700} and UBC-170_{550}. This marker was the starting point for subsequent detection of several *Foc* genes. Later the same RILs were used by Tullu et al. (1998) and these RAPD markers (CS-27_{700} and UBC-170_{550}) amplified fragments associated with race 4 resistance and were located 9 map units from the race 4 resistance. Resistance to Race 1 and 4 are 5 map units apart. Further, genetic analysis on the F_1s, F_2s and F_3 families from the cross of JG62 x Surutato-77 by Tullu et al. (1999) revealed that RAPD marker,

Table 8.4 Summary of the markers linked to *Fusarium* wilt resistance to different races in chickpea

Fusarium races	Gene tagged	Markers linked	Genetic distance (cM)	Linkage group (LG)	References
Race 1	*H1*	UBC-170_{550}	7.0	–	Mayer et al. (1997)
		CS-27_{700}	7.2		
		CS27A	4.9		
Race 4	*Foc4*	CS-27_{700}	9.0	–	Tullu et al. (1998, 1999)
Race 4 and Race 1	*Foc4*	Co-segregation of UBC-855_{500} with CS-27_{700}	5.2	6	Ratnaparkhe et al. (1998)
Race 4	*Foc4*	CS-27_{700}	3.7	2	Winter et al. (2000)
		TA96	3.4		
Race 5	*Foc5*	ECAMCTA07	6.4	2	Winter et al. (2000)
		TA27	3.5		
Race 0	*Foc01/ Foc01*	TR59 -$OPJ20_{600}$	2.0 3.0	3	Tekeoglu et al. (2000), Rubio et al. (2003), Cobos et al. (2005)
Race4	*Foc4*	CS-27-TA96	7.1	2	Huettel et al. (2002)
Race 5	*Foc5*	ECAMCTA07-TA27	9.9	2	Huettel et al. (2002)
Race 4	*Foc4*	CS-27_{700}	7.2	2	Tekeoglu et al. (2002)
Race 5	*Foc5*	CS-27_{700}	4.0	2	Tekeoglu et al. (2002)
Race 1	*H1*	CS-27_{700}	14	2	Tekeoglu et al. (2002)
Race 4	*Foc4*	R-2609-1	2.0	2	Benko-Iseppon et al. (2003)
Race 1	*Foc1* *Foc3*-	CS-27-*Foc*1 -TA96, TA27, CS27A *Foc*3-TA96, TA27, CS27A	7.1 9.2 8.1	2	Sharma et al. (2004)
Race 1	*Foc-1*	TA59	4.4	2	Iruela et al. (2006), Sharma and Muehlbauer (2007)
		TA96	4.9		
		TA27	4.9		
Race 1	*H2*	A07C	21.7	–	Soregaon et al. (2007)
Race 5	*Foc5*	TA110	6.5	2	Iruela et al. (2007)
		TA59	8.9		
Race 1	*Foc1*	TA110-*Foc*1-H3A12	6.0	2	Gowda et al. (2009)
Race2	*Foc2*	H3A12-*Foc*2-TA96	2.9		
Race3	*Foc3*	TA194-*Foc*3-H1BO6Y	0.9		
Race 3	*Foc3*	TA59-TA96	2.0	2	Millan et al. (2010)
Race 0	*Foc02*	TS47-TA27	1.5	2	Millan et al. (2010)
Race 0	*Foc02/ Foc02*	TS 47 – TA 59	<5.0	2	Halila et al. (2010)

(continued)

Table 8.4 (continued)

Fusarium races	Gene tagged	Markers linked	Genetic distance (cM)	Linkage group (LG)	References
Race 1	*Foc1*	TA37	0.2	2	Barman et al. (2014)
		TA200	1.0		
		TR2	3.0		

CS-27_{700} linked to race 4 and the genes for resistance to races 1 and 4 are closely linked.

Ratnaparkhe et al. (1998) for first time reported an ISSR marker UBC-855_{500} linked to the gene for resistance to race 4 at a distance of 5.2 cM. It co-segregated with CS-27_{700}, resistance to Fw race 1 and was mapped to linkage group 6 of the *cicer* genome. Further, Winter et al. (2000) reported genes for resistance to races 4 and 5 on linkage group (LG2) that includes a SCAR marker $CS27_{700}$ previously shown to be linked to Fw race 1, at distances of 3.7 and 18 cM, respectively, indicating a clustering of Fw resistance genes around this locus. Tekeoglu et al. (2000) indicated that resistance to *Foc* 0 may be controlled by only one recessive allele at the *Foc 01*/*Foc 01* locus, a second independent locus *Foc 02*/*Foc 02* was assigned by Rubio et al. (2003). Huettel et al. (2002) mapped *Foc4* and *Foc5* on LG 2 flanked by markers CS-27- TA96 and ECAMCTA07-TA27, respectively. Further, Tekeoglu et al. (2002) mapped resistance genes for FW race 1 (*Foc1*), race 4 (*Foc4*) and race 5 (*Foc5*) on the same LG 2 and were separated by 11.2 cM units.

Benko-Iseppon et al. (2003) identified DNA amplification finger printing (DAF) markers linked to races 4 and 5. The DAF marker R-2609-1 found closely linked (2.0 cM) to race 4 resistance and sequences of the linked markers were highly similar to genes encoding proteins involved in plant pathogen response, such as PR-5 thaumatin-like protein gene and another is homologous to the gene for anthranilate N-hydroxy cinnamoyl-benzoyltransferase, a regulator of the phytoalexin pathway, both important components of the plant's defense against pathogens. Sharma et al. (2004) reported that *Foc3* gene was mapped 0.6 cM from Sequence tagged microsatellite site (STMS) markers TA96 and TA27, and STS marker $CS27_{700}$. Another STMS marker TA194, at 14.3 cM, flanked the gene on the other side and linkage between *Foc3* and two other chickpea wilt resistance genes, *Foc 1* (syn *h1*) and *Foc 4* was established.

Rubio et al. (2003) demonstrated that resistance to race 0 is controlled by two genes, which segregate independently; one present in accession JG62 (*Foc 01*/*Foc 01*) mapping to LG5 and the second present in accession CA2139 (*Foc 02*/*Foc 02*) but remaining unmapped. Further, work of Halila et al. (2009) revealed that STMS markers sited on LG2 were strongly associated with *Foc 02*/*Foc 02* and in a region where genes for resistance to wilt races 1, 2, 3, 4 and 5 have previously been reported and which is highly saturated with tightly linked STMS markers that could be used in MAS. They also found a strong association of TR59 with wilt resistance gene *Foc-01*. Second resistance gene for race 0 (*Foc*-02) was found on LG2 closely linked with *Foc5* (Huettel et al. 2002; Flandez-Galvez et al. 2003; Cobos et al. 2009; Palomino et al. 2009).

Soregaon et al. (2007) identified an RAPD marker A07C amplifies 417 bp linked to *H2* locus of wilt susceptibility for race 1. Iruela et al. (2007) reported resistance to race 5 of *F. oxysporum* (*Foc* 5) was inherited as a single gene and mapped to LG2, flanked by the STMS markers TA110 (6.5 cM apart) and TA59 (8.9 cM apart). Gowda et al. (2009) mapped the three *Fusarium* wilt resistance genes *viz.*, *Foc 1*, *Foc 2* and *Foc 3* using RIL population derived from the cross JG62 x Vijay. They also identified SSR markers tightly linked and flanking the resistance genes on LG2 for *Foc* races 1, 2 and 3. The SSR markers H3A12 and TA110 flanked the *Foc1* locus at 3.9 and 2.1 cM, respectively, while *Foc* 2 was tagged with SSR markers TA96 and H3A12 at a distance of 0.2 and 2.7 cM, respectively. Similarly H1B06y and TA194 markers flanked the *Foc* 3 locus at 0.2 and 0.7 cM, respectively. Millan et al. (2010) found that TA96 and TA59 flanks *Foc3*, while *Foc02* flanked by TS47 and TA27 on the LG2. Halila et al. (2010) identified the markers (TS 47- TA 59) closely linked to races 0. Barman et al. (2014) mapped *Foc1* resistance loci flanked by markers TA37 and TA200. Thus, DNA markers identified tightly linked to Fw resistance genes would aid in screening resistant genotypes at early stage of plant growth without subjecting them to pathogens.

8.3.4 QTLs for Fusarium *Wilt in Chickpea*

Earlier studies of Fw resistance has been measured as qualitative traits. However studies using molecular markers illustrated the presence of QTLs influencing Fw reaction in chickpea rather than only major genes (Gowda et al. 2009; Sabbavarapu et al. 2013). Therefore, in present section the efforts made to identify QTLs associated with Fw resistance are discussed. The list of QTLs identified for Fw resistance to different races in chickpea have been presented in Table 8.5.

Cobos et al. (2009) reported a high percentage of the phenotypic variation for the resistance to *Foc*- 5 and *Foc*-0 (46.5 and 73%, respectively) was explained by a QTL on LG2 in a RIL derived from an interspecific cross. Sabbavarapu et al. (2013) identified two QTLs *viz.*, *FW-Q-APR-6-1* and *FW-Q-APR-6-2* on LG 6 with 10.4 and 18.8% phenotypic variation, respectively in RIL population of C-214 x WR-315 for resistance to *Foc*-1. Patil et al. (2014) identified QTLs associated with early and late wilt resistance on LG 2. QTLs for early wilt (*Wilt*-1) spanning through marker TA- 27 and TA-59 explained 36% of phenotypic variation and QTL for late wilt (*Wilt*-2) spanned through markers TA-27-TA110 and explained 16% of phenotypic variation. Jingade and Ravikumar (2015) identified five QTLs for *Foc1,* of which one was stable QTL (GSSR18-TC14801) across seasons. A total of seven QTLs were identified for race 1 by Raghu (2017) using two RIPs (JG62x WR315 and K850xWR315), of which four major QTLs (*q30pot-3-1, q60pot-3-1, qW30-07-3-1* and *qW60-07-1-1*) explained phenotypic variance ranging from 12.82 to 23.62% and three minor QTLs (*qW30-08-1-1, qW30-08-3-1* and *qWilt60-07-3-1*) explained

Table 8.5 List of QTLs identified for *Fusarium* wilt resistance to different races in chickpea

Fusarium races	QTL name	Population pedigree	Mapping population	Flanking markers	Position (cM)	LG	LOD	R2 (%)	Additive effect	References
Race 1	*Wilt 1*	JG 62 × WR 315	RI	TA27-TA59	9.0	2	9.19	36	−15.98	Patil et al. (2014)
	Wilt 2			TA27-TA110	22.2		3.53	16	−9.26	
Race 1	*QTLs between*	K850 × WR315	RI	TR 24 – EST SSR 21	63.50	2	4.43	56.80	21.98	Jingade and Ravikumar (2015)
				ESTSSR21 – ESTSSR65	63.60		3.13	56.20	21.94	
				GSSR18 – TC14801	158.80		8.90	69.80	−25.00	
				GSSR11 – ESTSSR 3	56.80		3.05	68.80	24.43	
Race 1	*FW-Q-APR-6-1*	C-214 × WR-315	F_2:F_3	CaM1402–CaM1101	36.3	6	8	10.4	–	Sabbavarapu et al. (2013)
	FW-Q-APR-6-2			CaM1125–TA22	41.4	6	7.6	18.8		
Race 1	*q30pot-3-1*	JG62 × WR315	RI	UASBC32-UASBC57	30	3	6.78	23.62	19.71	Raghu (2017)
	q60pot-3-1			UASBC48-UASBC50	32	3	7.93	25.97	19.08	
	qW30-07-3-1			UASBC48-UASBC50	32	3	4.94	17.31	10.84	
	qW60-07-1-1			UASBC79-UASBC92	11	1	3.55	12.82	7.87	
	qW30-08-1-1			UASBC79-UASBC92	11	1	2.50	8.56	7.74	
	qW30-08-3-1			UASBC48-UASBC50	32	3	32.00	2.59	8.64	

(continued)

Table 8.5 (continued)

Fusarium races	QTL name	Population pedigree	Mapping population	Flanking markers	Position (cM)	LG	LOD	R2 (%)	Additive effect	References
Race 1	*qWilt60-07-3-1*	K850 × WR315	RI	UASBC51 – UASBC53	4.88	3	2.69.	9.45	8.16.	Raghu (2017)
Race 1	*FW-Q-APR-2-1*	JG 62 × ICCV 05530	RI	TR19 – H2B061	10	2	18.54	24.56	27.4	Garg et al. (2018)
	FW-Q-APR-4-1			TA132 – TA46	16	4	5.6	7.94	15.56	
	FW-Q-APR-6-1			TA80 – CaM0594	37	6	4.38	6.63	9.79	
	FW-Q-APR-2-1			TR19 – H2B061	12	2	12.31	17.53	12.02	
	FW-Q-APR-4-1			CKAM1256 TS72	9	4	21.02	31.55	16.19	

RI recombinant inbred, *LGs* linkage groups, *LOD* logarithm of odd

phenotypic variation ranging from of 8.56 to 9.45%. Garg et al. (2018) detected five QTLs to Fw with phenotypic variance explained from 6.63 to 31.55%. Of the five QTLs, three QTLs including a major QTL on CaLG02 and a minor QTL each on CaLG04 and CaLG06 were identified for resistance to race 1. For race 3, a major QTL each on CaLG02 and CaLG04 were identified.

Comparison of different studies revealed that two major chromosomal loci on LG2 and LG5 govern resistance to the all reported races. However, the four Fw resistance gene (*Foc1*, *Foc3*, *Foc4* and *Foc5*) clusters on the LG 2 alone (Winter et al. 2000; Huettel et al. 2002; Tekeoglu et al. 2002; Sharma et al. 2004; Millan et al. 2006, 2010) therefore it has been speculated that CaLG02 is a hot spot for pathogen defence (Millan et al. 2006). Thus, it is important to fine map these loci to determine the candidate genes for resistance. Moreover, the genetically mapped candidate genes would be a more effective choice than the gene/QTL flanked anonymous markers for marker-assisted selection (MAS) as the candidate gene-based markers are derived directly from the protein-coding region of a gene and thus are in complete linkage with the allele associated with the trait of interest.

8.3.5 Fine Mapping

Candidate gene mapping in chickpea has been initiated in the early 2000s with the genetic mapping of resistance gene analogs (RGAs) as a candidate resistance gene and analyzed their association or co-localization with the QTLs for *Fusarium* wilt and *Ascochyta* blight (Huettel et al. 2002; Tekeoglu et al. 2002). Tekeoglu et al. (2002) integrated 50 SSR markers and a RGA locus into the linkage map reported by Santra et al. (2000) and improved the map resolution to an average inter-marker distance of 7.0 cM. This was the first time a candidate gene associated with disease resistance was genetically mapped in chickpea. Huettel et al. (2002) isolated a series of RGAs from both *C. arietinum* and *C. reticulatum* using two degenerate primer pairs targeting sequences in the NBS domain. Thirty of these RGAs were mapped on the reference genetic map of chickpea (Winter et al. 2000), where they could be located on principally five linkage groups, some of them as clusters on LGs 2 and 5, respectively. Rajesh et al. (2002) genetically mapped RGA protein kinase gene (RGAPto-kin1) on chickpea LG 5. While, Flandez-Galvez et al. (2003) mapped 12 RGA markers that clustered on three LGs. Gupta et al. (2013) described serine/threonine kinase, along with other proteins, as key molecular candidates controlling important hubs of the defensive network in the presence of *Foc*. Caballo et al. (2019) for the first time identified a candidate genes related to resistance to *Foc5* on LG 2. A putative region of 820 Kb delimited by SNP markers includes the gene, LOC101511605 (predicted to be a CBL-interacting serine/threonine-protein kinase 8-like) responsible for resistant reaction to *Foc 5*. These studies confirmed that potential resistance and pathogenesis-related genes for *Foc* races were localised on the CaLG02.

The integrated physical/genetic maps are very much needed for development of tightly linked markers to fine map the Fw resistance genes. In this context Varshney et al. (2014b) integrated physical, genetic and genome map of chickpea and located some of the Bacterial artificial chromosome (BACs) in the vicinity of some important QTLs for drought tolerance and resistance to *Fusarium* wilt and *Ascochyta* blight. Raghu (2017) used the *in silico* physical mapping approach to fine map the Fw resistance regions and identified the candidate genes. He redefined QTL positions of 44 Fw resistance linked markers covering 32 Fw resistance loci and nine QTLs from earlier studies. *In silico* physical mapping in this study revealed that the QTL-hotspot region/Fw resistance region on LG2 in genetic map for different races (*Foc02, Foc1, Foc3, Foc4, Foc5)* was physically mapped on chickpea Ca2. Further, QTL hot spot has been redefined to 3.98 Mb with closely linked markers UASBC48 and UASBC50 through genetic mapping that delineated to small regions harbouring 75 putative candidate genes for Fw resistance in chickpea. The candidate genes identified encodes protein, which are potentially involved in plant defence responses especially two genes encoding proteins annotated as cell wall remodulation (extensions and expansions).

8.3.6 Marker Assisted Backcross (MABC)

Molecular mapping studies revealed the clustering of Fw resistance genes on LG 2 and validated in different genetic backgrounds (Castro et al. 2010). Thus, identification of genomic regions associated with Fw resistance would serve as the basis to initiate marker-assisted transfer of resistant genes from *Fusarium* wilt resistant cultivars carrying individual genes into agronomically superior chickpea varieties. Further, the available DNA-based genetic markers linked to *Fusarium* wilt resistance would allow breeders to use them in MAS to directly infer the genotype of very young plants that avoids time delays caused by phenotype testing, and simultaneously avoid confounding effects of genotype x environment interactions. Marker-assisted backcrossing aims at conversion of targeted lines with respect to one or two traits without disturbing remaining all other native traits of target cultivar. However, MABC applying FGS (foreground) using QTL linked markers, and BGS (background selection) using genome-wide SSR markers for recovery of RPG (Recurrent Parent Genome) is an environment-independent, precise, and quick approach for the development of cultivars for the trait of interest (Varshney et al. 2010).

Marker-assisted backcrossing (MABC) has been successfully employed in chickpea to introgress *Ascochyta* blight (Ab) resistance with double-podding traits in chickpea cultivars CDC Xena, CDC Leader, and FLIP98-135C (Taran et al. 2013). A genomic region (known as QTL-hotspot) harbouring several QTLs for drought component traits was identified (Varshney et al. 2014a) and successfully introgressed initially into an elite cultivar JG 11 (Varshney et al. 2013a). MABC approach was adopted to develop dual resistance *Fusarium* wilt (Fw) and *Ascochyta* blight (Ab) C 214, an elite chickpea cultivar. To develop resistant lines, *Foc*1 locus for Fw and two quantitative trait loci (QTL) regions, ABQTL-I and ABQTL-II, for Ab were targeted for introgression. After three backcrosses and three rounds of selfing, 22 BC_3F_4 lines were generated for Fw and 14 MABC lines for Ab (Varshney et al. 2013b). Phenotyping of these lines has identified three resistant lines for Fw and seven resistance lines for Ab. Recently, Mannur et al. (2019) has successfully introgressed resistance to Fw race 4 (*Foc*4) into Annigeri 1 and JG 74 through marker-assisted backcrossing. Multiplication evaluation of improved version of Annigeri 1 (Super Annigeri 1) recorded 8% mean yield advantage and enhanced disease resistance over Annigeri 1. JG 74315-14 (the superior line in JG 74) recorded 125% and 25% advantage over recurrent parent and local check, respectively. These improved lines with enhanced Fw resistance and high yield performance demonstrate the successful examples of deployment of molecular breeding to develop superior lines for Fw resistance in chickpea.

8.4 Transcriptome Study on Chickpea – *Foc* Interaction

Several omics approaches have been applied to understand the molecular mechanism of plant-pathogen interactions. It has identified a set of unique genes, transcripts, proteins and metabolites via genomics, transcriptomics, proteomics and metabolomics, respectively. Chickpea-*Fusarium* interaction has been analyzed at different omic level for better understanding; in this section the work on transcriptome level is reviewed to understand the functional role of mRNA.

8.4.1 Transcriptome Analysis

A cell undergoing a functional or developmental process has a specific set of genes undergoing transcription at a particular time and is collectively called the 'transcriptome'. Thus, a transcriptome represents up to an extent physical, biochemical and developmental status of a cell. A transcriptome represents a pool of protein coding as well as non protein-coding RNAs; moreover, there may be the presence of variants of genes originating from alternative splicing and RNA editing, making the transcriptome more complex than a genome. Study of transcriptome may reveal information regarding spatial and temporal expression patterns of genes, and therefore it is possible to generate global expression profiles of genes representing developmental stages of an organism (Garg et al. 2011).

Transcriptome analysis was initiated with the generation of expressed sequence tags (ESTs) that are 200–800 nucleotide long cDNA sequences, synthesized from mRNA through reverse transcription. ESTs represent the expressed part of an organism's genome and hence are an excellent resource for the study of gene expression at a genome-wide level. Conventionally, EST resources have been developed through Sanger sequencing. Although this process is used to generate and sequence longer fragments of cDNA, it is tedious and labour intensive and offers poor coverage of the transcriptome. These limitations of EST-based transcriptome analysis inspired scientists to develop microarray and other tag-based methods for gene expression analysis. Therefore, tools such as microarrays and serial analysis of gene expression (SAGE) continued to be used for several years for analysis of global gene expression patterns. However, with the advent of NGS and the simultaneous development of *in silico* analytical tools, global genome and transcriptome analysis has become a standard practice for deriving information to relate genotype to phenotype. However, it is not possible to sequence the transcripts to the full length due to technological limitations. Transcriptome analysis is based on the principle that the depth of coverage of a sequence is proportional to the level of expression of the corresponding gene. Therefore, by mapping and counting the sequenced reads onto the given transcript, expression can be measured, thereby translating sequence information to some biologically significant information. A host of NGS technologies such as sequencing by synthesis (Illumina Inc., USA), SOLiD (ThermoFisher Scientific) and pyrosequencing (454 biosciences/Roche) has provided unprecedented opportunities for high-throughput functional genomic research. Moreover, a number of technologies for transcriptome sequencing are emerging such as The Ion Torrent (ThermoFisher Scientific), single-molecule real time (SMRT) (Pacific Biosciences, USA) and Nanopore (Oxford Nanopore Technologies, UK) (Mardis 2008; Morozova and Marra 2008; Schuster 2008).

Over the last few decades, several laboratories have used a wide range of transcriptomics techniques encompassing cDNA-AFLP, cDNA-RAPD, LongSAGE, expressed sequence tag (EST) analysis, serial analysis of gene expression (SAGE), microarray and next-generation sequencing (NGS) to detect several defense related genes of chickpea as well as virulence related genes of *Foc*. Transcriptome analysis has been utilized extensively to identify non-coding RNAs such as miRNAs and long intergenic non-coding (LINC) RNAs. Transcriptome analysis has facilitated the development of molecular markers such as simple sequence repeats (SSRs), single-nucleotide polymorphisms (SNPs) and potential intron polymorphisms (PIPs) that are being used to expedite the chickpea breeding programmes (Kant et al. 2017).

8.4.2 cDNA-RAPD and cDNA-AFLP

The potential of the AFLP technique for generating mRNA fingerprints was first recognized by Bachem et al. (1996) for the study of differential gene expression during potato tuber formation. Since then it has been used to profile genes in a range of different systems including plants and microbes. cDNA-AFLP remains a useful technique for several reasons; it is versatile, easy, inexpensive, robust and quantitative (Nimbalkar et al. 2006). Similarly cDNA-RAPD is also useful technique for profiling of genes, where random decamer primers are used. Differentially expressed genes in chickpea, (*Cicer arietinum* L.) during root infection by *Fusarium oxysporum* f sp. *ciceri* Race1, were identified using cDNA-RAPD and cDNA-AFLP approaches. Under cDNA-RAPD technique decamer primers were employed on cDNA template and revealed nine differentially expressed transcripts in the resistant-infected chickpea variety. Among the 2000 transcript-derived fragments (TDFs) screened by cDNA-AFLP, 273 were differentially expressed in chickpea roots during *Fusarium* infection. Only 13.65% of the TDFs were differentially regulated during pathogen challenge, while the other 86% were expressed non-differentially during the process of pathogen infection in chickpea roots. Nineteen TDFs, which expressed differentially in the resistant-infected chickpea variety were cloned and sequenced. Two of these TDFs were similar to transcription factors like WRKY proteins and 14-3-3 proteins, while three TDFs represented the NBS-LRR-type gene sequences. Two TDFs had sequence identity to genes known to have function in defence. The RAPD TDF CaFRi60 showed sequence identity to gamma-glutamyl-cysteine synthetase. Among the TDFs examined by cDNA-AFLP, 19 were confirmed by reverse Northern blot to be differentially expressed. The data confirmed the effectiveness of the cDNA-AFLP technique in detecting differentially expressed genes during pathogenesis (Nimbalkar et al. 2006). Early events of chickpea (*Cicer arietinum*) and *Fusarium oxysporum* f. sp. *ciceri* (race 1) interaction through cDNA-AFLP analysis was studied through triggering the pathogen-challenged disease responses in both susceptible and resistant plants followed by monitoring the expression of stress induced genes or gene fragments at the transcript level. cDNA-AFLP followed by homology search helped in differentiating and analyzing the up- and down-regulated gene fragments. Several detected DNA fragments appeared to have relevance with pathogen-mediated defence. Some of the important transcript-derived fragments were homologous to genes for sucrose synthase, isoflavonoid biosynthesis, drought stress response, serine threonine kinases, cystatins, arginase, and so on. Gupta et al. (2009) performed reverse-transcriptase polymerase chain reaction with samples collected at 48 and 96 h post infection confirmed a similar type of differential expression pattern. Based on the above results the interacting pathways of cellular processes during chickpea (*Cicer arietinum*) and *Fusarium oxysporum* f. sp. *ciceri* (race 1) interaction were generated. Gurjar et al. (2012) used cDNA-RAPD and semi-quantitative RT-PCR to study gene expression analysis of chickpea roots during *Foc* infection. cDNAs derived from total mRNA during infection process of susceptible (JG62) and resistant (Digvijay) cultivars, were amplified using random oligonucleotides. Sequence characterization of differentially expressed transcripts revealed their homology with many plant genes essential for various metabolic functions including defence. Further, expression patterns of specific candidate gene transcripts were analyzed in the *Foc* inoculated and uninoculated resistant and susceptible chickpea cultivars, on day 6 of infection. Semiquantitative RT-PCR analysis of defence related genes was performed using gene specific oligonucleotides in resistant and susceptible chickpea cultivars. The expression of fungal pathogenesis related genes and their race specific response was determined throughout the course of chickpea-*Foc* interaction. Temporal expression and race specific response of plant defence related and fungal virulence genes were studied in the resistant and susceptible cultivars of chickpea inoculated with three races of *Foc* highlighting the host-pathogen interactions. Few genes, involved in chickpea defence against *Fusarium* wilt which was not reported previously were unveiled in the study.

8.4.3 Suppression Subtractive Hybridization (SSH)

The suppression subtractive hybridization (SSH) allows rapid identification and isolation of differentially expressed genes in a single experiment. The SSH is a powerful technique widely used in comparing two populations of mRNA and to isolate clones of differentially expressed genes (Diatchenko et al. 1996). It has been utilized widely to isolate and characterize differentially expressed genes during defense response against various pathogens in different crops (Fernandez et al. 2004; Coram and Pang 2005; Jaiswal et al. 2011). By applying SSH method chickpea–*Fusarium* interaction at the transcriptome level was studied to identify and validate differentially regulated genes predicted to be involved in defense response between resistance (WR315) and susceptible (JG62) genotypes of chickpea against *Fusarium oxysporum* f. sp. *ciceris* (*Foc*) race 4 during compatible and incompatible interaction at early stage of the infection. Altogether, 162 chickpea-expressed sequence tags (ESTs) were identified from two libraries and analyzed to catalog eight functional categories. These ESTs could be assembled into 18 contigs and 144 singletons with 10 contigs and 68 singletons from compatible and 8 contigs and 70 singletons from incompatible interaction. The largest category consisted of ESTs which encode for proteins related to hypothetical proteins (22.8%), followed by energy and metabolism (20.3%)-related genes, defense and cell rescue-related genes (17.9%) and signal transduction-related genes (16%). Among them, 17.1 and 18.7% were defense-related genes in compatible and incompatible interaction, respectively. These ESTs mainly includes various putative genes related to oxidative burst, pathogenesis and secondary metabolism. Induction of putative superoxide dismutase, metallothionein, 4-coumarate-CoA ligase, heat shock proteins and cysteine proteases indicated oxidative burst after infection. The ESTs belonged to various functional categories which were directly and indirectly associated with defense signaling pathways. Quantitative and semi-quantitative polymerase chain reaction exhibited differential expression of candidate genes and detected higher levels in incompatible interaction compared to compatible interaction (Saabale et al. 2018).

8.4.4 Serial Analysis of Gene Expression (SAGE)

Serial analysis of gene expression (SAGE) is a transcriptomic technique used by molecular biologists to produce a snapshot of the messenger RNA population in a sample of interest in the form of small tags that correspond to fragments of those transcripts. Several variants have been developed since, most notably a more robust version, LongSAGE, RL-SAGE and the most recent SuperSAGE. Many of these have improved the technique with the capture of longer tags, enabling more confident identification of a source gene. Transcriptome profiling using LongSAGE provided a valuable insight into the molecular interactions between chickpea and *Foc*. Upasani et al. (2017) revealed several known as well as novel genes with differential or unique expression patterns in chickpea contributing to lignification, hormonal homeostasis, plant defense signaling, ROS homeostasis, R-gene mediated defense, etc. Similarly, several *Foc* genes characteristically required for survival and growth of the pathogen were expressed only in the susceptible cultivar with null expression of most of these genes in the resistant cultivar. Overall, higher number of stress responsive transcripts was observed in JGI. However, up-regulated genes in response to fungal attack were more in DVI. The transcripts associated with abiotic stress, ROS homeostasis, heat-shock proteins, secondary metabolism and PR proteins were noticeably up-regulated in DVI as compared to those in DVC and JGI. Interestingly, reticulon-like protein B2 (RTNLB2), showed highest expression (15.36 fold) in DVI. Transcripts belonging to ROS homeostasis were peroxidase, glutathione-S transferase, catalase, glutaredoxin and thioredoxin family proteins. The PR protein category particularly included disease resistance proteins (DRR) SR1, DRR 206 and TMV resistance protein. Likewise, several heat shock proteins (HSPs) were highly expressed in DVI. Moreover, lignin biosynthetic enzymes such as 4-coumarate–CoA ligase (4CL), cinnamoyl-CoA reductase 1 (CCR1) and caffeoyl-CoA O-methyltransferase (CCoAOMT) were also up-regulated in DVI. *Foc* transcript (identified as TKL protein kinase) expressing only in DVI, was recognized as a virulence factor. Three of the 18 *Foc* transcripts (ATP synthase subunit alpha and two hypothetical proteins) and 85 *Foc* transcripts that expressed only in JGI, showed homology to experimentally proven virulence factors. Transcripts for several pathogenesis related proteins, major latex proteins 40, auxin binding proteins 41, etc. from 'biotic and abiotic stress response' category were detected with higher abundance and elevated expression in DVI. These proteins provided R gene mediated response, strong lignification, proteinase inhibitory activities and chaperon like functions enriching the defense response of resistant cultivar against *Foc*. In addition, the expression of signaling components and pathogenesis related proteins in resistant cultivar gives the plant an advantage in mounting prompt defense. Conversely, up-regulation of certain factors such as actin depolymerization factor (ADF), aquaporins tetrapyrrole synthesis, etc. in JGI adds to its susceptibility. Thus, the key difference between the resistant and the susceptible cultivars was timely detection of the invading pathogen and rapid and immediate activation of defense responses in response to pathogen effectors in the resistant cultivar.

8.4.5 Microarray

Microarray is a method of transcriptome analysis that is widely used for identification of transcripts under biotic and abiotic stress condition. In order to identify genes involved in wilt resistance in chickpea, EST sequencing followed by microarray analysis of chickpea wilt susceptible genotype (JG-62) and resistant genotype (WR-315) the tranascriptome analysis through microarray was performed after infecting them with *F. oxysporum ciceri*. They reported 6272 gene sequences of immune-response pathway that provided genotype-dependent spatial information on the presence and relative abundance of each gene. The sequence assembly led to the identification of a *Ca* Unigene set of 2013 transcripts comprising of 973 contigs and 1040 singletons, two-third of which represent new chickpea genes hitherto undiscovered. They identified 209 gene families and 262 genotype-specific SNPs. Further, several novel transcription regulators were identified indicating their possible role in immune response. The transcriptomic analysis revealed 649 non-cannonical genes besides many unexpected candidates with known biochemical functions, which have never been associated with pathostress-responsive transcriptome (Ashraf et al. 2009).

8.4.6 RNA Sequencing Technique

RNA-Sequencing (RNA-Seq) technique uses next-generation sequencing (NGS) to reveal the presence and quantity of RNA in a biological sample at a given moment, analyzing the continuously changing cellular transcriptome (Chu and Corey 2012; Wang et al. 2009). Specifically, RNA-Seq facilitates the ability to look at alternative gene spliced transcripts, post-transcriptional modifications, gene fusion, mutations/SNPs and changes in gene expression over time, or differences in gene expression in different groups or treatments (Maher et al. 2009). In addition to mRNA transcripts, RNA-Seq can look at different populations of RNA to include total RNA, small RNA, such as miRNA, tRNA, and ribosomal profiling (Ingolia et al. 2012). RNA-Seq can also be used to determine exon/intron boundaries and verify or amend previously annotated 5′ and 3′ gene boundaries. Recent advances in RNA-Seq include single cell sequencing and in situ sequencing of fixed tissue (Lee et al. 2014). Molecular complexities in chickpea during *Fusarium oxysporum* f. sp. *ciceri* Race 1 (*Foc*1) infection was dissected using transcriptomic analysis by RNA-seq revealed significantly differential expression of molecular transcripts between infected and control plants of both susceptible and resistant genotypes (Gupta et al. 2017). Radar plot analyses showed maximum expressional undulations after infection in both susceptible and resistant plants. Gene ontology and functional clustering showed large number of transcripts controlling basic metabolism of plants. Network analyses demonstrated defense components like peptidyl cis/trans isomerase, MAP kinase, beta 1,3 glucanase, serine threonine kinase, patatin like protein, lactolylglutathione lyase, coproporphyrinogen III oxidase, sulfotransferases; reactive oxygen species regulating components like respiratory burst oxidase, superoxide dismutases, cytochrome b5 reductase, glutathione reductase, thioredoxin reductase, ATPase; metabolism regulating components, myo-inositol phosphate, carboxylate synthase; transport related gamma tonoplast intrinsic protein, and structural component, ubiquitins to serve as important nodals of defense signaling network. These nodal molecules probably served as hub controllers of defense signaling. Sharma et al. (2016) revealed new insights into infection related genes by genome wide transcriptome profiling of *Fusarium oxysporum* f sp. *ciceris* conidial germination. The infection was initiated by conidia that invaded the host tissue often by penetration of intact epidermal cells. They reported the characterization of the transcriptome of *Foc* sequenced using Illumina Hiseq technology during its conidial germination at different time points. Genome-wide expression profiling revealed that genes linked to fungal development were transcribed in successive ways. Analysis showed that *Foc* have large sets of germination-related genes and families of genes encoding secreted effectors, cell wall/pectin-degrading enzymes, metabolism related enzymes, transporters and peptidases. They found that metabolism related enzymes were up-regulated at early time point whereas most transporters and secondary metabolites important for tissue colonization and pathogenicity were up-regulated later as evident from the qRT-PCR. The study demonstrated that early conidial germination in *Foc* was accompanied by rapid shifts in gene expression that prepared the fungus for germ tube outgrowth, host cell invasion and pathogenesis. Jain et al. (2015) studied the differential response of susceptible and tolerant/resistant chickpea varieties to *F. oxysporum*, transcriptomes of wilt susceptible (JG62) and wilt-tolerant/wilt-resistant (ICCV2, K850 and WR315) chickpea varieties using the Illumina platform. Comparison among the transcriptomes led to identification of 303 polymorphic SSRs, 14,462 SNPs and 1864 insertions/deletions (InDels). Moreover, a large number of SNPs and/or InDels were found to be present in defense-related genes. Kohli et al. (2014) constructed libraries from fungal-infected (*F. oxysporum* f.sp. *ciceris*), salt-treated and untreated seedlings of chickpea and sequenced using the Illumina GAIIx platform. After analysis they identified 122 conserved miRNAs belonging to 25 different families along with 59 novel miRNAs. miR156, miR396 and miR319 were upregulated in response to salt stress. miR156 and miR396 expression was found to be 1.5 times upregulated in both wilt and salt stresses, indicating a common mechanism implied by chickpea involving these miRNAs to cope up with both the stresses. miR530 was found to be significantly upregulated during wilt stress and may be involved in defense to fungal infection. Three legume-specific miRNAs, miR2111, miR2118 and miR5213, were also indicated to play a critical role in defense to pathogen attack. Targets of miR2111 included F-box protein and TIR (Toll/Interleukin-1 Receptor) domain-containing NBS-LRR disease-resistance proteins, and miR2118 and miR5213 also targeted the same class of R genes. Interestingly, miR2118 was up-regulated following wilt infection and down-regulated following salt stress. Functional characterization of these hub molecules would not only help in developing better understanding of chickpea- *Foc*1 interaction but also place them as promising candidates for resistance management programs against vascular wilt of legumes.

8.5 Concluding Remarks

The chickpea – *Foc* interaction is extensively studied, further *Foc* infection process is well understood at microbial, biochemical and molecular level. The prevailing races diversity in *Foc* is well documented along with its symptomatology, further molecular markers have stimulated this process. From the plant side the race specific resistance genes have been deciphered and most saturated generic maps are available in chickpea among grain legumes. The markers linked to resistance gene along with major QTLs are exploited to develop resistance genotypes with sustainability. This strategy is useful for quick improvement of known genetic background where disease resistance varieties such as Super Annigeri 1 and JG 74315-14 are developed. The transcriptome studies have identified molecular mechanisms; several biotic stress related key genes, transcription factors and miRNAs that play significant role in offering sustainable *Foc* resistance in chickpea.

References

Agrios GN (2005) Environmental effects on the development of the infectious disease. In: Plant pathology, 5th edn. Elesvier Academy Press, Burlington, Mass, pp 251–262

Al-Taae AK, Hadwan HA, Al-Jobory SAE (2013) Physiological races of *Fusarium oxysporum* f. sp. *Ciceris in* Iraq. J Life Sci 7:1070–1075

Anupama K (2001) Inheritance of shoot and root characters and molecular markers in chickpea (*Cicer arietinum* L.) Doctoral dissertation, Acharya NG Ranga Agricultural University, India

Arvayo-Ortiz RM, Esqueda M, Acedo-Felix E, Sanchez A, Gutierrez A (2011) Morphological variability and races of *Fusarium oxysporum* f. sp. *ciceris* associated with chickpea (*Cicer arietinum*) crops. Am J Agric Biol Sci 6:114–121

Ashraf N, Ghai D, Barman P, Basu S, Gangisetty N, Mandal MK, Chakraborty N, Datta A, Chakraborty S (2009) Comparative analyses of genotype dependent expressed sequence tags and stress-responsive transcriptome of chickpea wilt illustrate predicted and unexpected genes and novel regulators of plant immunity. BMC Genomics 10:415. https://doi.org/10.1186/1471-2164-10-415

Ayyar RV, Iyer BR (1936) A preliminary note on the mode of inheritance to *Fusarium* wilt in *Cicer arietinum*. Proc Indian Acad Sci 3:438–443

Bachem CW, Van der Hoeven RS, de Bruijn SM, Vreugdenhil D, Zabeau M, Visser RG (1996) Visualization of differential gene expression using a novel method of RNA fingerprinting based on AFLP: analysis of gene expression during potato tuber development. Plant J 9:745–753

Backman PA, Turmer JT (1989) Plant response and disease control following seed inoculation with *Bacillus subtilis*. In: Brown JM (ed) Proceedings beltwide cotton products research conference (Book 2). National Cotton Council of American, Memphis, pp 16–17
Barman P, Handique AK, Tanti B (2014) Tagging STMS markers to *Fusarium* wilt race-1 resistance in chickpea (*Cicer arietinum* L.). Indian J Biotechnol 13:370–375
Bayraktar H, Dolar FS (2012) Pathogenic variability of *Fusarium oxysporum* f. sp. *ciceris* isolates from chickpea in Turkey. Pak J Bot 44:821–823
Beckman CH (1987) The nature of wilt diseases of plants. American Phytopathology Society Press, University of California, St Paul, p 175
Benko-Iseppon AM, Winter P, Huettel B, Staginnus C, Muehlbauer FJ, Kahl G (2003) Molecular markers closely linked to *Fusarium* resistance genes in chickpea show significant alignments to pathogenesis-related genes located on Arabidopsis chromosomes 1 and 5. Theor Appl Genet 107:379–386
Bishop CD, Cooper RM (1983) An ultrastructural study of root invasion in three vascular wilt diseases. Physiol Plant Pathol 22:15–27
Bouhadida M, Jendoubi W, Gargouri S, Beji M, Kharrat M, Chen W (2017) First report of *Fusarium* redolens causing *Fusarium* yellowing and wilt of chickpea in Tunisia. Plant Dis 1038:101–106
Butler EJ (1918) Fungi and diseases of plants. In: Singh Bishen, Singh Mahendra Pal, New Connaugh Place, Dehradun Periodical Experts. 42-D, Vivek Vihar, Delhi, 32–547
Caballo C, Madrid E, Gil J, Chen W, Rubio J, Millan T (2019) Saturation of genomic region implicated in resistance to *Fusarium oxysporum* f. sp. *ciceris* race 5 in chickpea. Mol Breed 39:1–11
Castro P, Piston F, Madrid E, Millan T, Gil J, Rubio J (2010) Development of chickpea near-isogenic lines for *Fusarium* wilt. Theor Appl Genet 121:1519–1526
Chaube HS, Pundhir VS (2005) Crop diseases and their management. Prentice-Hall of India, Private Limited, New Delhi, p 724
Chu Y, Corey DR (2012) RNA sequencing: platform selection, experimental design, and data interpretation. Nucleic Acid Ther 22(4):271–274
Cobos MJ, Fernandez MJ, Rubio J, Kharrat M, Moreno MT, Gil J, Millan T (2005) A linkage map of chickpea (*Cicer arietinum* L.) based on populations from Kabuli × desi crosses: location of genes for resistance to *Fusarium* wilt race 0. Theor Appl Genet 110:1347–1353
Cobos MJ, Winter P, Kharrat M, Cubero JI, Gil J, Millan T, Rubio J (2009) Genetic analysis of agronomic traits in a wide cross of chickpea. Field Crop Res 111:130–136
Coram TE, Pang ECK (2005) Isolation and analysis of candidate *Ascochyta* blight defence genes in chickpea. Physiol Mol Plant Phathol 66:201–210
Correll JC (1991) The relationship between *formae speciales*, races and vegetative compatibility groups in *F. oxysporum*. Phytopathology 81:1061–1064
Couteaudier Y, Alabouvette C (1990) Quantitative comparison of *Fusarium oxysporum* competitiveness in relation with carbon utilization. FEMS Microbiol Ecol 74:261–268
Demers JE, Garzon CD, Jiménez-Gasco MM (2014) Striking genetic similarity between races of *Fusarium oxysporum* f. sp. *ciceris* confirms a monophyletic origin and clonal evolution of the chickpea vascular wilt pathogen. Eur J Plant Pathol 139:303–318
Diatchenko L, Lau YF, Campbell AP, Chenchik A, Moqadam F, Huang B, Lukyanov S, Lukyanov K, Gurskaya N, Sverdlov ED, Siebert PD (1996) Suppression subtractive hybridization: a method for generating differentially regulated or tissue-specific cDNA probes and libraries. Proc Natl Acad Sci 93:6025–6030
Dolar FS (1997) Determination of the races of *Fusarium oxysporum* f. sp. *ciceris* in Ankara province. Turk J Phytopathol 26:11–15
Fang X, Kuo J, You MP, Finnegan PM, Barbetti MJ (2012) Comparative root colonisation of strawberry cultivars camarosa and festival by *Fusarium oxysporum* f. sp. *fragariae*. Plant Soil 358:75–89
Fernandez D, Santos P, Agostini C, Bon MC, Petitot AS, Silva MC, Guerra-Guimaraes L, Ribeiro A, Argout X, Nicole M (2004) Coffee (*Coffea arabica* L.) genes early expressed during infection by the rust fungus (*Hemileia vastatrix*). Mol Plant Pathol 5:527–536

Flandez-Galvez H, Ford R, Pang EC, Taylor PW (2003) An intraspecific linkage map of the chickpea (*Cicer arietinum* L.) genome based on sequence tagged microsatellite site and resistance gene analog markers. Theor Appl Genet 106:1447–1456

Garg R, Patel RK, Jhanwar S, Priya P, Bhattacharjee A, Yadav G, Bhatia S, Chattopadhyay D, Tyagi AK, Jain M (2011) Gene discovery and tissue-specific transcriptome analysis in chickpea with massively parallel pyrosequencing and web resource development. Plant Physiol 156(4):1661–1678

Garg T, Mallikarjuna BP, Thudi M, Samineni S, Singh S, Sandhu JS, Kaur L, Singh I, Sirari A, Basandrai AK, Basandrai D, Varshney RK, Gaur PM (2018) Identification of QTLs for resistance to *Fusarium* wilt and *Ascochyta* blight in a recombinant inbred population of chickpea (*Cicer arietinum* L.). Euphytica 214(3):1–11

Girase VS, Deshmukh RB (2002) Inheritance of resistance to wilt (*Fusarium oxysporum* f.sp. *ciceri*) in chickpea. J Maharashtra Agric Univ 27(1):10–12

Gordon TR (2017) *Fusarium oxysporum* and the *Fusarium* wilt syndrome. Annu Rev Phytopathol 55:23–39

Gowda SJM, Radhika P, Kadoo NY, Mhase LB, Gupta VS (2009) Molecular mapping of wilt resistance genes in chickpea. Mol Breed 24:177–183

Gumber RK, Kumar J, Haware MP (1995) Inheritance of resistance to *Fusarium* wilt in chickpea. Plant Breed 114:277–279

Gupta S, Chakraborti D, Rangi RK, Basu D, Das S (2009) A molecular insight into the early events of chickpea (*Cicer arietinum*) and *Fusarium oxysporum* f. sp. *ciceri* (race 1) interaction through cDNA-AFLP analysis. Phytopathology 99:1245–1257

Gupta S, Bhar A, Chatterjee M, Das S (2013) *Fusarium oxysporum* f.sp. *ciceri* race 1 induced redox state alterations are coupled to downstream defense signaling in root tissues of chickpea (*Cicer arietinum* L.). PLoS One 8:73163

Gupta S, Bhar A, Chatterjee M, Ghosh A, Das S (2017) Transcriptomic dissection reveals wide spread differential expression in chickpea during early time points of *Fusarium oxysporum* f. sp. *ciceri* Race 1 attack. PLoS One 12(5):e0178164

Gurjar G, Barve M, Giri A, Gupta V (2009) Identification of Indian pathogenic races of *Fusarium oxysporum* f. sp. *ciceris* with gene specific, ITS and random markers. Mycologia 101(4):484–495

Gurjar GS, Giri AP, Gupta VS (2012) Gene expression profiling during wilting in chickpea caused by *Fusarium oxysporum* f. sp. *Ciceri*. Am J Plant Sci 3:190–201

Halila HM, Strange R (1996) Identification of the causal agent of wilt of chickpea in Tunisia as *Fusarium oxysporum* f. sp. ciceris race 0. Phytopathol Mediterr 35:67–74

Halila I, Cobos MJ, Rubio J, Millan T, Kharrat M, Gil J (2009) Tagging and mapping a second resistance gene for *Fusarium* wilt race 0 in chickpea. Eur J Plant Pathol 124:87–92

Halila I, Rubio J, Millan T, Gil J, Kharrat M, Marrakchi M (2010) Resistance in chickpea (*Cicer arietinum*) to *Fusarium* wilt race 0. Plant Breed 129:563–566

Haware MP, Nene YL (1982) Races of *Fusarium oxysporum* f. sp. *ciceri*. Plant Dis 66:809–810

Haware MP, Jiménez-Díaz RM, Amin KS, Phillips JC, Halila H (1990) Integrated management of wilt and root rots of chickpea. In: Chickpea in the nineties: proceedings of the second international workshop on chickpea improvement, 4–8 December 1989. Patencheru, India: ICRI-SAT Center, p129–133

Haware MP, Nene YL, Pundir RPS, Narayana RJ (1992) Screening of world chickpea germplasm for resistance to *Fusarium* wilt. Field Crop Res 30:147–154

Honnareddy N, Dubey SC (2006) Pathogenic and molecular characterization of Indian isolates of *Fusarium oxysporum* f. sp. *ciceris* causing chickpea wilt. Curr Sci 91:661–666

Huettel B, Santra D, Muehlbauer FJ, Kahl G (2002) Resistance gene analogues of chickpea (*Cicer arietinum* L.): isolation, genetic mapping and association with a *Fusarium* resistance gene cluster. Theor Appl Genet 105:479–490

Huisman OC (1982) Interactions of root growth dynamics to epidemiology of root-invading fungi. Annu Rev Phytopathol 20:303–327

Ingolia NT, Brar GA, Rouskin S, McGeachy AM, Weissman JS (2012) The ribosome profiling strategy for monitoring translation in vivo by deep sequencing of ribosome-protected mRNA fragments. Nat Protoc 7(8):1534–1550

Iruela M, Rubio J, Barro F, Cubero JI, Millán T, Gil J (2006) Detection of two quantitative trait loci for resistance to ascochyta blight in an intra-specific cross of chickpea (*Cicer arietinum* L.): development of SCAR markers associated with resistance. Theor Appl Genet 112(2):278–287

Iruela M, Castro P, Rubio J, Cubero JI, Jacinto C, Millán T, Gil J (2007) Validation of a QTL for resistance to ascochyta blight linked to resistance to *Fusarium* wilt race 5 in chickpea (Cicer arietinum L.). Eur J Plant Pathol 119:29–37

Jain M, Pole AK, Singh VK, Ravikumar RL, Garg R (2015) Discovery of molecular markers for *Fusarium* wilt via transcriptome sequencing of chickpea cultivars. Mol Breed 35(10):198–204

Jaiswal P, Cheruku RJ, Kumar K, Yadav S, Singh A, Kumari P, Dube SC, Upadhyaya KC, Verma PK (2011) Differential transcript accumulation in chickpea during early phases of compatible interaction with a necrotrophic fungus *Ascochyta rabiei*. Mol Biol Rep 39:4635–4646

Jendoubi W, Bouhadida M, Boukteb A, Béji M, Kharrat M (2017) *Fusarium* wilt affecting chickpea crop. Agriculture 7:23. https://doi.org/10.3390/agriculture7030023

Jiménez-Díaz RM, Basallote-Ureba MJ, Rapoport H (1989) Colonization and pathogenesis in chickpea infected by races of *Fusarium oxysporum* f. sp. *ciceris*. In: Tjamos EC, Beckman C (eds) Vascular wilt diseases of plants, vol H28. Springer, Berlin, pp 113–121

Jiménez-Díaz RM, Alcalá-Jiménez AR, Hervás A, Trapero-Casas JL (1993) Pathogenic variability and host resistance in the *Fusarium oxysporum* f. sp. *ciceris/Cicer arietinum* pathosystem. Proceedings of the 3rd Eur. Semin. *Fusarium* mycotoxins, taxonomy, pathogenicity and host resistance. Plant Breeding and Acclimatization Institute, Radzikov (Poland), pp 87–94

Jiménez-Díaz RM, Castillo P, Del Mar Jiménez-Gasco M, Landa BB, Navas-Cortés JA (2015) *Fusarium* wilt of chickpeas: biology, ecology and management. Crop Prot 73:16–27

Jiménez-Gasco MM, Jiménez-Díaz RM (2003) Development of a specific polymerase chain reaction-based assay for the identification of *Fusarium oxysporum* f. sp. *ciceris* and its pathogenic races 0, 1A, 5 and 6. Phytopathology 93:200–209

Jiménez-Gasco M, Pérez-Artés E, Jiménez-Díaz RM (2001) Identification of pathogenic races 0, 1B/C, 5, and 6 of *Fusarium oxysporum* f. sp. *ciceris* with random amplified polymorphic DNA (RAPD). Eur J Plant Pathol 107(2):237–248

Jiménez-Gasco MM, Milgroom MG, Jiménez-Díaz RM (2002) Gene genealogies support *Fusarium oxysporum* f. sp. *ciceris* as a monophyletic group. Plant Pathol 51:72–77

Jiménez-Gasco MM, Milgroom MG, Jiménez-Díaz RM (2004a) Stepwise evolution of races in *Fusarium oxysporum* f. sp. *ciceris* inferred from fingerprinting with repetitive DNA sequence. Phytopathology 94:228–235

Jiménez-Gasco MM, Navas-Cortés JA, Jiménez-Díaz RM (2004b) The *Fusarium oxysporum* f. sp. *ciceris/Cicer arietinum* patho system: a case study of the evolution of plant-pathogenic fungi into races and pathotypes. Int Microbiol 7:95–104

Jingade P, Ravikumar RL (2015) Development of molecular map and identification of QTLs linked to *Fusarium* wilt resistance in chickpea. J Genet 94(4):723–729

Jorge IJI, Navas J, Jimenez R, Tena M (2006) Cell wall degrading enzymes in *Fusarium* wilt of chickpea: correlation between pectinase and xylanase activities and disease development in plants infected with two pathogenic races of *Fusarium oxysporum* f. sp. *ciceris*. Can J Bot 84(9):1395–1404

Kant C, Pandey V, Verma S, Tiwari M, Kumar S, Bhatia S (2017) Transcriptome analysis in chickpea (*Cicer arietinum* L.): applications in study of gene expression, non-coding RNA prediction, and molecular marker development. In: Applications of RNA-seq and omics strategies – from microorganisms to human health. INTECH, pp 245–263. https://doi.org/10.5772/intechopen.69884

Kohli D, Joshi G, Deokar AA, Bhardwaj AR, Agarwal M, Katiyar-Agarwal S, Srinivasan R, Jain PK (2014) Identification and characterization of wilt and salt stress-responsive microRNAs in chickpea through high-throughput sequencing. PLoS One 9(10):e108851

Kumar S (1998) Inheritance to *Fusarium* wilt (race 2) in chickpea. Plant Breed 117:139–142
Kumar J, Haware MP (1982) Inheritance of resistance to *Fusarium* wilt in chickpea. Phytopathology 72:1035–1036
Ladizinsky G, Adler A (1976) Genetic relationships among the annual species of *Cicer* L. Theor Appl Genet 48(4):197–203. https://doi.org/10.1007/BF00527371
Lee JH, Daugharthy ER, Scheiman J, Kalhor R, Yang JL, Ferrante TC, Yang JL, Terry R, Jeanty SSF, Li C, Amamoto R, Peters DT, Turczyk BM, Marblestone AH, Inverso SA, Bernard A, Mali P, Rios X, Aach J, Church GM (2014) Highly multiplexed subcellular RNA sequencing in situ. Science 343(6177):1360–1363
Leroch M, Kleber A, Silva E, Coenen T, Koppenhofer D, Shmaryahu A, Valenzela PD, Hanh M (2013) Transcriptome profiling of *Botrytis cinerea* conidial germination reveals upregulation of infection-related genes during the prepenetration stage. Eukaryot Cell 12(4):614–626
Leslie JF, Summerell BA (2006) The *Fusarium* laboratory manual, 1st edn. Blackwell Publishing Ltd, Oxford, London
Liu S, Zhou X, Liu B, Pan K (2011) Sugars in watermelon root exudates and their effects on *Fusarium oxysporum* f. sp. *niveum*. Allelopath J 28(1):21–28
Lockwood JL (1977) Fungistasis in soil. Biol Rev 51:1–43
Maher CA, Kumar SC, Cao X, Kalyana Sundaram S, Han B, Jing X, Sam L, Barrette T, Palanisamy N, Chinnaiyan AM (2009) Transcriptome sequencing to detect gene fusions in cancer. Nature 458(7234):97–101
Mannur DM, Babbar A, Thudi M, Sabbavarapu MM, Roorkiwal M, Yeri SB, Bansal V, Jayalakshmi SK, Yadav S, Rathore A, Chamarthi SK, Mallikarjuna BP, Gaur PM, Varshney RK (2019) Super Annigeri 1 and improved JG 74: two *Fusarium* wilt-resistant introgression lines developed using marker-assisted backcrossing approach in chickpea (*Cicer arietinum* L.). Mol Breed 39(1):2–8
Mardis ER (2008) Next-generation DNA sequencing methods. Annu Rev Genomics Hum Genet 9:387–402
Mayer MS, Tullu A, Simon CJ, Kumar J, Kaiser WJ, Kraft JM, Muehlbauer FJ (1997) Development of a DNA marker for *Fusarium* wilt resistance in chickpea. Crop Sci 37:1625–1629
Michielse CB, Rep M (2009) Pathogen profile update: *Fusarium oxysporum*. Mol Plant Pathol 10:311–324
Millan T, Clarke HJ, Siddique KHM, Buhariwalla HK, Gaur PM, Kumar J, Gill J, Kahl G, Winter P (2006) Chickpea molecular breeding: new tools and concepts. Euphytica 147:81–103
Millan T, Winter P, Jungling R, Gil J, Rubio J, Cho S, Cobos M, Iruela M, Rajesh P, Tekeoglu M (2010) A consensus genetic map of chickpea (*Cicer arietinum* L.) based on 10 mapping populations. Euphytica 175(2):175–185
Morozova O, Marra MA (2008) Applications of next-generation sequencing technologies in functional genomics. Genomics 92(5):255–264
Nimbalkar SB, Harsulkar AM, Giri AP, Sainani MN, Franceschi V, Gupta VS (2006) Differentially expressed gene transcripts in roots of resistant and susceptible chickpea plant (*Cicer arietinum* L.) upon *Fusarium oxysporum* infection. Physiol Mol Plant Pathol 68(2006):176–188
Padwick GW (1940) The genus *Fusarium*, III. A- critical study of the fungus causing wilt of gram (*Cicer arietinum* L.) and the related species of the subsection *Orthocera* with special relation to the variability of key characteristics. Indian J Agric Sci 10:241–284
Palomino C, Fernández-Romero M, Rubio J, Torres A, Moreno M, Millán T (2009) Integration of new CAPS and dCAPS-RGA markers into a composite chickpea genetic map and their association with disease resistance. Theor Appl Genet 118:671–682
Pathak MM, Singh KP, Lal SB (1975) Inheritance of resistance to wilt (*Fusarium oxysporum* f.sp. *ciceri*) in gram. Indian J Farm Sci 3:10–11

Patil BS, Ravikumar RL, Bhat JS, Soregaon CD (2014) Molecular mapping of QTLs for resistance to early and late *Fusarium* wilt in chickpea. Czech J Genet Plant Breed 50:171–176

Poornima KN, Sabaale PR, Prakash GP, Das A, Soren KR, Singh NP (2017) PCR based race identification of *Fusarium oxysporum* f. sp. *ciceri* using molecular markers. J Food Legumes 30(1):6–10

Raghu R (2017) Genetic mapping of functional markers located in QTL region for *Fusarium* wilt resistance in chickpea (*Cicer arietinum* L.). Ph D (thesis) University of Agricultural Sciences Bengaluru, India

Rajesh PN, Tekeoglu M, Gupta VS, Ranjekar PK, Muehlbauer FJ (2002) Molecular mapping and characterization of an RGA locus RGA Ptokin1-2171 in chickpea. Euphytica 128(3):427–433

Ratnaparkhe MB, Tekeoglu M, Muehlbauer FJ (1998) Intersimple- sequence-repeat (ISSR) polymorphisms are useful for finding markers associated with disease resistance gene clusters. Theor Appl Genet 97(4):515–519

Rubio J, Haji-Moussa E, Kharrat M, Moreno MT, Millant Gil J (2003) Two genes and linked RAPD markers involved in resistance to *Fusarium oxysporum* f.sp. *ciceris* race 0 in chickpea. Plant Breed 122:189–191

Saabale PR, Dubey SC, Priyanka K, Sharma TR (2018) Analysis of differential transcript expression in chickpea during compatible and incompatible interactions with *Fusarium oxysporum* f. sp. *ciceris* Race 4. Biotech 8:111. https://doi.org/10.1007/s13205-018-1128-z

Sabbavarapu MM, Sharm M, Chamarthi SK, Swapna N, Rathore A, Thudi M, Gaur MP, Pande S, Singh S, Kaur L, Varshney RK (2013) Molecular mapping of QTLs for resistance to *Fusarium* wilt (race 1) and *Ascochyta* blight in chickpea (*Cicer arietinum* L.). Euphytica 193:121–133

Santra DK, Tekeoglu M, Ratnaparkhe M, Kaiser WJ, Muehlbauer FJ (2000) Identification and mapping of QTLs conferring resistance to Ascochyta blight in chickpea. Crop Sci 40:1606–1612

Saxena MC, Singh KB (1987) The chickpea published by C. A. B. International ICARDA, pp 250–252

Schroth MN, Hildebrand DC (1964) Influence of plant exudates on root infecting fungi. Annu Rev Phytopathol 2(1):101–132

Schuster SC (2008) Next-generation sequencing transforms today's biology. Nat Methods 5(1):16–18

Sharma KD, Muehlbauer F (2007) *Fusarium* wilt of chickpea: physiological specialization, genetics of resistance and resistance gene tagging. Euphytica 157:1–14

Sharma KD, Winter P, Kahl G, Muehlbauer FJ (2004) Molecular mapping of *Fusarium oxysporum* f. sp. *Ciceris* race 3 resistance gene in chickpea. Theor Appl Genet 108:1243–1248

Sharma KD, Chen W, Muehlbauer FJ (2005a) Genetics of chickpea resistance to five races of *Fusarium* wilt and a concise set of race differentials for *Fusarium oxysporum* f. sp. *ciceris*. Plant Dis 89:385–390

Sharma M, Rajeev Kumar V, Narayan Rao J, Seetha K, David H, Suresh P (2005b) Genetic diversity in Indian isolates of *Fusarium oxysporum* f. sp. *ciceris*, chickpea wilt pathogen. Afr J Biotechnol 8(6):1016–1023

Sharma M, Nagavardhini A, Thudi M, Ghosh R, Pande S, Varshney RK (2014) Development of DArT markers and assessment of diversity in *Fusarium oxysporum* f. sp. *ciceris*, wilt pathogen of chickpea (*Cicer arietinum* L.). BMC Genomics 15:454

Sharma M, Sengupta A, Ghosh R, Agarwal G, Tarafdar A, Nagavardhini A, Pande S, Varshney RK (2016) Genome wide transcriptome profiling of *Fusarium oxysporum* f sp. *ciceris* conidial germination reveals new insights into infection related genes. Sci Rep 6:37353. https://doi.org/10.1038/srep37353

Shehabu M, Ahmed S, Sakhuja K (2008) Pathogenic variability in ethiopian isolates of *Fusarium oxysporum* f. sp. *ciceris* and reaction of chickpea improved varieties to the isolates. Int J Pest Manag 54:143–149

Sindhu J, Singh K, Slinkard A (1983) Inheritance of resistance to *Fusarium* wilt in chickpeas. J Hered 74(1):68–76

Singh H, Kumar J, Haware MP, Smithson JB (1987a) Genetics of resistance to *Fusarium* wilt in chickpeas. In: Day PR, Jellis GJ (eds) Genet plant pathogenesis. Blackwell Scientific Publications, Oxford, pp 339–342

Singh H, Kumar J, Smithson JB, Haware MP (1987b) Complementation between genes for resistance to race 1 of *Fusarium oxysporum* f.sp. *ciceri* in chickpea. Plant Pathol 36:539–543

Singh R, Hasan A, Chaudhary R (2010) Variability in *Fusarium oxysporum* f. sp. *ciceri* causing vascular wilt in chickpea. Arch Phytopathol Plant Protect 43:987–995

Smithson JB, Kumar JB, Haware MP, Singh HP (1983) Complementation between genes for late wilting of race 1 of *Fusarium oxysporum* f. sp. *ciceri* in chickpea. Proceedings of the International Genetic Congress, IARI, New Delhi, India

Soregaon CD, Ravikumar RL, Thippeswamy S (2007) Identification of DNA marker linked to *H2* locus of wilt resistance of chickpea. Indian J Genet Plant Breed 67(4):323–328

Synder WC, Hansen HN (1940) The species in *Fusarium*. Am J Bot 27:64–67

Talboys PW (1957) The possible significance of toxic metabolites of *Verticillium* albo-atrum in the development of hop wilt symptoms. Trans Br Mycol Soc 40:415–427

Taran B, Warkentin TD, Vandenberg A (2013) Fast track genetic improvement of ascochyta blight resistance and double poding in chickpea by marker assisted backcrossing. Theor Appl Genet 126(6):1–9

Tekeoglu M, Tullu A, Kaiser WJ, Muehlbauer FJ (2000) Inheritance and linkage of two genes that confer resistance to *Fusarium* wilt in chickpea. Crop Sci 40:1247–1251

Tekeoglu M, Rajesh PN, Muehlbauer FJ (2002) Integration of sequence tagged microsatellite sites to the chickpea genetic map. Theor Appl Genet 105:847–854

Tullu A, Muehlbauer F, Simon C, Mayer M, Kumar J, Kaiser W, Kraft J (1998) Inheritance and linkage of a gene for resistance to race 4 of *Fusarium* wilt and RAPD markers in chickpea. Euphytica 102(2):227–267

Tullu A, Kaiser WJ, Kraft JM, Muehlbauer FJ (1999) A second gene for resistance to race 4 of *Fusarium* wilt in chickpea and linkage with a RAPD marker. Euphytica 109:43–50

Upadhyaya HD, Haware MP, Kumar J, Smithson JB (1983a) Resistance to wilt in chickpea. Inheritance of late wilting in response to race1. Euphytica 32:447–452

Upadhyaya HD, Smithson JB, Haware MP, Kumar J (1983b) Resistance to wilt in chickpea. II. Further evidence for two genes for resistance to race 1. Euphytica 32:749–755

Upasani ML, Limaye BM, Gurjar GS, Kasibhatla SM, Joshi RR, Kadoo NY, Gupta VS (2017) Chickpea-*Fusarium oxysporum* interaction transcriptome reveals differential modulation of plant defense strategies. Sci Rep 7:7746. https://doi.org/10.1038/s41598-017-07114-x

Varshney RK, Thudi M, May GD, Jackson SA (2010) Legume genomics and breeding. Plant Breed Rev 33:257–304

Varshney RK, Gaur PM, Chamarthi SK, Krishnamurthy L, Tripathi S, Kashiwagi J, Samineni S, Singh VK, Thudi M, Jaganathan D (2013a) Fast-track introgression of BQTLhotspot for root traits and other drought tolerance traits in JG 11, an elite and leading variety of chickpea. Plant Genome 6:1–9

Varshney RK, Mohan S, Gaur PM, Chamarthi S, Singh V, Srinivasan S, Swapna N, Sharma M, Pande S, Singh S (2013b) Marker-assisted backcrossing to introgress resistance to *Fusarium* wilt race 1 and ascochyta blight in C 214, an elite cultivar of chickpea. Plant Genome 7(1):278–289

Varshney RK, Thudi M, Nayak SN, Gaur PM, Kashiwagi J, Krishnamurthy L, Jaganathan D, Koppolu J, Bohra A, Tripathi S, Rathore A, Jukanti AK, Jayalakshmi V, Vemula A, Singh S, Yasin M, Sheshshayee MS, Viswanatha KP (2014a) Genetic dissection of drought tolerance in chickpea (*Cicer arietinum* L.). Theor Appl Genet 127:445–462

Varshney RK, Mir RR, Bhatia S, Thudi M, Hu Y, Azam S, Zhang Y, Jaganathan D, You FM, Gao J, Rieralizarazu O (2014b) Integrated physical, genetic and genome map of chickpea(*Cicer arietinum* L.). Funct Integr Genomics 14(1):59–73

Wang Z, Gerstein M, Snyder M (2009) RNA-Seq: a revolutionary tool for transcriptomics. Nat Rev Genet 10(1):57–63

Watt M, Silk WK, Passioura JB (2006) Rates of root and organism growth, soil conditions, and temporal and spatial development of the rhizosphere. Ann Bot 97:839–855

Winter P, Benko-Iseppon A-M, Hüttel B, Ratnaparkhe M, Tullu A, Sonnante G, Pfaff T, Tekeoglu M, Santra D, Sant JV, Rajesh NP, Kahl G, Muehlbauer JF (2000) A linkage map of the chickpea (*Cicer arietinum* L.) genome based on recombinant inbred lines from a *C. arietinum*×*C. reticulatum* cross: localization of resistance genes for *Fusarium* wilt races 4 and 5. Theor Appl Genet 101:1155–1163

Chapter 9
Biotechnological Approaches for Enhancing Stress Tolerance in Legumes

P. S. Chandrashekharaiah, Vishal Paul, Shivbachan Kushwaha, Debanjan Sanyal, and Santanu Dasgupta

Abstract Legumes are the major sources of protein, dietary fibre, high-quality food & feed. In the world crop production, legumes occupy third place only after cereals and oilseeds. Leguminous crops are known to fix the atmospheric nitrogen for their growth through bacterial association and can be grown in low fertile soils. The requirement of synthetic nitrogenous fertilizers for leguminous crops is far lesser than the other crops, there by reduces the emission of greenhouse gases. Because of these qualities, legumes are considered as potential crops for sustainable agriculture. Due to continuous change in climatic conditions and modern agriculture systems, leguminous crops are frequently exposed to biotic and abiotic stresses. These stresses are considered as major constraints which mainly affects the physiology, metabolism and finally the cultivation and yield of leguminous crops. The yield losses in leguminous crops were accounted for 30–100% based on the magnitude and severity of biological and physical stresses. Hence, it's required to improve productivities of leguminous plants to harness the potential nutritional values. Developing stress tolerant varieties is one of the strategies to overcome this. So far, traditional breeding approaches have been exploited to develop stress tolerant varieties, however, these are laborious and time consuming. The recent establishments of biotechnological tools in model leguminous crops such as *Medicago truncatula* and *Lotus japonicas,* have helped in understanding the regulation and mechanism of action of stress related genes. In the current book chapter, we have discussed about the major biotic and abiotic stresses and the application of various biotechnological tools in developing stress tolerant leguminous plants for improved productivity.

P. S. Chandrashekharaiah · V. Paul · S. Kushwaha · D. Sanyal (✉)
Reliance Industries Ltd., Jamnagar, Gujarat, India
e-mail: Chandra.Shekharaiah@ril.com; Vishal.Paul@ril.com; Shivbachan.K@ril.com; Debanjan.Sanyal@ril.com

S. Dasgupta
Reliance Industries Ltd., Navi Mumbai, Maharashtra, India
e-mail: Santanu.Dasgupta@ril.com

P. Guleria et al. (eds.), *Sustainable Agriculture Reviews 51*, Sustainable Agriculture Reviews 51, https://doi.org/10.1007/978-3-030-68828-8_9

Keywords Legumes · Sustainability · Stress tolerance · Biotic and abiotic stresses · Biotechnology · Gene discovery

9.1 Introduction

Legumes belongs to fabaceae or leguminosae family, which includes ~800 genera and 20,000 species (Lewis et al. 2005) and it is one of the 3rd largest family after orchidaceae and compositae. Legumes are distributed around the world across various climatic conditions from extreme cold to tropical desert and includes plants with wide range of morphological characteristics. Legumes act as valuable sources of food (protein), feed and forage. They are the source of wide range of nutrients that include protein with essential amino acids, complex carbohydrates (~60%), fibres (5–37%), essential minerals and vitamins (Yvonne and Jideani 2017). Based on the season of cultivation, legumes are broadly classified into cool season and warm (tropical) season food legumes (Miller et al. 2002; Toker and Yadav 2010). Some of the examples for cool season legumes includes, lentil (*Lens culinaris*), broad bean (*Vicia faba*), dry pea (*Pisum sativum*), lupin (*Lupinus spp.*), chickpea (*Cicer arietinum*), grass pea (*Lathyrus sativus*) and common vetch (*Vicia sativa*) (Andrews and Hodge 2010). Whereas, cow pea (*Vigna unguiculata*), pigeon pea (*Cajanus cajan*), mung bean (*Vigna radiata*) common bean (*Phaseolus spp.*) and urad bean (*Vigna mungo*) are some of the warm season food legumes (Singh and Singh 2011). Globally around $31 billion tons of legumes are produced annually, (Popelka et al. 2004; Varshney and Dubey 2009) and India is the largest producer of pulses and contributes 25% of global production (http://www.fao.org/india/fao-in-india/india-at-a-glance/en/).

Suboptimal environmental conditions prevent the plants from attaining their maximum growth and yield potentials. (Bray et al. 2000; Rockstrom and Falkenmark 2000). The lower yield during stress is because of the uncongenial climatic conditions and changes in the physiological status of the plants (Shao et al. 2008). Various abiotic stresses are impacting the agriculture productivity worldwide and reported to reduce the yield by >50% in most of the important pulse crops (Wang et al. 2003). Though the legumes are widely grown across the globe and are tolerant to wide range of climatic conditions, still the productivity is majorly influenced by increased biotic and abiotic constraints. The biotic stresses include high pressure of disease, insect and pests (Scheelbeek 2018). Whereas, abiotic stresses include temperature (heat & cold), drought salinity and flood. The abiotic stresses further increase the susceptibility of plants to diseases and pests and causes significant yield loss (Reddy et al. 2004).

Whenever pathogen/insect attacks the plants, the plant defence mechanisms are triggered. These mechanisms are called immune responses, which leads to change in the host gene expression (Kareem and Taiwo 2007). The host defence responses against biotrophic pathogen is mediated by salicylic acid pathway and for necrotrophs and hemibiotrophs it involves the production of ethylene and jasmonic acid (Glazebrook 2005). Various biotechnological tools such as RNA sequences, micro arrays, subtractive hybridization, amplified fragment length polymorphism (AFLP) and complementary deoxy ribonucleic acid (cDNA) AFLP, have been used to identify the genes responsible for tolerance to biotic and abiotic stresses in various crops (Mehrtens et al. 2005). The understanding of metabolism and physiology of legumes along with mechanisms of resistance are required for successful application of these biotechnological tools. The bigger genome size and polyploidy nature in some of the leguminous crops have restricted the developmental approaches. However, the recent biotechnological developments in model leguminous plants (*Medicago truncatula* and *Lotus japonicus*) helped to understand the resistance or tolerance mechanisms to various stresses (Zhu et al. 2005). The development of power full genetic tool box such as genome sequencing (Kato et al. 2003), genetic and physical maps (Pedrosa et al. 2002; Thoquet et al. 2002), isolation of Expressed Sequence Tags (ESTs) (Kulikova et al. 2001; Asamizu et al. 2004) in model plants make these plants valuable for conducting molecular studies related to biotic and abiotic stresses. These two model plants are autogamous, diploids, have shorter generation time, prolific seed production and small genome size (Cook 1999). Besides model leguminous plants, the genome sequences of various leguminous crop plants are also available, and this makes the research easier and more effective. Examples of the pulse crops for which the full gene sequences available includes, cultivated soybean (*Glycine max*), wild soybean (*Glycine soja*), chickpea (*Cicer arietinum*), pigeon pea (*Cajanus Cajun*), adzuki bean (*Vigna. angularis*), mung bean (*Vigna radiata*), wild peanut (*Arachis duranensis*), cultivated peanut (*Arachis hypogaea*), cowpea (*Vigna unguiculate*), common bean (*Phaseolus vulgaris*), lentil (*Lens culinaris*), lupin (*Lupinus angustifolius*) and red clover (*Trifolium pretense*) (https://legumeinfo.org/genomes).

With the advancement of affordable gene sequencing techniques such as Next-Generation sequencing (NGS) and availability of various bioinformatics tools enable the extensive study of genome wide expression and changes during the interactions of the crop with the external factors. These technologies helped to understand the interactions and to identify the pathways involved in defense (Zhu et al. 2005) and in applying the findings from model legumes to other legumes. In this book chapter we are discussing about the various biotic and abiotic stresses observed in leguminous crops, their effects and the management using various biotechnological approaches.

9.2 Biotic Stress in Leguminous Crops

The damages caused by fungi, bacteria, viruses, insects, nematodes and weeds etc., are called as biotic stress (Newton et al. 2011). Among these, fungi, bacteria and viruses affect leguminous crops at various stages of plant growth (Pande et al. 2009). The intensity of the biotic stress depends on the plant resistance, location of

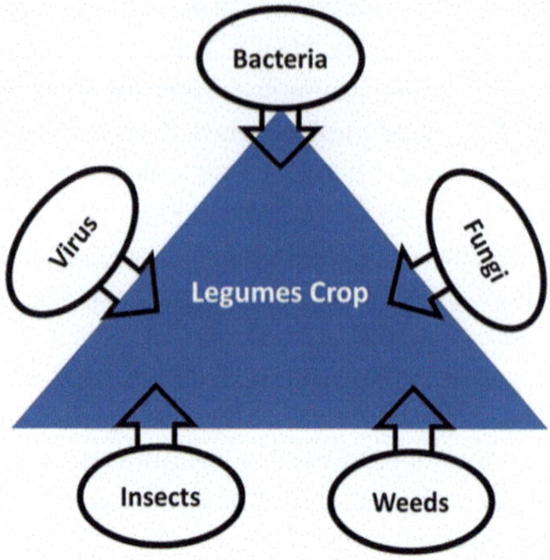

Fig. 9.1 Major biological agents causing biotic stress in leguminous crops

the crop, the disease and pest control measures. The intensive agriculture mostly depends on the use of chemicals for control of pest and diseases. In the modern agriculture system due to extensive use of chemicals (fertilizers, pesticide and insecticides) pathogens and pests have developed resistance and found to cause huge yield losses. The changing climatic conditions and exposure to altered abiotic stresses made leguminous plants more susceptible for biotic stresses (Reddy et al. 2004). The major biotic stresses observed in leguminous crops are briefly discussed and shown in Fig. 9.1.

9.2.1 Fungi

The severity of fungal diseases varies with the years and regions. Few of the fungi were known to affect large cropping areas and found to cause huge productivity and quality loss. The fungal pathogens causing rusts, powdery and downey mildew disease are mainly biotrophs (Sillero et al. 2005). These biotrophic fungi derive required nutrients from the leguminous plants and don't cause much destruction to the plants. However, these biotrophic fungi are prevalent in almost all regions of legume cultivation. (Sillero et al. 2006). The fungal species belongs to *Uromuces* and *Puccinia* causes rust disease in grain and forage legumes. The faba bean rust caused by *Uromyces viciae-fabae* is most important rust disease in European countries, Middle east, China and North Africa and causes considerable yield losses (Sillero et al. 2010). The *U. viciae-fabae* is also known to causes rust disease in other leguminous crops such as vetch, lentil and pea. (Emeran et al. 2005, 2008; Barilli et al. 2011). It has been reported that the *U. viciae-fabae and U. pisi were* known to infect pea in sub tropics and in cool season respectively (Kushwaha et al. 2006). The *U. appendiculatus* is known to cause rust disease in beans all over the world, but more common in humid and sub trophic regions (Miklas et al. 2006).

U. ciceris-arietini causes rust in chick pea, the pathogen was found to cause significant yield losses, when it infects the plant at early stage (Sillero et al. 2012). The soybean rust caused by *P. pachyrhizi* is wide spread in nature and most commonly found in almost all parts of the world (Russia, Japan, Australia, India, China, Africa, Hawaii, South America and USA). The downey mildew diseases are widespread in nature and specifically found in cool climates. Example., *Peronospora viciae* and *P. fabae* causes downey mildew of peas and faba bean respectively. *Erysiphae polygony* causes powdery mildew in most of the legumes. Examples., *E. pisi, E. medicaginis*, *E. vicia-sativa, E. diffusa,* and *Podosphaera phaseoli* causes powdery mildew disease in pea, alfa-alfa, vetches, soybean and cow pea respectively (Falloon and Viljanen-Rollinson 2001; Trabanco et al. 2012).

The fungi which kills the plant and feeds on the plant dead material called necrotrophs. The necrotrophic fungi which causes diseases in legumes are anthracnose, *Ascochyta blight* and chocolate spot (Tivoli et al. 2006). The botrytis gray mold is a serious and destructive disease of chick pea and lentil and is caused by *Botrytis cinereal* (Davidson et al. 2004; Pande et al. 2005). This, fungi had caused epidemics in several counties such as, Bangladesh, Australia, Argentina, Nepal, India, and Pakistan (Stoddard et al. 2010). The anthracnose diseases caused by *Colletotrichum truncatum* in lentil, *Colletotrichum lupini* in lupin are economically important diseases and were found to mainly appear in warm and humid climates (Buchwaldt et al. 2004; Mackie et al. 2007). The lesions and crown rot caused by these fungus are major bottlenecks in these crops.

The soil borne diseases of fungi are mainly classified into stem wilt, vascular wilt and damping-off/root rot (Infantino et al. 2006). The effects of these soil borne diseases are influenced by environmental conditions. The cooler climatic conditions mainly favour the root rots and damping-off (*Rhizoctonia solani, Aphanomyces euteiches*, *Phytophthora spp.* and *Verticillium spp.*) and these are mostly seen in Northern European regions and North American latitude (Kraft and Pfleger 2001; Chen et al. 2011 Vandemark et al. 2006). The wilt and root rot diseases are caused by *Fusarium* spp. and *Macrophomina phaseolina* respectively. These pathogens are favoured by dry or warm climatic conditions (Chaudhary et al. 2009; Afouda et al. 2009) or drought conditions. The intensity of these diseases would vary with the changing climatic conditions and needs to be addressed properly to avoid the huge yield losses. Some of the economically important fungal diseases and yield losses observed in various leguminous crops are shown in Table 9.1.

9.2.2 Bacteria

The bacteria are third most important pathogenic microbes causing significant yield losses after fungi and viruses. Bacteria causes various destructive diseases in legumes such as wilts, cankers, soft rots, stem and leaf spots. The Gram-negative bacteria (*Ralstonia solanacearum, Pseudomonas* sp., *Erwinia* sp.,) mainly causes these diseases and use type III secretion system to release effector protein into the

Table 9.1 The major fungal diseases of some important leguminous crops

Leguminous crop	Name of the disease and causative agent	Effect of biotic stress caused by fungi	References
Chick pea	Fusarium wilt: *F. oxysporum* Ascochyta blight: *Ascochyta rabiei* Botrytis grey mold: *Bortytis cinerea*	Wilting affect seed setting and pod formation. Sometimes causes 50–100% yield loss	Nene and Reddy (1987)
Field pea	Powdery mildew: *Erysiphe pisi* Rust: *Uromyces* sp. Downey mildew: *Pernospora viciae*	Yield loss up to 10–30%	Sillero et al. (2005)
Lentil	Fusarium root rot: *Fusarium* spp.	Affect seedling and crop growth. Causes 20–70% yield loss	Schneider et al. (2001) and Hamwieh et al. (2005)
Pigeon pea	Wilt: *Fusarium* oxysporum, Stem blight: *Phytophthora* sp. Leaf spot: *Alternaria* sp. Powdery mildew: *Oidiopsis taurica*	Causes wilt, yellowing of plant and affect the yield by 20–70%	Reddy (2006)
Soybean	Asian rust: *Phakopsora pachyrhizi*	This disease observed all over the world and causes severe yield loss	Carmona et al. (2005)
Cowpea, urd bean and mung bean	Leaf spot: *Cercospora* sp. Powdery mildew: *Erysiphe polygoni* Root rot: *Rhizoctonia solani*	10–100% yield loss	Rubiales et al. (2002)

host cells (He and Jin 2003). The bacterial diseases found in legumes are broadly categorized into leaf spots, leaf blights, wilts, dwarfing and rots. The bacteria such as *Xanthomonas axonopodis* or *Pseudomonas syringae* mostly causes the blight or spots in legumes and found to infect >1 species of legumes across the world. Example: *X. axonopodis* pv. *glycines, Xanthomonas axonopodis* pv. *phaseoli, X. axonopodis* pv. *vignicola* causes blight diseases in soybean, common bean and cow pea respectively. These diseases were found more severe under high rainfall and high temperature conditions. Higher yield losses were observed when infection occurs at early stage of plant growth due to defoliation. The pea leaf blight caused by *Pseudomonas syringae* pv. pisi is a seed borne disease, the different pathovars were found to cause different diseases for example, *P. syringae pv. phaseolicola* causes halo blight and *P. syringae pv. syringae* causes brown spot in common bean (Arnold et al. 2011). Some of the pathogens such as *P. syringae* pv. *phaseolicola* is a broad-spectrum pathogen and can infect several different legumes such as pigeon pea, mung bean and many wild beans (Taylor et al. 1996). These reports indicate that the same pathogen can cause diseases in many other crops and pathogen was not found host specific. The wilt diseases are mainly caused by Gram-positive bacteria. The *Curtobacterium flaccumfaciens* pv. *flaccumfaciens* causes wilts in cow pea, soybean and common bean. The main symptoms of the wilt include irregular broad yellowing from leaf margin followed by necrosis. The pathogen was found to colonize the seed through vascular tissue (Tegli et al. 2002). Bacterial wilt caused by *Ralstonia solanacearum* is another important bacterial wilt disease which affects the yield of peanut in various countries (China, Vietnam, Uganda and Indonesis) (Jiang et al. 2007). The *R. solanacerum* pathogen takes an entry into plants through wounds formed by nematodes, insects or through the secondary roots. The pathogen can also colonize in the root and alter the root tip of *M. truncatula* (Turner et al. 2009). After successful entry of bacteria into plant cell, the bacterial cells multiply at faster rate and produces slime which blocks the plants phloem and cause wilt. The *Clavibacter michiganensis* ssp. *Insidiosus,* causes wilt in alfalfa and enters the plant root/stem wounds. The infected Alfalfa plant shows spindly stems, stunted growth, distorted leaves (Eichenlaub and Gartemann 2011) and transfer to other healthy plants through seed materials (Cormack and Moffatt 1956; Samac et al. 1998). Some of the biotic stress causing bacteria in some of the important pulse crops are shown in Table 9.2.

Table 9.2 The major bacterial diseases of some important leguminous crops

Leguminous crop	Name of the disease and causative agent	Effect of biotic stress caused by bacteria	References
Green gram and black gram	Mung bean phyllody-16Sr II-D phytoplasma	Plants commonly show symptoms that include phyllody, yellowing virescence and stunted growth.	Akhtar et al. (2010) and Bertaccini et al. (2005)
Lentil	Ascochyta blight: *Ascochyta lentis*	Irregular lesions on petioles, leaves and stem. Darker brown spots on pods	Tullu et al. (2010)
Soybean	Bacterial blight: *Pseudomonas syringae pv.* glycinea	Brown spots on the margins of the cotyledons, affects seed quality. Yield losses of 4–40% has been recorded	Jagtap et al. (2012)
Dry bean	Common bacterial blight (CBB): *Xanthomonas axonopodis pv.* phaseoli	Affects yield and seed quality. Yield losses of 22–40% has been reported	Gillard et al. (2009) and Fininsa and Tefera (2001)
Cow pea	Bacterial blight: *Xanthomonas campestris* pv. vignicola	Necrotic, light brown lesions of irregular spots appear on the leaf. Disease affects the crop growth and finally the yield	Constantin et al. (2016)

9.2.3 *Viruses*

Viruses are obligate parasites and multiply only inside the host cells. The diagnosis of viral disease is very difficult because of the symptoms are very similar to those which occur due to abiotic stress (nutrient, temperature and herbicide) (Burrows 2012). There are two groups of viruses which affects the legumes, the first group mainly causes mosaic or mottle symptoms on the leaves of legumes. Examples includes, Cucumber mosaic virus (CMV), Bean yellow mosaic virus (BYMV), Pea seed borne mosaic virus (PSbMV). The other groups of viruses affect the growth, induces yellow or reddening of leaves and affects the pod and seed setting. Most of these viruses causes major economic losses in some parts of the world and complete crop failure in some cases. Example. Bean leaf roll virus, chickpea chlorotic stunt virus (Makkouk and Kumari 2009; Makkouk et al. 2014). The virus diseases affect significantly the legumes production in tropics and subtropics regions (Sastry and Zitter 2014). The food legumes are known to get infected by >150 viruses of different genera (ICTV 2012). The impact of virus diseases depends on the crop cultivar, strain or pathotype of virus and influenced by climatic conditions. The time and duration of infection also influences the yield loss and severity. The sterility mosaic (SMV) virus infestation in pigeon pea before 45 days of crop growth causes 95–100% yield loss, whereas in older plants (>45 days) the losses were accounted up to 26–97% (Guptha et al. 2014). In cowpea there are > 140 viruses are reported and among them 20 have wide spread occurrence. Due to viral diseases, the losses in cow pea were accounted up to 10–100% (Kareem and Taiwo 2007).

The peanut plant was known to get infected by > 31 types of viruses. However, out of which only few were reported to cause economic damages (Nigam et al. 2012) across the globe. Virus diseases of peanut varies from region to region. In Africa the groundnut rosette disease is most common while in United states of America, tomato spotted wilt virus is most prevalent. The economically important peanut viruses such as bud necrosis, peanut stripe virus, cucumber mosaic virus was found specifically in South Asia, East and Southeast Asia, Argentina and china respectively (Nigam et al. 2012).Some important viruses found in leguminous crops are, cowpea chlorotic mottle virus (CPCMV), cucumber mosaic virus (CMV), cowpea severe mosaic virus (CPSMV), cowpea mosaic virus (CPMV), cowpea aphid-borne mosaic virus (CABMV), cowpea mild mottle virus (CPMMV), cowpea chlorotic mosaic virus (CCMV), mung bean yellow mosaic India virus (MYMIV), bean common mosaic virus (BCMV), soybean mosaic virus (SbMV), peanut mottle virus (PeMoV), alfa-alfa mosaic virus (AMV), cucumber mosaic virus (CMV). For most of these viruses, insects like aphids, white fly, thrips, jassids acts as a vector (Loebenstein and Thottappilly 2003; Alegbejo and Kashina 2001). The arthropod insects were reported to transmit ~75% of the 700 types of viruses (International Committee on Taxonomy of Viruses, Hogenhout et al. 2008). Sometimes the outbreak of viral diseases can be forecasted by monitoring the aphid/insect populations.

9.2.4 Insect Pests

Insects are considered as most important biotic stresses of legumes all over the world as compared to other biotic stresses (Edwards and Singh 2006). Apart from feeding they also act as vectors for many pathogens (Edwards and Singh 2006). The type of insects varies with climatic conditions (dry and cool season). The insects such as pod borers (*Helicoverpa armigera* and *H. punctigera*) and aphids (*Aphis glycine*) majorly appear in cool season legume crops (Yoshida et al. 1997). The weevils (*Zabrodes subfasciatus* and *Apion godmani*) appear in warm season legumes (Garza et al. 1996). The pod borer and cut worms known to cause 10–90% yield loss in chick pea. In pigeon pea the pod borer and leaf roller infestation cause yield losses up to ~70–80%. White fly, jassids and pod borer are known to cause 25–55% yield loss in cow pea, urad bean and mung bean (Das 2008; Pande et al. 2009; Chandrashekar et al. 2014). Chewing insects are one of the major class of insect pests, in this the most important group is the storage insects (bruchid beetles) such as *Callosobruchus maculatus*, *Callosobruchus chinensis*, *Callosobruchu analis*, *Bruchus incarnates* and *Acanthoscelides obtectus* (Keneni et al. 2011) cause serious damages to the stored legume seeds. Due to their short generation time and high fertility rates (Southgate 1979) the losses caused by storage pests are very significant. The damages caused by these storage insects affects the seed germination, quality and make these seeds unfit for consumption purpose. Another group of chewing pest targets the pod in the field and have reported to cause major losses in various legumes for example, *Nezara viriolvla* in *Glycine max* (Soybean), *Helicoverpa armigera* in chick pea and pigeon pea (War et al. 2013; Acharjee et al. 2010) and *Maruca vitrata* in cowpea (Higgins et al. 2012). Sucking pests mainly feeds on xylem or phloem and other various parts of plant cells. The major sucking pests of legumes includes leafhoppers, psyllids, whiteflies, aphids and mirids. Out of these, the aphids are observed as major sucking pests and damage the crop by directly feeding and act as potential virus transmitters. Examples of major aphids includes *Aphis craccivora* in cow pea, *Aphis glycine* in soybean, *Therioaphis trifoli* in spotted alfalfa, *Acyrthosiphon pisum* in pea (Kamphuis et al. 2012). The soybean aphid observed for the first time in North America was reported to cause $1.6 billion loss in 2008 (Kim et al. 2008).

9.2.5 Nematodes

Nematodes are described as biotrophic obligate parasites and polyphagous in nature. Nematodes modify cellular responses of crop plants and coexist with the hosts. The plant parasitic nematodes (PPNs) cause various diseases in leguminous crops and know to degrade the plant parts (De Coninck et al. 2015). The major parasitic nematodes of pulses are cyst and root-knot nematodes (*Meloidogyne incognita* and *Meloidogyne javanica*), pigeon pea cyst nematode (*Heterodera cajani*), reniform nematode (*Rotylenchulus reniformis*), and root lesion nematode (*Pratylenchus thornei*) (Rubiales et al. 2015). Overall, nematode infection is known to reduce the yield of legumes by 11–12% annually. The cyst and root knot nematodes were found to cause losses up to > 90 billion dollars worldwide annually (Dhandaydham et al. 2008). In chick pea PPNs were found to reduce yields up to ~14% (Castillo et al. 2008). The root not nematode *Heterodera cicero* was known to cause significant yield losses in semi-arid pulse cultivation. The eggs of this nematodes get hatched in the presence of root exudates of chick pea without undergoing hibernation (Castillo et al. 2008) and causes new infection. Another root knot nematode, *Meloidogyne* spp. are endoparasites, obligate in nature and present in various climatic conditions. There are > 70 species of *Meloidogyne* have been reported, out of these only four species. (*M. incognita, M. arenaria, M. javanica and M. hapla)* were found to infect >95% of plants (Sasser et al. 1983). The root knot nematode infection is sometimes associated with fungal wilt (*Fusarium* sp.) (France and Abawi 1994; Castillo et al. 2003). *Heterodera glycine* is the most important nematode pest of soybean which is found to affects the production across the globe (Koenning et al. 1999) and because of this yield losses up to ~$430 million was observed per year in the USA (Wrather et al. 1997). Apart from soybean, this pathogen was found to have broad host range such as peanuts, pigeon pea, pea, and bean (Koenning et al. 1999; Dhanbaydham et al. 2009). The clover cyst nematode *Heterodera trifoli*, is most commonly found in North America and was found to reduce clover yield by ~12–34% (Dhandaydham et al. 2008).

9.2.6 Weeds

Weeds are problematic and unchecked weeds cause yield losses up to 20–90% in many pulse crops (Pooniya et al. 2015). Weeds compete for the nutrient and acts as hosts for many insect-pests, nematodes and pathogens. For example, *Chenopodium album* L, is an alternative host for caterpillars in peas and pigeon peas and is a main host for *Agrotis ipsilon* (greasy cut worm) in peas and grams (Das 2008). Weeds like *Melilotus indica, Chenopodium album, Avena ludoviciana* are reported to cause 20–25% yield loss in chick pea. In cowpea, mung bean and urdbean the weeds such as *Cynodon dactylon, Cyprus rotundus, Amaranthus* spp., *Bedens Pilosa, Physalis minima* etc., were known to cause yield losses of 50–90% (Das 2008; Pande et al. 2009; Chandrashekar et al. 2014). The broomrape species *Orobanche* and *Phelipanche* were observed as dominant parasitic weeds in cultivation of various leguminous crops. *Orobanche crenata* was found as major limitation in Mediterranean basin and Middle East region, for grain and forage legume cultivation and was found to cause major economic losses in clover seed production. Whereas, the *Phelipanche aegyptiaca* is an important weed in legume as well as in some other vegetable crops in Asia and Western Mediterranean region and reported as very problematic to faba bean and common vetch in the regions of Tunisia and Morocco respectively (Parker 2009; Rubiales and Fernández-Aparicio 2012). In

Table 9.3 The major weeds observed in some of the important leguminous crops and the extent of yield loss

Leguminous crop	Most commonly found weeds	Extent of yield loss (%)	References
Field pea	*Chenopodium album* *Circium arvense*, *Anagalis arvensis* *Avena* spp.,	15–67	Satyagopal et al. (2014a, b), Pande et al. (2009), Das (2008), and Chandrashekar et al. (2014)
Lentil	*Chenopodium spp.*, *Avena spp.*, *Guizotia scabra* , *Phalaris spp.*, *Fumaria parviflora*	50	
Pigeon pea	*Eclipta alba*, *Euphorbia parviflora* *Portulaca oleracea*, *Celosia argentea*, *Commelina benghalensis*	30–90	

grain legumes such as cowpea, considerable yield loss was observed due to *Striga gesnerioides* and *Alectra vogelii* in sub-Saharan area of Africa. The complete parasite *Cuscuta* spp. was found to affect growth of some legumes in some areas (Riches et al. 1992; Parker 2009; Rubiales et al. 2006) (Table 9.3).

9.3 Management of Biotic Stress by Biotechnological Approaches

The biotechnological approaches are very much necessary to resolve the issues caused by biotic and abiotic stresses to minimize the losses. This would require extensive biological information of leguminous crops. The application of the biotechnological tools to improve the tolerance or resistance of most of the leguminous crop is limited owing to the large genome size and polyploidy. The identification of model leguminous plants helped to apply the genetic tools and to develop the improved varieties (Cook 1999; Handberg and Stougaard 1992). Some of the synthetic biology approaches which are applied for legume crop improvement are briefly discussed below.

9.3.1 Molecular Markers

The markers (DNA) or quantitative trait loci (QTL) which are linked with the agronomic traits would be identified by genome analysis. The molecular techniques such as amplified fragment length polymorphism (AFLP), random amplified polymorphism (RAPD), restriction fragment length polymorphism (RFLP), simple sequence repeat (SSR) and derivatives have been reported both for identification of markers or QTL related to biotic (Roman et al. 2002; Ouedraogo et al. 2002) and abiotic (Lee et al. 2004; Kassem et al. 2004) stresses. The identification and analysis of these markers provides various information about the presence of target gene on the chromosome, loci and the gene interactions. The marker-based techniques are much faster and very effective when the trait is controlled by many genes. Identification of molecular markers for diseases and pest resistance will reduce the dependency on phenotypic data. Identification of pest or insect resistance by phenotypic observation would be laborious and time consuming in large scale trials (Chen et al. 2007). Molecular markers can be used to transfer insect or pest resistance from one plant species to other. The trait which is linked with the marker could be precisely transferred to the cultivar through marker assisted breeding programme (MAS) (Yu et al. 2004). MAS programme was successfully applied in soybean to develop the nematode (Diers 2004) resistant varieties and in pinto bean for bacterial blight resistance (Mutlu et al. 2005). The RAPD markers linked, 11 QTL genes (Ur-1-11) were used to develop the rust resistance and other key diseases (Bean common mosaic, Bean golden mosaic virus, bacterial blight and anthracnose) in common bean. In some cases, the MAS was combined with the new genes from other species to induce the resistance. This was demonstrated in soybean to develop the resistance to soybean looper (*Pseudoplusia includens*) and to corn earworm (*Helicoverpa zea*). In this technique the insect resistance QTL and synthetic Bt genes (Cry 1 AC) were combined to develop the resistant varieties. The development of resistant varieties through gene pyramiding by MAS approach could provide a long-term durable resistance. The molecular marker techniques have been successfully applied to develop bruchid resistant varieties in mung bean. The molecular techniques were used to identify the bruchid resistant wild type mung bean plants, genes responsible for resistance and developing molecular markers (Schafleitner et al. 2016; Somta et al. 2007).

The QTLs for Ascochyta blight disease caused by *Didymella pinodes* and *Phoma medicaginis* var *pinodella* were identified. These QTLs were mapped into two populations such as A26 × Rovar and A88 × Rovar (Timmerman-Vaughan et al. 2016). The UpDSII and UpDSIV QTLs were mapped for *Uromyces pisi* which causes rust diseases in pea. These two QTLs were identified in linkage groups of II and IV which control resistance to *Uromyces*. (Barilli et al. 2018). The soil borne fungi *Verticillium alfalfae* causes wilt disease in alfalfa, using molecular markers such as SSRs, SNPs, 17SNP, the marker linked to pathogen wilt resistance was identified. (Zhang et al. 2014). The oomycete fungi, *Phytophthora sojae* causes root rot disease in soybean which results in yellowing of leaves, damping-off and wilting of plants. In the cultivar Qichadou, the QTL which confers résistance to *Phytophthora sojae* (RpsQ) was identified using SSR marker and mapped to 118-kb region of chromosome no 3 of soybean. This mapped region consists 11 disease resistance genes, out of these the gene Glyma.03g027200, was found to encode threonine receptor-like kinase (RLK) and later it was proved to induce resistance to phytophthora root rot (RpsO) (Lee et al. 2017). Root- knot nematodes (RKN) are the major biotic stress in cow pea. Two resistance genes such as Rk and Rk2 were previously reported for RKN in cowpea. The QTL study using populations of resistance inbred line (524BxIT84S-2049) resulted in identification of QRK-vu9.1 QTL associated with *Meloidogyne javanica* and it was mapped on to LG9 linkage group at 13.37 cM position (Santos et al. 2018).

9.3.2 *Somaclonal Variations and Mutagenesis*

The development of plants through tissue culture approach (embryo cultivation or callus regeneration) has the chances of generating agronomically superior traits (Larkin and Scowcroft 1981) and was reported in pea (Griga et al. 1995) and pigeon pea (Chintapalli et al. 1997). The variations generated through somatic approaches may not be useful for massive multiplication through micropropagation but can be used as base material in classical breeding programmes. The application of chemical (ethyl methane sulfate) and physical (UV and X ray) mutagenesis approaches are more common in non-pulse crop like cereals for developing abiotic (Bhagwat and Duncan 1998) and biotic (Khan et al. 2001; Fuller and Eed 2003) stress tolerant varieties. Due to difficulty in regeneration efficiency in pulse crops, only the mutagenesis approach has been explored to improve the nitrogen fixation (Sagan et al. 1994) rather than disease and pest resistance. The tissue culture approaches must be improved specially in leguminous crops for developing the disease resistance varieties.

9.3.3 *Transformation*

The use of *Agrobacterium tumefaciens* (De Clercq et al. 2002; Li et al. 2004) and particle bombardment techniques for transformation of targeted genes into embryogenic or organogenic cultures of leguminous crops for inducing resistance is an important breakthrough. The transfer of insect resistant genes from bacteria (*Bacillus thuringiensis*) (Walker et al. 2004), virus (Aragao et al. 2002) and genes encoding proteins (phytoalexins) were transferred into leguminous crops using *A. tumefaciens* transformation techniques (He and Dixon 2000; Samac et al. 2004). The technique has been demonstrated in soybean to develop the insect resistance lines by transferring *cry A* gene from *B. thuringiensis.*

9.3.4 Gene Expression Analysis

Control of biotic stress in plants can be achieved by transcriptional expression or repression of target genes. For this differential expression of genes in response to stress needs to be identified and studied. There are many techniques were developed to achieve this for example., microarrays (Schena et al. 1995), suppression subtractive hybridization library (Diatchenko et al. 1996), serial analyse gene expression (Velculescu et al. 1995), quantitative measurement of transcription factors (TF) and expression patterns in addition to Northern blotting. The expression studies in legumes during pathogen attack revealed that there were large number of genes that are differentially expressed and more precisely certain family of defense genes were expressed irrespective of pathogens. Ex., PR-proteins (PR-10), glucanase, chitinase (Salles et al. 2002), lipoxygenases (Lox genes) (Cho and Muehlbauer 2004;) and phytoalexins such as medicarpin (He and Dixon 2000).

9.3.5 Transcriptional Factors

Transcriptional factors (TF-s) controls the biochemical and stress response gene expression (Eulgem 2005). The TF-s were extensively studied in *Arabidopsis thaliana* for various stress factors. Dr. M. Udvardi from Max Planck Institute for Plant Physiology, Golm, Germany has established the platform for TF-s in leguminous crops. The DREB/CREB proteins (Yamaguchi-Shinozaki and Shinozaki 2005; Singh et al. 2002) and ethylene-responsive element-binding factors (ERF) (Anderson et al. 2005) are the major TF-s reported in *Medicago truncatula*. Each family of these genes has specific roles. The ERF family TF-s are responsible for both biotic and abiotic stresses. The WRKY family involves in regulation of receptor protein kinases (Asai et al. 2002; Robatzek and Somssich 2002), bZIP family members regulate PR-1 and glutathione S23Tranferase genes (Fan and Dong 2002; Chen and Singh 1999). The genetic transformation of TF-s of different family is attractive to achieve the resistance or tolerance to certain stresses (Eulgem 2005; Yamaguchi Shinozaki and Shinozaki 2005). Over expression or knock out of TF-s also have been studied in many crops for achieving specific phenotypes (Kim et al. 2004; Onate-Sanchez and Singh 2002).

9.3.6 Gene Knock out Approaches

Gene knock out or suppression of the gene expression techniques are powerful tools for determining the function of gene. The several techniques used are targeted gene replacement, anti-sense RNA suppression, insertional mutagenesis, targeted induced local lesion and gene silencing. Post transcriptional gene silencing would be done by using RNA silencing (Baulcombe 2000). It has been used to study the selective gene knockouts. Phenotypic instability in future generations and the requirement of reliable transformation systems (Hannon 2002) are the main limitations of this technology. In plant system, RNA silencing is a natural process against a virus infection. Based on this principle the virus induced gene silencing technique (VIGS) has been developed to inhibit the expression of host gene through virus vector infection (Baulcombe 2000; Britt and May 2003). VIGS based technique was used in pea for control of early browning virus (Constantin et al. 2004; Van den Boogaart et al. 2004)

9.3.7 Gene Editing Approaches

Transfer of genes from one variety to other through breeding approaches always associated with the inheritance of undesirable traits. Hence the plant breeding associated with precise gene editing technologies helps in achieving resistance to various pathogens very precisely without any other undesirable traits. The technologies such as, activator-like effector nucleases (TALENs), zinc finger nucleases (ZFNs), clustered regularly interspaced short palindromic repeat (CRISPR)-associated protein9 (Cas9)-based genome editing tools (CRISPR/Cas9) are becoming popular to generate target specific mutations. These tools create mutations based on the pairing of single guider RNA (ssg RNA) to the target DNA site (Cong et al. 2013; Jiang et al. 2013). In legume using CRISPR/Cas9 (Michno et al. 2015) the methodology was developed to edit gene in *M. truncatula* (Curtin et al. 2018). CRISPR/Cas 9 approaches were used to study the virulence of some important pathogens (*Sclerotinia sclerotiorum* and *P. sojae*) in pulses (Li et al. 2018a; Fang and Tyler 2016). Rodriguez-Leal et al. (2017) designed a genetic scheme which exploits heritability of Cas9 activity in heterozygous mutant's background. This kind of system can be used to screen the QTLs for disease resistance and it would be a valuable tool in breeding program. Even though the gene editing tools are highly precise much work has not been done with leguminous crops to develop a disease resistance, we anticipate these tools would be used more frequently in near future.

9.3.8 Proteomic and Metabolomic Study

The qualitative and quantitative protein profiling with respect to stress is very much essential to identify the specific type of protein production (Gygi and Aebersold 2000). The translation of mRNA does not always produce the targeted protein and its affected by post translational modification (Canovas et al. 2004). Both transcriptomic and proteomic data are essentially required to decode a complex biological process along with metabolomic studies (Sumner et al. 2003, Dixon 2001). This kind of holistic approach was studied in *M. truncate* for various stresses (Bell et al. 2001) and in *L. japonica*, to study the host and symbiotic nitrogen fixing bacterial interaction (Rispail 2005; Desbrosses et al. 2005). Even though the large-scale metabolic study is difficult, the targeted approach was applied to study the production of various substances during the stress. The specific metabolites accumulation was observed in response to pathogen and elicitor treatment in various plants EX., medicarpin, pisatin, glyceolin or sativan in alfalfa, pea, soybean and *L. japonicus* respectively (Saunders and O'Neill 2004; Shimada et al. 2000; Borejsza Lozovaya et al. 2004). The spider mite infection in *L. japonicus* has induced the production of terpenoids which are volatile in nature. These compounds are believed to attract natural predators (Ozawa et al. 2000) to controls the mite. As a defence reaction accumulation of triterpene saponins in response to insect attack was studied in alfalfa (Agrell et al. 2004) and these compounds were also known to have anti-insect, antimicrobial, and allelopathic properties (Dixon and Sumner 2003). The metabolic approaches in combination with transgenic are necessary since most of the metabolic pathways are connected to each other and are highly complex. The biotechnological modification of any one pathway affects the other pathways and can create synergistic or antagonistic impacts. Therefore, the metabolic analysis is very much important for plant growth and development under various environment and stress conditions. (He and Dixon 2000; Wu and Van Etten 2004).

9.4 Functions of R Gene and Receptors in Disease Resistance

From the different plants system there were ~40 R genes have been identified and isolated so far (Martin et al. 2003), out of these only two are from legumes. The R genes isolation from legumes is difficult due to lack of good transformation system, unavailability of genetic map and the polyploidy nature of legumes. The identification of model leguminous crops and development of genetic toolbox has improved the isolation of more R genes. The sequence similarity study using known R genes with the conserved motifs resulted in the identification of many R genes in legumes. (Zhu et al. 2002; Yu et al. 1996) but the biggest constraint is mapping of disease resistance gene to pathogen. For, example the Rpg1-b gene in soybean confers resistance to pathogen *Pseudomonas syringae pv glycinea* which carries virulent gene (avrB gene) and causes blight disease (Ashfield et al. 2004). The Rpg1-b genes belongs to the nucleotide binding Leu-rich repeat (CC-NB-LRR) class of R genes and does not share much similarity with RPM1 R genes found in *Arabidopsis* which is also conferring resistance to *P. syringae* expressing avr B gene. Phylogenetic analysis of these two genes suggested that, these two genes are not orthologus but have the specificity to avrB genes. The study and isolation of four (Rps1-k-1, Rps1-k-3 and Rps1-k-4) highly similar R genes from Rps1-k locus found to confer resistance to *Phytophthora sojae* causing root and stem rot (Gao et al. 2005). The gene Rps1-k-1 shares 89.9% amino acid sequence similarity with other three genes. Characterization of R genes provided the information on interaction of R genes with pathogen and symbiont, for example the Rj2 locus which is responsible for nodulation in soybean was found to induce resistance to a pathogen *P. sojae* and powdery mildew. The sequence analysis of 120 kb region from nodulating genotype of soybean shown no evidence of nodulating genes in the plant, instead it was entirely composed of R genes (Graham et al. 2006). These results prove that R genes not only induce resistance but also involve in recognizing beneficial symbionts.

Zou et al. (2005) studied the response of soybean 'Williams 82' cultivar to virulent and a-virulent strains of *Pseudomonas syringae pv glycinea* differing in avrB gene. The pathogenicity was mainly due to the presence of avrB gene in the bacteria. They have used 27648 cDNA arrays for studying the response of plant to the pathogen. In the study, they found that 45 genes were mapped for susceptible response. Where, in case of resistance response 2000 and 1000 genes have differentially expressed after 8 and 24 h of inoculation of pathogen respectively. Many of these genes were also found to express in biotic and abiotic stresses. In the study they found that, the genes associated with chloroplast and phenylpropanoid pathway were down or up regulated. The chloroplast associated genes were down regulated in response to oxidative stress which has resulted in production of reactive oxygen species and shown the hypermediated response in plants. The genes which regulate the production of flavone and isoflavone compounds from the phenylpropanoid pathway were upregulated during the resistance response and these compounds were believed to have antioxidant and antimicrobial activity (Zabala et al. 2006).

Iqbal et al. (2005) studied the response of soybean plant to the fungal infection *Solani* f. sp. glycines, which causes sudden death syndrome (SDS). The susceptible variety (EssexX) shown the death symptom, whereas the inbred line which was partially resistant and carrying six quantitative trait loci shown resistance to SDS. During the inoculation study with leguminous plant containing Response Induced Gene (RIL123), there were total 81, 88 and 129 genes have upregulated by at least 2 or sometimes many folds after 3, 7 and 10 days of inoculation of pathogen respectively. In the inbred line, after 10 days of disease infestation, up-regulations of several genes involved in phenylpropanoid pathway was noticed. This study indicates that the involvement of pathogen in inducing resistance to SDS.

Moy et al. (2004) studied the transcriptomic profile of the susceptible soybean plant to the *P. sojae*. In the infection study it was found that after 3 and 6 h of inoculation of pathogen, 22 and 97 genes were upregulated respectively. The genes which have upregulated after 6 h of infection were belongs to 1a (PR), which is a pathogenesis related protein produced in phenylpropanoid metabolism and oxidative stress. Between 12 and 48 h of inoculation, genes which are involved in defence mechanism, phytoalexin metabolism and signalling proteins were up regulated. The genes such as peroxidases and lipoxygenase were down regulated, whereas genes involved in terpenoid metabolism were found unaffected. Isoflavones synthesis was found more dominant in R-gene mediated resistance. Subramanian et al. (2005) down regulated the isoflavone synthase genes in roots of soybean by RNAi approach and decreased the isoflavone synthesis by ~95%. The decreased isoflavone was found to increase the susceptibility of the host to the pathogen.

The infection of *Medicago truncate* plant with *Erysiphe pisi* causing powdery mildew shown the expression of genes involved in basal defence and hyper reaction

(Foster-Hartnett et al. 2007). The genes involved in PR protein belongs to PR10 family (PR1 EDS1) and b-1,3-glucanase glutathione S-transferase. These genes were upregulated in resistant and moderately resistant genotypesand plant shown salicylic acid response to pathogen infection. The legumes are the main sources for several subclasses of PR-10 family proteins, some of these proteins have ribonuclease and antifungal activities (Bantignies et al. 2000; Chadha and Das 2006) and confer resistance against the pathogen.

9.5 Abiotic Stress in Leguminous Crops

The major abiotic stresses (Fig. 9.2) comprise of salinity, drought, water logging, nutritional deficiency, high temperature and low temperature. In addition to the major stresses, there are some minor abiotic stresses which are sporadic and highly limited in occurrence. Minor stresses includes shade, ultra-violate light (UV), photo inhibition, air pollution, windstorm and hail storm.

9.6 The Complex Plant Response to Abiotic Stress

Abiotic stress tolerance is a complex mechanism in plants. It comprises of multiple gene working together to cope with a stress. The complexity further increases when same or different stress is present at different stages of plant. The story doesn't stop here, and the complexity of stress response increases again with multiple stress at the same time. Plants has also evolved to simplify the stress response by imparting same signalling response pathways to different and multiple stress.

The plant response to abiotic stress is a multifaceted response involving multiples genes, biochemical reactions and molecular pathways working together (Seki

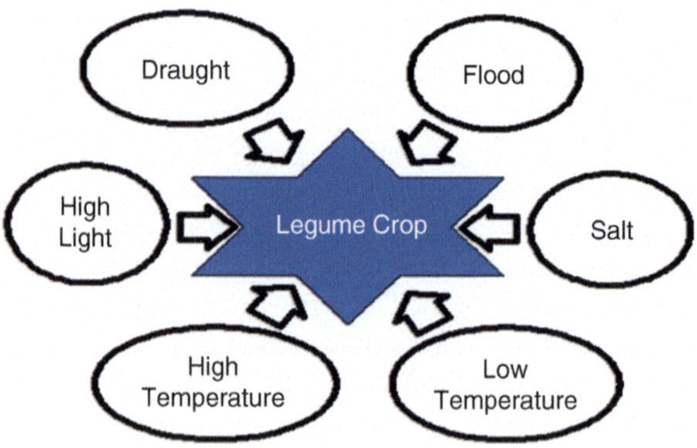

Fig. 9.2 Major physical factors causing abiotic stress in leguminous crops

et al. 2003; Avnioktem et al. 2008). To understand the mechanisms behind stress tolerance, it is important to analyse stress-inducible genes so that it can be improved by gene manipulation through synthetic biology approach. Stress may occur at different and/or multiple stages of development and at the same time more than one stress can also occur. Acquiring the signal of abiotic stress and then transducing the signal to switch on the plant response are the critical steps responsible for determining the survival as well as reproduction of plants exposed to the stress (Chinnusamy et al. 2004). Studies in the past years has indicated that different stresses can induce the same molecular response by plant. This overlapping stress response can be explained as cross-tolerance phenomenon. Cross tolerance is the capability of plants to protect itself from damage imposed by different stresses associated with a primary stress. For example, several drought-inducible genes are also upregulated by cold and salt stresses, suggesting the presence of one primary stress resulting same stress response pathways to combat different abiotic stresses (Xu and Zhou 2006). Genes involved in abiotic stress response are classified in three major groups and shown in Table 9.4.

9.7 Types of Abiotic Stresses and tolerance mechanism

9.7.1 *Drought Stress.*

Legumes are among the important crops in the world, which are affected by water shortage. The extent of draught impact varies with legumes species and the phenological state. However, it is unclear how the effects of drought vary with species, climate, soil, and water shortage timing. Recent studies have shown that the quantity of water reduction has direct effect on yield reduction in legumes (Wang et al. 2018b; Hao et al. 2013). Legumes on exposure to drought condition during their reproductive stage have shown higher yield loss as compared to vegetative stage, whereas climatic region and the associated factors were found to contribute insignificantly to yield loss. (Daryanto et al. 2015). Recent research has identified that the root characteristics play a significant role in drought and nutrient stress tolerance (Bengough et al. 2011). Plants with deep root system were able to perform better in drought condition by acquiring water from deeper soil domains (Hammer et al. 2009). According to recent study under different environmental conditions soybean genotypes with better root traits were found more tolerant to drought stress during reproductive stage.(Prince et al. 2013). Based on above results, the root architecture is considered as a key trait to enhance crop productivity in the coming years of climate change (Den Herder et al. 2010).

Development of drought-resistant genotypes is important to maintain and improve the productivity of leguminous crops in today's changing environmental conditions. Recent advancements in the field of biotechnology allows us to identify the genes that provide resistance to drought and related stresses. With those

Table 9.4 classification of molecules responsible for abiotic stress in legumes

Classification (Gene product)	Direct: protecting molecules	Indirect: signalling molecules	Transport molecule
Response	Protects molecules in the stress conditions	Induces the response to stress through a cascade of signalling pathways	Induces active transport water or ion molecules across membrane to balance osmotic stress
Examples of stress	heat, cold, salt, free radicals, etc.	Cold, heat, heavy metals, UV light, etc	High salinity, flooding, etc.
Examples	1. **Osmoprotectants**: *Medicago truncatula* (Trehalose); Glycine max (Amino acids: proline) 2. **Antioxidant Enzymes:** ***Lotus japonicas:*** scavenging enzymes (Ascorbate-glutathione cycle enzymes, Catalase, superoxide dismutase, glutathione peroxydase and) ***Medicago truncatula:*** total peroxidase, superoxide dismutase, ascorbate peroxidise, glutathione reductase, guaicacol peroxidise 3. Ion/proton transporters (SOS-mediated signalling pathways: SOS1 and SOS2	1. Transcription Factors *M. truncatula* (CBF genes, AP2-like transcription factors, 2. Kinases: *M. truncatula*: salt-induced receptor-like kinase MtSrlk 3. Calcium-dependent protein kinase 4. SOS kinase 5. Phospholipases	1. Aqua porins 2. Na^+/K^+ transporters ***Lotus japonicas:*** Putative chloride channel (Ljwgs 016759.2) and Putative K^+ transporter (chr5. CM0911.54.1) Glycine max: 3. Ca^{++} transporters
References	Elmaghrabi et al. (2013), Kang et al. (2010), Lopez et al. (2008), Aghaei et al. (2008), Bray et al. (2000), Rubio et al. (2009), Bianco and Defez (2009), Kang et al. (2010), and Mhadhbi et al. (2011)	Ludwig et al. (2004), Frank et al. (2000), Merchan et al. (2007), Zhang et al. (2010), and Li et al. (2011)	Blumwald et al. (2000) and Sanchez et al. (2008)

methods, legumes can be transformed to alter the genetic makeup for protecting against devastating drought conditions. Here we discuss the genes discovered to impart stress tolerance to drought and various other related stresses in legumes. As explained earlier, transcription factor plays a key role in imparting stress tolerance by inducing signaling cascade. In recent years, many leguminous plants were engineered with single-gene alterations to identify the specific genes involved in conferring resistance to drought conditions. Transgenic chickpea plants with DREB1A transcription factor and Arabidopsis rd29A promoter has shown increased expression of the DREB1A gene before 50% soil dehydration conditions (Anbazhagan et al. 2014). AtDREB1A Expression in transgenic peanut by water stress showed increased transpiration efficiency under drought conditions. Earlier studies on rd29A DREB1A mechanisms has revealed that it controls stomatal opening, water intake and root development under dry environments of plants and thereby improve drought tolerance of plants as compared to wild type (WT). Transcription factors DREB2A, WRKY, bZIP, MyB, AP2/ERF, XPB1 are among the promising candidate genes involved in dehydration tolerance in chickpea. (Nayak et al. 2009; Yoo et al. 2005; Hiremath et al. 2011; Savitri and Fauziah 2018; Deokar et al. 2011). Transcription factors such as GmNAC, GmDREB, GmZIP, ERF089 were found to increase root development and drought tolerance in soybean (Tran et al. 2004; Hu et al. 2006). Peanut DREB1A, rd29A transcription factors overexpression has shown drought tolerance in transgenic peanut (Bhatnagar et al. 2007). GmFDL19 is a basic leucine zipper (bZIP) family of factors showed enhanced drought and salt tolerance in soybean at seedling stage (Li et al. 2017). The agronomic characteristics of transgenic plants expressed with AtABF3 gene were compared to WT plants. In normal condition, transformants were dwarf, however, under stress conditions (Drought and salt) the seed productivity in terms of weight was significantly higher than WT plants. With these results, it was concluded that overexpression of the AtABF3 gene hadimproved salt and drought tolerance in soybean (Kim et al. 2018). Expression of multiple transcription factors (AtDREB2A, AtHB7 and AtABF3) at the same time had shown significant improvement of Peanut (*Arachis hypogaea* L.) to salt and drought tolerance (Pruthvi et al. 2014).

A group of protein kinase enzymes that modifies other proteins by phosphorylation are also known to impart stress tolerance in the legumes by indirect mechanism. In soybean GmBIN2 gene from protein kinase family GSK3 on overexpression in transgenic roots had shown relatively higher root growth rate as compared to control when subjected to drought and salt stress. Studies of physiological indicators (SOD activity, proline content and EC) has also supported the conclusion further. Furthermore, itwas also found that GmBIN2 overexpression up-regulate the stress-related genes (Wang et al. 2018c). Transgenic AtEDT1 alfalfa plants revealed larger root systems in terms of lengths, weight and thickness than wild type plants. The AtEDT1 transgenic alfalfa plants have shown enhanced expression of stress inducible genes, especially drought-responsive genes, in comparison with WT plants in 2 years of field trials (Zheng et al. 2017). Osmosensors or aquaporins like PgTIP1 confers drought and salt tolerance by sustaining salt ions, ROS and water homeostasis when the gene PgTIP1 was transformed into soybean. Transgenic soybean PgTIP1 lines have shown (1) better stomatal movement with greater leaf water-retention capacity (2) improved water-gas exchange capacity; (3) less transportation and buildup of Na^+, Cl− ions; (4) lower Na^+/K^+ ratio; and (5)boosted antioxidase activities (6) reduced membrane damage in roots and leaves than WT. With these improved performances, the transgenic- PgTIP1 soybean lines showed drought as well as salt tolerance (An et.al. 2018). GmHK, GmCLV1A, GmCLV1B, GmRLK1, GmRLK2, G mRLK3, GmRLK4 were found to work as osmosensor in soybean (Yamamoto et al. 2000). VfPIP1 transgenic broad bean plants showed reduced transpiration rate resulted in higher growth rate when compared to wild type in drought stress (Cui et al. 2008). In addition to these, the stomata of the transgenic plants were found operating considerably faster than the WT plants under ABA or dark treatment. These observations predict that the improved drought resistance by VfPIP1 expression could be due to closure of stomata to avoid water loss (Cui et al. 2008). CDPK nodulin 26 was expressed and targeted to form an aquaporin channelin permeability (Guenther et al. 2003). It stimulated its intrinsic water transport rate and enhanced the water permeability across the membrane in soybean. Transgenic chickpea and soybean with overexpressed aquaporin (AQPs) resulted inincreased transport of water and other solutes across membranes (Azeem et al. 2019).The genes are up-regulated for achieving the stress tolerance in plants, but there are some genes which needs to be down regulated for increasing the drought tolerance level. The receptor for activated C kinase 1 (GmRACK1) silenced soybean lines with RNAi technology have shown high drought tolerance where as, the overexpressed seedlings were found highly sensitive to drought in comparison with wild type (Li et al. 2018a). Silencing of SPL13 (SQUAMOSA promoter binding-like transcription factor protein-13) has shown improved drought tolerance in alfalfa plants (*Medicago sativa*) by by RNAi silencing with MicroRNA156 (Arshad et al. 2017).

9.7.2 *Flooding/Waterl ogging Stress*

Asian countries are annually affected by water logging condition on large rain fed areas. Flooding conditions are going to worsen in near future due to frequent extreme precipitation events. Flooding negatively affects plant growth by causing continued exposure to hypoxia stress (Sasidharan et al. 2017a, b) resulting wilting and physiological damage to the plants and eventually death. Legumes are very much affected by flooding (Githiri et al. 2006). Crops like chickpea, pea lentil and lupin are susceptible to flooding during vegetative stages (Siddique and Sykes 1997; Yu and Rengel 1999; Palta et al. 2010). To avoid yield loss in legumes, improvement for waterlogging tolerance is greatly needed. The nitrogen fixation process of legumes from air is a highly energy demanding process and nitrogen fixation stops immediately in water submerged roots as the gas exchange stops even before roots become completely hypoxic (Amarante and Sodek 2006). Different legumes were having varied tolerance to waterlogging during vegetative growth. Faba bean (*Vicia faba*) was found to be the most tolerant legume for waterlogging. Legumes such as yellow lupin, grasspea, chickpea, narrow-leafed lupin and lentil were observed as somewhat tolerant while field pea is the least tolerant (Solaiman et al. 2007). Within the species, variation was observed in waterlogging tolerance due to differential genetic makeup in faba bean, soya bean and chickpea (Solaiman et al. 2007; Henshaw et al. 2007; Palta et al. 2010). Zombi pea (*Vigna vexillata*) is a leguminous crop known for waterlogging tolerance. In the tolerant variety of zombie pea, auxin-regulated lateral rooting genes were induced. In addition to lateral root formation, cell wall modification genes, peroxidase genes and aquaporin were highly expressed, imparting waterlogging tolerance in zombie pea. The legume genus *Trifolium*, is known for waterlogging-tolerance with the help of higher root porosity and lateral root formation (Gibberd et al. 2001).

Most existing studies in legumes for molecular basis of water logging tolerance were focused in soybean. Energy crisis in root-zone resultant from hypoxic conditions was found as an important factor in waterlogging condition. Tolerant soybean varieties were found with improved root growth and aerenchyma as compared to sensitive varities under flooding stress (Valliyodan et al. 2014; Jitsuyama 2015; Kim et al. 2015; Sakazono et al. 2014). Recently, a major QTL, qWT_Gm_03, was also identified regulating root development, root plasticity and auxin pathways during waterlogging stress in soybean (Ye et al. 2018). Recent studies in Glycine max, also found various quantitative trait locus (QTL) associated with water logging tolerance namely Chr-3, Chr-5 (satt_269), Chr-6 (satt_100), Chr-10, Chr-11, Chr-12 (satt_052, satt_302), Chr-13 (satt_269), Chr-18 (sat_64) (Githiri et al. 2006; Van Toai et al. 2001; Cornelious et al. 2005; Valliyodan et al. 2017; Nguyen et al. 2012; Ye et al. 2018; Nguyen et al. 2017).

9.7.3 Temperature Stress

Temperature stress during reproductive stages is a serious concern for production of grain legumes. The temperature is increasing abruptly year after year due to climate change. Huge genetic variations exist in legumes for temperature tolerance that can be further exploited for development of heat/cold tolerant cultivars.

9.7.3.1 High Temperature Stress

Many food legumes were evaluated for heat tolerance under heat stress, and was in the following order of most tolerant to least tolerant: groundnut > soybean > pigeon pea > chickpea based. The heat tolerance in these crops was assayed based on membrane stability analysis and photosystem (PSII) function (Srinivasan et al. 1996). Higher temperature harms seed vigour, seed germination, seedling development and finally survival of the plant (Wahid et al. 2007) in alfalfa, mung bean and chickpea (Kumar et al. 2011; Mingpeng et al. 2010; Piramila et al. 2012). Heat tolerance was investigated in soybean, bean and pea at different (Day/Night) temperature with control (at 10 °C). The moderate stressat 25 °C and severe stress at 30 °C with highest heat strain injuries were observed under severe heat stress as expected (Nemeskèri 2004). Chimeric translational fusion of TPS1-TPS2 genes expression by rd29A a stress regulated promoter has displayed promisingly increased temperature (heat and freezing) tolerance in alfalfa. Yeast gene encoding TPS1-TPS2 expression with rd29A promoter has enhanced growth and triggered increased trehalose accumulation in different lines under stress conditions (Ramón et al. 2009). The above results suggest that the TPS-TPP fusion protein from yeast could be used for developing stress-tolerance in legume crops. The over expression of VrWRKY11 in mung bean

(*Vigna radiate*) under heat stress suggest the importance of gene for heat tolerance(Srivastava et al. 2018). In mung bean AtWRKY26 was triggered by drought & heat and shown ethylene-dependent resistance response (Fu and Yu 2010). Transcriptional factor DREB2 (Dehydration-Responsive Element-Binding Protein 2) is particularly responsible for plants response to heat and cold stress by regulating the expression of stress inducible genes. The DREB2 genes have been identified in diverse plant types including desert legume (*Eremosparton songoricum*) and is a promising candidate gene for the improvement of stress tolerance (Xiaoshuang et al. 2014).

9.7.3.2 Low Temperature Stress

Limited research has been done on the various factors responsible for low temperature tolerance in legumes. Low temperatures damages plants both by a chilling and by freezing. Chilling causes physiological and developmental abnormalities whereas freezing results in cellular damage. Usually symptoms of low-temperature include wilting, bleaching, browning, and finally necrosis of leaves leading plant death (Levitt et al. 1980). It was predicted that around 15% of global arable land is affected by freezing stress (Dudal 1976). The mechanisms of cold tolerance generally act by blocking the induction of tissues damaging factors. Cold acclimation is the first line of defense and it works by stabilizing plasma membranes with the help of hydrophilic and LEA proteins. Low temperature reduces the water absorption by roots and water transportation to shoots causing desiccation shock to the plant and leading to wilting of the plant (Uemura et al. 1995; Thomashow 1999). The mechanisms of tolerance to subnormal temperatures consist multiple genes and multiple regulatory pathways, and it is the biggest bottleneck in development of transgenic plants with higher levels of subnormal temperature tolerance. Another big challenge is the lack of in-depth understanding of temperature stress tolerance mechanism in legumes. Various studies with legume have suggested that the stable sucrose metabolism in anthers is critical during cold stress for pollen development. Studies opened up that the main category of genes central to cold tolerance in anthers are related to carbohydrate metabolism, signal transduction and pollen development were up-regulated during cold stress (Sharma and Nayyar 2014).

In a comparative study ABA-induced and cold-induced freezing tolerance in alfalfa cultivars, it was observed that ABA alone could give freezing tolerance to plants for some extent, but to achieve maximum tolerance it is essential to club cold acclimation induced tolerance. DREB1C (Dehydration-responsive element-binding protein 1C) is found to mediate cold-inducible genes transcription and plays a key role in freezing tolerance and cold acclimation in *Medicago truncatula*. (Chen et al. 2010). The Yamasaki and Randall (2016) confirmed that the CBF/DREB1 expression in soybean imparted the cold tolerance . The DRE element binding site is suggested as the common cross-talk switch between low temp stress, drought stress and salinity stress signalling. DREB1 family genes are suggested as a probable "master-switch" for induction of cold responsive (COR) genes and cold acclimation during cold stress (Thomashow et al. 2001). Over-expression of GsZFP1 in *Medicago sativa* (Alfalfa) caused increased tolerance to cold and dehydration stress. The over-expression of GsZFP1was found to decrease the water loss by inducing expression of multiple stress-responsive gene markers suggesting its crucial role in plant responses to cold and drought conditions. (Tang et al. 2013)

9.7.3.3 Salinity Stress

Soil salinization is one of the most common cause of land degradation. It was found predominantly in arid and semiarid areas, where the evapotranspiration is more than rainfall and insufficient rain to leach soluble salts (Miller and Donahue 1990). As per the Food and Agriculture Organization (FAO) study, around 7% of the total worlds land is affected with salinity problem (Massoud, 1974), and it is increasing rapidly (Pessarakli 1999), and numbers are doubled in the past two decades. Ironically, majority of the land loss is due to salinization by improper irrigation (Flowers and Yeo 1995). The presence of excessive amounts of soluble salts in the soils hamper normal plant growth. Saline soils have Electrical conductivity (EC) of >4 dSm^{-1}, ESP < 15%, and pH below 8.5 (Abrol 1986). The capability of legumes to survive at high salt stress is important for future agriculture (Flowers and Yeo 1989). In salinity stress, growth limitations are generally due to the decline in water potential of tissue and less availability of water (Ashraf 2004) resulting closure of stomata, dropped photosynthesis and inhibited growth (Robinson et al. 1997). Normally the chlorophyll content is reduced (Del Rio et al. 2005), photosynthesis is declined, and growth is retarded (Taïbi et al. 2016) under salinity stress in soybean plant. A salinity stress also causes damage to ultrastructure of the cell and tissue (Barba-Espín et al. 2011).In a study, Glycine max SKP1-like 1 (GmSK1), with similar features of SKP1 protein was induced by abscisic acid (ABA), jasmonic acid (JA), salicylic acid (SA), salt, freezing and water deficit suggesting central roles of GmSK1 in plant (*Glycine max*) responses to hormones and abiotic stress. SKP1 homologue in soybean was also found imparting tolerance to salt (Chen et al. 2018). The PgTIP1-transgenic soybean showed better water-gas exchange capacity, reduced leaf water evaporation, enhanced antioxidase activities,reduced absorption & accumulation of Na^+, Cl^- ions as well as reduced cell membrane damage in roots and leaves (An et al. 2018). Overexpression of GmFDL19 (flowering promoter) in transgenic soybean, caused early flowering, enhanced salt and drought tolerance at the seedling stage (Li et al. 2017). Overexpression of GmBIN2 gene, a protein kinase in soybean shown significantly higher root growth rate than the control under salt and mannitol stress (Wang et al. 2018c). VrbZIP, a basic leucine zipper (bZIP) genes was found to play a key role in responses to various abiotic stresses like salt and drought by ABA dependent response. In a study, 54 and 50 bZIP proteins were identified from whole-genome sequences of *V. radiata* and *V. angularis*, respectively which were found responsible for resistance to salt and drought in mung bean (Wang et al. 2018a).

Transgenic alfalfa plants with CsLEA expression has showed a relatively lower Na+ content in the leaves. The transgenic CsLEA alfalfa plants also maintained higher relative water content (RWC), higher shoot biomass with decreased membrane damage & osmotic stress injury and shown enhanced tolerance to drought and salt tolerance (Zhang et al. 2016). GsZFP1 encodes a zinc-finger protein was also found to confer tolerance to stresses like salt, drought, cold in Alfalfa plants (Tang et al. 2013). AVP1 transgenic alfalfa had shown increased water retention, potassium uptake & root activity and steady ion homeostasis with phenotype of salt and drought tolerance. The AVP1 transgenic alfalfa had also showed increased photosynthetic efficiency and decreased cell membrane damage during salt and water-deficit stress (Bao et al. 2009). The over expression of codA had shown improved abiotic stress tolerance in mung bean which is generally very sensitive to salt, water logging and temperature stress during the flowering and reproductive stages (Baloda et al. 2017). The same gene has also shown salinity and drought tolerance in transgenic alfalfa. Homeodomain-leucine zipper (HD-Zip) was cfound to confer tolerance to water deficit and salinity in transgenic alfalfa plants (Li et al. 2014; Cabello et al. 2017).

9.7.4 High Light Stress

Light is an important parameter and was found to affect the growth and development of plant. Plants utilize light as the primary energy source and convert light into chemical energy through the process of "photosynthesis". Plant exposure to light beyond what is required for photochemistry causes inactivation of photosynthetic molecules and induces production of highly reactive intermediate molecules as by-products called as reactive oxygen species (ROS). ROS can cause photoinhibition as well as photo-oxidative damage and inhibits photosynthesis. ROS can oxidise proteins, lipids and enzymes required for the normal functioning of the organelles and the cell. On response to high light stress, plants modulate the expression levels of light-responsive genes. Light responsive genes regulate the processes like seed germination, photo morphogenesis, chloroplast development, stem growth, pigmentation and flowering (Ciarmiello et al. 2011).The plants grown under extreme light, shown thicker leaves, higher photosynthetic rate higher, respiration ratio , and the leaves were covered with a waxy/ cuticle layer(s). Generally, morphological impacts of intense light and UV exposure are growth reduction, leaves thickening and wax/cuticle layers formation over leaves (Saleh et al. 2007). Light signals are detected by different families of photoreceptors, which include phytochromes, cryptochromes, phototropins and UVB photoreceptors. Light can modify the photoreceptor activity that alter their form and cellular localisation. High light affects the redox potential of photoreceptors and regulates the expression (Karpinski et al. 1997). Furthermore, the redox state of plastoquinone (PQ) has been found to involve in expression of chloroplast genes (Pfannschmidt et al. 1999). Various regulators downstream of photoreceptors have been identified through molecular approaches.

These regulators are known to encode various protein like transcription factors (TFs), kinases, phosphatases and catabolic pathway proteins (Klimczak et al. 1995). Recently, some light-responsive transcriptional factors have been recognized for light-responsive element (LRE)-binding proteins through genetic analyses of mutants that are deficient for their response to specific light. Such analyses have been successfully executed to identify cis-acting elements involved in the light response of PhANGs, such as G-box, I-box and GATA motifs (Giuliano et al. 1988; Gidoni et al. 1989). Although many LREs are known, the single element was not found in all light-regulated promoters, suggesting presence of a very complex light regulation network without any universal switch (Jiao et al. 2007).

9.8 Management of Abiotic Stress by Biotechnological Approaches

The threat represented by various abiotic stresses can be addressed by numerous approaches, right from classical breeding to modern genetic approaches. However, understanding the mechanisms of stress management is first step to make the feasible strategy. Biotechnological approaches can contribute to speed up classical breeding and overcome major problems. Some of the majorly useful biotechnological tools for abiotic stress management in legumes are discussed.

Molecular marker-assisted breeding Molecular markers are mostly useful while targeting the stress tolerance mechanism controlled by multiple genes together. We can map different Quantitative Trait Loci (QTL) contributing to a agronomical trait and identify linked molecular markers to transfer several QTLs in one improved cultivar. Several molecular marker-based techniques such as RAPD, RFLP, AFLP, and SSR are being used in legumes to make a strategy to overcome various abiotic stresses (Kassem et al. 2004; Lee et al. 2004). Genetic maps for many species are established and potential QTL have been located. This has improved the knowledge of the chromosomal locations controlling the specific stress in legumes. This knowledge may be useful in the breeding programs through Marker-Assisted Selection (MAS) to select abiotic stress tolerant varieties of legumes. Conversely, the use of MAS in legumes for the genetic improvement to stress have been limited due to vast genetic complexity of most stress related traits and lack of funding. However, MAS was shown useful to select drought tolerant common bean (Schneider et al. 1997). The knowledge of QTLs in relation with abiotic stress in legume is still at an early stage to apply for gene pyramiding. Advances achieved in non-legume crops are promising to obtain an improved stress resistant cultivar (Foolad 2004). In vitro selection has already been used for improvement of legumes although they have not been sufficiently explored in legumes. This method can also be combined with other approaches and it is attractive for abiotic stresses, where appropriate screening methods are unavailable or have low efficiency. For example, salinity is one of the abiotic stress that has been addressed by in vitro selection (Flowers 2004; Zair et al.

2003) and zinc tolerance (Samantaray et al. 1999) among others. Somaclonal variation and invitro mutagenesis following by invitro selection offers an alternative way for breeding.

Double Haploids and Wide Hybridization (DH) This method is tissue culture-based method and it has been tried in *Lupinus* spp. (Bayliss et al. 2004), alfalfa (Zagorska et al. 1997) and soybean (Rodrigues et al. 2005) but no commercial legume varieties have been produced so far. Successful application of DH technology coupling with MAS technology will be more efficient to select individuals carrying desirable traits in legumes.With the improved wide hybridization techniques, Stress tolerance traits from wild germplasms can be transferred into target species. Due to effective embryo rescue techniques it was possible to produce interspecific hybrids in legumes such as faba bean from *Vicia faba* x *V. narbonensis* (Lazaridou et al. 1993), grass pea from *Lathyrus odoratus* x *L. belinensis* (Hammett et al. 1994), and pigeon pea from *Cajanus cajan* x *C. platycarpus* (Mallikarjuna and Moss 1995). Though the hybrids were made but stress responses were not assessed. Inspite of the huge potential, very little research has been done with these techniques to overcome abiotic stresses in legumes. The potential of Double haploids and wide hybridization has been demonstrated in non-legumes crops and stress tolerance to abiotic (Arumugam et al. 2002) stresses together with advances in tissue culture was achieved.

Genetic Transformation
Crop improvement through genetic manipulations is possible now (Dunwell 2000) as transformation of grain legumes is proven (Chandra and Pental 2003; Somers et al. 2003). Both micro-particle bombardment (Gulati et al. 2002) and *A. tumefaciens* (De Clercq et al. 2002; Li et al. 2004) have been proven for DNA delivery in legumes. Transformation has been generally based on the infection of *A. tumefaciens* to produce composite plants with hairy roots (Boisson-Dernier et al. 2001; Wu and Van Etten 2004). By genetic engineering approach specific gene can be inserted for specific biochemical function, regulatory control, or multiple genes for a more complex pathway. Here we will descuss few examples related abiotic stress in legumes (Chandra and Pental 2003; Popelka et al. 2004; Somers et al. 2003). The response of legumes to various abiotic stresses are shown in Table 9.5.

Addressing the abiotic stress with conventional breeding strategies is difficult due to its poorunderstanding and complexity along with the lack of good sources of natural tolerance. Abiotic stresses in general causes disruption of various cellular functions and triggers activation of complex metabolic pathways (Kassem et al. 2004; Lee et al. 2004; Popelka et al. 2004). The recent progresses in genetic engineering approaches suggest that there are good prospects for developing transgenic legumes with enhanced salt tolerance in the near future (Foolad 2004; Sharma and Lavanya 2002). At the same time, increased tolerance to aluminium and cyanamide toxicity have been demonstrated in transgenic alfalfa and soybean revealing the potential of these approaches in legumes to acquire abiotic stress tolerance (Morphew et al. 2004; Zhang et al. 2005).

Table 9.5 Response of legumes to abiotic stress and biotechnological approaches to overcome the stress

Stress	Physiological effect	Legume response	Target genes for biotechnological approaches	References
Drought Stress	Photooxidative damage in chloroplast structures, cell growth reduction, changes in the pool of sugars used for signalling	Accumulation of osmo-protectants like proline (Pro) is a common response	Expression of photo-protective proteins (ELIPs) in legumes. Accumulation of proline in soybean has improved drought tolerance. In *M. truncatula*, manipulation of trehalose metabolism enhanced resistance and recovery from severe water deficit	Chaves et al. (2009), Liu et al. (2013), Hermans, (2008), Kim and Nam (2013), and Duque et al. (2013)
Salt Stress	Impairs plant growth, reduced water availability, ion toxicity, cytosolic solute imbalance, osmotic stress in *M. truncatula,* L. japonicus and G. max.	Increased root growth and accumulation of more ions (Na^+ & Cl^-) in shoots in soybean. Compartmentalization of Na^+ into root cell vacuoles	The vacuolar Na^+/H^+ antiporter NHX1 (for Na^+/H^+ exchanger) is involved in Na^+ compartmentalization. *M. truncatula,* MtZPT2-1, a TFIIIA transcription factor is involved in modulating root adaptation. Helix-Loop-Helix transcription factor (MtbHLH–658) increased adaptation of root growth to salt stress	Conde et al. (2011), Obhanian et al. (2011), Zahaf et al. (2012), Dabuxilatu and Ikeda, (2005), Luo et al. (2005), Zahaf et al. (2012), Li et al. (2006), Zahran et al. (2007), Teakle et al. (2010), and Merchan et al. (2007)
Cold Stress	*M. truncatula* under cold acclimation increased stem and root dry matter without any changes in leaf dry matter	Increased levels of sucrose, fructose, lactose, proline, osmolality	Over-expression of MtCBF4 in *M. truncatula* improved cold tolerance (4°C). CBF genes are involved in freezing tolerance in *M. truncatula*	Zhang et al. (2011) and Li et al. (2011)

(continued)

Table 9.5 (continued)

Stress	Physiological effect	Legume response	Target genes for biotechnological approaches	References
Heat stress	Affects vegetative growth in peanut, pea and chickpea water loss, reduced cell size, reduced leaf area and biomass. Reduced Fertility	Observed the expression of HSP-interacting proteins for improved heat stress tolerance in soybean.	Accumulation of ClpB/ HSP100 increasedthe pollen viability in faba bean reported increased levels of Ef-Tu protein in soybean in protecting key enzymes and proteins from heat stress. Application of ABA increased growth reduced oxidative damage in chickpea	Bolhuis and De Groot (1959), Poehlman (1991), Singh and Dhaliwal (1972), Zhu et al. (2006), Nayyar et al. (2005), Kumar et al. (2012b, 2015b), and Das et al. (2016)
Flooding	Produces hypoxia condition	During flooding, energy deficiency and during recovery plants experience oxidative stress	During flooding, alcohol dehydrogenase (ADH2) accumulates in roots of soybean to cope with the hypoxic condition. Limitedlight availability increases succinate-semialdehyde dehydrogenase and γ-amino butyrate (GABA) shunt increases replenishment energy during recovery, SOD was found increased, suggesting that the antioxidative system may play a crucial role in protecting cells from oxidative damage.	Sasidharan et al. (2017a, b), Jackson and Ram (2003), Yeung et al. (2018), and Komatsu et al. (2011)

So far, we know only a soybean line with cryA gene from *B. thuringiensis* as a commercial variety (Babu et al. 2003). This low number has been influenced by regeneration recalcitrance of most legumes along with social and political reasons. However, the recent advancements in legume transformation technology and the increasing political interest for high protein content should see an upsurge in the development of genetically modified legumes in near future.

9.9 Conclusion

Biotic and abiotic stresses are among the major constraints which reduces legume yield. Biotechnological approaches are being intensively used in identifying the genes and molecular mechanisms involved in stress response. Identification of makers and mapping of biotic and abiotic stresses have helped us to find desired and elite genotypes which has superior agronomic traits in given stress conditions. Many of these traits are quantitative and are influenced by genetic as well as environmental factors. Though the extensive studies in model crops has been done, our knowledge about the molecular mechanisms operational during stress tolerance under in vivo conditions in legumes is limited. Many studies have concluded that a single mechanism does not impart much improvement in growth and quantitative yield, suggesting incorporation of multiple stress resistant mechanisms so that the effect is cumulative and enhanced. In the nature, generally the legumes are exposed to multiple stress at different growth stages of plant. In this perspective, the plant must have resistance to multiple stresses at a time, which can be achieved by intelligently transferring the genes or pathways which are known to provide resistance to multiple stresses with the same primary signalling mechanism. Biotechnological approaches are very much essential, through which we can identify, transfer and express the genes which induces the resistance and can monitor the post effects of transgene by advanced proteomic and metabolomics approaches.

References

Abrol IP (1986) Salt-affected soils: an overview. In: Chopra VL, Paroda SL (eds) Approaches for incorporating drought and salinity resistance in crop plants. IBH Publishing Company, New Delhi/Oxford, pp 1–23

Acharjee S, Sarmah BK, Kumar PA, Olsen K, Mahon R, Moar WJ, Moore A, Higgins T (2010) Transgenic chickpeas (Cicer arietinum L.) expressing a sequence-modified cry2Aa gene. Plant Sci 178:333–339

Afouda L, Wolf G, Wydra K (2009) Development of a sensitive serological method for specific detection of latent infection of Macrophomina phaseolina in cowpea. J Phytopathol 157:15–23

Aghaei K, Ehsanpour AA, Shah AH, Komatsu S (2008) Proteome analysis of soybean hypocotyl and root under salt stress. Amino Acids 36:91–98

Agrell J, Anderson P, Oleszek W, Stochmal A, Agrell C (2004) Combined effects of elevated CO_2 and herbivore damage on alfalfa and cotton. J Chem Ecol 30:2309–2324

Akhtar KP, Dickinson M, Hodgetts J, Abbas G, Asghar MJ, Shah TM, Atta B, Ahmad M, Haq MA (2010) The phytoplasma disease 'mung bean phyllody' is now present in Pakistan. Plant Pathol 59:399

Alegbejo MD, Kashina BD (2001) Status of legume viruses in Nigeria. J Sustain Agric 18:55–69

Amarante L, Sodek L (2006) Waterlogging effect on xylem sap glutamine of nodulated soybean. Biol Plant 50:405–410

An J, Cheng C, Hu Z, Chen H, Cai W, Yu B (2018) The Panax ginseng *PgTIP1* gene confers enhanced salt and drought tolerance to transgenic soybean plants by maintaining homeostasis of water, salt ions and ROS. Environ Exp Bot 155:45–55

Anbazhagan K, Bhatnagar-Mathur P, Vadez V, Dumbala SR, Kishor PK, Sharma KK (2014) DREB1A overexpression in transgenic chickpea alters key traits influencing plant water budget across water regimes. Plant Cell Rep 34:199–210

Anderson JP, Thatcher LF, Singh KB (2005) Plant defence responses: conservation between models and crops. Funct Plant Biol 32:21–34

Andrews M, Hodge S (2010) Climate change, a challenge for cool season grain legume crop production. In: Yadav S, Redden R (eds) Climate change and management of cool season grain legume crops. Springer, Dordrecht, pp 1–9

Aragao FJL, Vianna GR, Albino MMC, Rech EL (2002) Transgenic dry bean tolerant to the herbicide glufosinate ammonium. Crop Sci 42:1298–1302

Arnold DL, Lovell HC, Jackson RW, Mansfield JW (2011) Pseudomonas syringae pv. phaseolicola: from "has bean" to supermodel. Mol Plant Pathol 12:617–627

Arshad M, Feyissa BA, Amyot L, Aung B, Hannoufa A (2017) *MicroRNA156* improves drought stress tolerance in alfalfa (*Medicago sativa* L.) by silencing SPL13. Plant Sci 258:122–136

Arumugam N, Mukhopadhyay A, Gupta V, Sodhi YS, Verma JK, Pental D, Pradhan AK (2002) Synthesis of somatic hybrids (RCBB) by fusing heat-tolerant Raphanus sativus (RR) and Brassica oleracea (CC) with Brassica nigra (BB). Plant Breed 121:168–170

Asai T, Tena G, Plotnikova J, Willmann MR, Chiu WL, Gomez-Gomez L, Boller T, Ausubel FM, Sheen J (2002) MAP kinase signalling cascade in Arabidopsis innate immunity. Nature 415:977–983

Asamizu E, Nakamura Y, Sato S, Tabata S (2004) Characteristics of the lotus japonicus gene repertoire deduced from large-scale expressed sequence tag (EST) analysis. Plant Mol Biol 54:405–414

Ashfield T, Ong LE, Nobuta K, Schneider CM, Innes RW (2004) Convergent evolution of disease resistance gene specificity in two flowering plant families. Plant Cell 16:309–318

Ashraf M (2004) Some important physiological selection criteria for salt tolerance in plants. Flora 199:361–376

Avnioktem H, Eyidogan F, Selçuk F, Tufanoz M, da Silva JAT, Yucel M (2008) Revealing response of plants to biotic and abiotic stresses with microarray technology. Genes Genomes Genomics 2:14–48

Azeem F, Bilal A, Rana M, Muhammad AA, Habibullah N, Sabir H, Sumaira R, Hamid M, Usama A, Muhammad A (2019) Drought affects *aquaporins* gene expression in important pulse legume chickpea (*Cicer arietinum* L.). Pak J Bot 51:81–88

Babu RM, Sajeena A, Seetharaman K, Reddy MS (2003) Advances in genetically engineered (transgenic) plants in pest management – an over view. Crop Prot 22:1071–1086

Baloda A, Madanpotra S, Aiwal PK (2017) Transformation of mung bean plants for salt and drought tolerance by introducing a gene for an osmo-protectant glycine betaine. J Plant Stress Physiol 3:5

Bantignies B, Seguin J, Muzac I, Dedalde champ F, Gulick P, Ibrahim R (2000) Direct evidence for ribonucleolytic activity of a PR-10-like protein from white lupin roots. Plant Mol Biol 42:871–881

Bao AK, Wang SM, Wu GQ, Xi JJ, Zhang JL, Wang CM (2009) Overexpression of the *Arabidopsis* H^+-PPase enhanced resistance to salt and drought stress in transgenic alfalfa (*Medicago sativa* L.). Plant Sci 176:232–240

Barba-Espin G, Clemente-Moreno MJ, Alvarez S, García-Legaz MF, Hernández JA, Díaz-Vivancos P (2011) Salicylic acid negatively affects the response to salt stress in pea plants: effects on PR1b and MAPK expression. Plant Biol 13:909–917

Barilli Cobos MJ, Carrillo E, Kilian A, Carling J, Rubiales D (2018) A high-density integrated DArTseq SNP-based genetic map of *Pisum fulvum* and identification of QTLs controlling rust resistance. Front Plant Sci 9:167

Barilli E, Satovic Z, Sillero JC, Rubiales D, Torres AM (2011) Phylogenetic analysis of Uromyces species infecting grain and forage legumes by sequence analysis of nuclear ribosomal internal transcribed spacer region. J Phytopathol 159:137–145

Baulcombe DC (2000) Unwinding RNA silencing. Science 290:1108
Bayliss KL, Wroth JM, Cowling WA (2004) Pro-embryos of Lupinus spp. Produced from isolated microspore culture. Aust J Agric Res 55:589–593
Bell CJ, Dixon RA, Farmer AD, Flores R, Inman J, Gonzales RA, Harrison MJ, Paiva NL, Scott AD, Weller JW, May GD (2001) The Medicago Genome Initiative: a model legume database. Nucleic Acids Res 29:114–117
Bengough AG, McKenzie BM, Hallett PD, Valentine TA (2011) Root elongation, water stress, and mechanical impedance: a review of limiting stresses and beneficial root tip traits. J Exp Bot 62:59–68
Bertaccini A, Franova J, Botti S, Tabanelli D (2005) Molecular characterization of phytoplasmas in lilies with fasciation in the Czech Republic. FEMS Microbiol Lett 249:79–85
Bhagwat B, Duncan EJ (1998) Mutation breeding of banana cv. Highgate (Musa spp., AAA Group) for tolerance to *Fusarium oxysporum* f. sp. cubense using chemical mutagens. Sci Hortic 73:11–22
Bhatnagar MP, Devi MJ, Reddy DS, Lavanya M, Vadez V, Serraj RK (2007) Stress inducible expression of *AtDREB1A* in transgenic peanut (*Arachis hypogaea* L.) increases transpiration efficiency under water-limiting conditions. Plant Cell Rep 26:2071–2082
Bianco C, Defez R (2009) Medicago truncatula improves salt tolerance when nodulated by an indole-3-acetic acid-overproducing Sinorhizobium meliloti strain. J Exp Bot 60:3097–3107
Boisson-Dernier A, Chabaud M, Garcia F, Becard G, Rosenberg C, Barker DG (2001) Agrobacterium rhizogenes-transformed roots of Medicago truncatula for the study of nitrogen-fixing and endomycorrhizal symbiotic associations. Mol Plant-Microbe Interact 14:695–700
Bolhuis GG, De Groot W (1959) Observations on the effect of varying temperatures on the flowering and fruit set in three varieties of groundnut. Neth J Agric Sci 7:317–326
Bray EA, Bailey SJ, Weretilnyk E (2000) Responses to abiotic stresses. In: Gruissem W, Buchannan B, Jones R (eds) Biochemistry and molecular biology of plants. American Society of Plant Physiologists, Rockville, pp 1158–1203
Britt AB, May GD (2003) Re-engineering plant gene targeting. Trends Plant Sci 8:90–95
Buchwaldt L, Anderson KL, Morrall RAA, Gossen BD, Bernier CC (2004) Identification of lentil germ plasm resistant to Colletotrichum truncatum and characterization of two pathogen races. Phytopathology 94:236–243
Burrows M (2012) Diseases of cool season legumes (Pulse Crops: dry pea, lentil and chickpea). Montana State University Extension. EB0207, p 10
Cabello JV, Giacomelli JI, Gómez MC, Chan RL (2017) The sunflower transcription factor HaHB11 confers tolerance to water deficit and salinity to transgenic *Arabidopsis* and alfalfa plants. J Biotechnol 257:35–46
Canovas F, Dumas-Gaudot E, Recorbet G, Jorrin J, Mock HP, Rossignol M (2004) Plant proteome analysis. Proteomics 4:285–298
Carmona MA, Gally ME, Lopez SE (2005) Asian soybean rust: incidence, severity, and morphological characterization of *Phakopsora pachyrhizi* (uredinia and telia) in Argentina. Plant Dis 89:109–109
Castillo P, Navas-Cortés JA, Gomar-Tinoco D, Di Vito M, Jiménez-Díaz RM (2003) Interactions between Meloidogyne artiellia, the cereal and legume root-knot nematode, and Fusarium oxysporum f. sp. ciceris race 5 in chickpea. Phytopathology 93:1513–1523
Castillo P, Navas-Cortés JA, Landa BB, Jiménez-Díaz RM, Vovlas N (2008) Plant-parasitic nematodes attacking chickpea and their in planta interactions with rhizobia and phytopathogenic fungi. Plant Dis 92:840–853
Chadha P, Das RH (2006) A pathogenesis related protein, AhPR10 from peanut: an insight of its mode of antifungal activity. Planta 225:213–222
Chandra A, Pental D (2003) Regeneration and genetic transformation of grain legumes: an overview. Curr Sci 84:381–387
Chandrashekar N, Guptha O, Yelshetty S, Sharma OP, Bhagat S, Chattopadhay C, Sehgal M, Kumari A, Amerasan N, Sushil SN, Sinha AK, Asre R, Kapoor KS, Sathyagopal K, Jeyakumar

P (2014) Integrated pest management for chickpea. National Centre for Integrated Pest Management, New Delhi, p 43
Chaves MM, Flexas J, Pinheiro C (2009) Photosynthesis under drought and salt stress: regulation mechanisms from whole plant to cell. Ann Bot 103:551–560
Chen WQ, Singh KB (1999) The auxin, hydrogen peroxide and salicylic acid induced expression of the Arabidopsis GST6 promoter is mediated in part by an ocs element. Plant J 19:667–677
Chen HM, Liu CA, Kuo CG, Chien CM, Sun HC, Huang CC (2007) Development of a molecular marker for a bruchid (*Callosobruchus chinensis* L.) resistance gene in mung bean. Euphytica 157:113–122
Chen JR, Lu JJ, Liu R, Xiong XY, Wang TX, Chen SY (2010) DREB1C from *Medicago truncatula* enhances freezing tolerance in transgenic *M. truncatula* and China rose (*Rosa chinensis* Jacq.). Plant Growth Regul 60:199–211
Chen W, Sharma HC, Muehlbauer FJ (2011) Compendium of chickpea and lentil diseases and pests. APS Press, St Paul
Chen Y, Chi Y, Meng Q, Wang YD (2018) *GmSK1*, an *SKP1* homologue in soybean, is involved in the tolerance to salt and drought. Plant Physiol Biochem 127:25–31
Chinnusamy V, Schumaker K, Zhu JK (2004) Molecular genetics perspectives on cross-talk and specificity in abiotic stress signalling in plants. J Exp Bot 55:225–236
Chintapalli PL, Moss JP, Sharma KK, Bhalla JK (1997) In vitro culture provides additional variation for pigeon pea *Cajanus cajan* (L) Millsp crop improvement. In Vitro Cell Dev Biol Plant 33:30–37
Cho SH, Muehlbauer FJ (2004) Genetic effect of differentially regulated fungal response genes on resistance to necrotrophic fungal pathogens in chickpea (*Cicer arietinum* L.). Physiol Mol Plant Pathol 64:57–66
Choudhary K, Singh M, Rathore MS, Shekhawat NS (2009) Somatic embryogenesis and in vitro plant regeneration in moth bean [Vigna aconitifolia (Jacq.) Marechal]: a recalcitrant grain legume. Plant Biotechnol Rep 3:205–211
Ciarmiello LF, Woodrow P, Fuggi A, Pontecorvo G, Carillo P (2011) Plant genes for abiotic stress. IntechOpen. https://doi.org/10.5772/22465
Conde A, Chaves MM, Geros H (2011) Membrane transport, sensing and signaling in plant adaptation to environmental stress. Plant Cell Physiol 52:1583–1602
Cong L, Ran FA, Cox D, Lin S, Barretto R, Habib N (2013) Multiplex genome engineering using CRISPR/Cas systems. Science 339:819–823
Constantin GD, Krath BN, MacFarlane SA, Nicolaisen M, Johansen IE, Lund OS (2004) Virus-induced gene silencing as a tool for functional genomics in a legume species. Plant J 40:622–631
Constantin EC, Cleenwerck I, Maes M, Baeyen S, Van Malderghem C, De Vos P, Cottyn B (2016) Genetic characterization of strains named as Xanthomonas axonopodis pv. dieffenbachiae leads to a taxonomic revision of the X. axonopodis species complex. Plant Pathol 65:792–806
Cook DR (1999) *Medicago truncatula* – a model in the making! Commentary. Curr Opin Plant Biol 2:301–304
Cormack MW, Moffatt JE (1956) Occurrence of the bacterial wilt organism in alfalfa seed. Phytopathology 46:407–409
Cornelious B, Chen P, Chen Y, de Leon N, Shannon JG, Wang D (2005) Identification of QTLs underlying water-logging tolerance in soybean. Mol Breed 16:103–112
Cui XH, Hao FS, Chen H, Chen J, Wang XC (2008) Expression of the *Vicia faba VfPIP1* gene in *Arabidopsis thaliana* plants improves their drought resistance. J Plant Res 121:207–214
Curtin SJ, Xiong Y, Michno JM, Campbell BW, Stec AO, Cermak T (2018) CRISPR/Cas9 and TALENs generate heritable mutations for genes involved in small RNA processing of Glycine max and *Medicago truncatula*. Plant Biotechnol J 16:1125–1137
Dabuxilatu MI, Ikeda M (2005) Distribution of K, Na and Cl in root and leaf cells of soybean and cucumber plants grown under salinity conditions. Soil Sci Plant Nutr 51:1053–1057
Daryanto S, Wang L, Jacinthe PA (2015) Global synthesis of drought effects on food legume production. PLoS One 10(6):e0127401

Das TK (2008) Weed Science: Basics and applications. Jain Brothers, New Delhi, pp 666–679
Davidson JA, Pande S, Bretag TW, Lindbeck KD, Krishna-Kishore G (2004) Biology and management of Botrytis spp. in legume crops. In: Elad Y, Williamson B, Tudzynski P, Denle N (eds) Botrytis: biology, pathology and control. Kluwer Academic Publishers, Dordrecht, pp 295–318
De Clercq J, Zambre M, Van Montagu M, Dillen W, Angenon G (2002) An optimized Agrobacterium-mediated transformation procedure for Phaseolus acutifolius A. Gray. Plant Cell Rep 21:333–340
De Coninck B, Timmermans P, Vos C, Cammue BP, Kazan K (2015) What lies beneath: belowground defense strategies in plants. Trends Plant Sci 20:91–101
Del Rio D, Stewart AJ, Pellegrini N (2005) A review of recent studies on malondialdehyde as toxic molecule and biological marker of oxidative stress. Nutr Metab Cardiovasc 15:316–328
Den Herder G, Van Isterdael G, Beeckman T, De Smet I (2010) The roots of a new green revolution. Trends Plant Sci 15:600–607
Deokar AA, Kondawar V, Jain PK, Karuppayil SM, Raju NL, Vadez V (2011) Comparative analysis of expressed sequence tags (ESTs) between drought-tolerant and -susceptible genotypes of chickpea under terminal drought stress. BMC Plant Biol 11:70
Desbrosses GG, Kopka J, Udvardi MK (2005) Lotus japonicus metabolic profiling. Development of gas chromatography-mass spectrometry resources for the study of plant16 microbe interactions. Plant Physiol 137:1302–1318
Dhandaydham M, Charles L, Zhu H, Starr JL, Huguet T, Cook DR, Prosperi JM, Opperman C (2008) Characterization of root-knot nematode resistance in Medicago truncatula. J Nematol 40:46–54
Diatchenko L, Lau YFC, Campbell AP, Chenchik A, Moqadam F, Huang B, Lukyanov S, Lukyanov K, Gurskaya N, Sverdlov ED, Siebert PD (1996) Suppression subtractive hybridization: a method for generating differentially regulated or tissue. Proc Natl Acad Sci U S A 93:6025–6030
Diers B (2004) Soybean genetic improvement through conventional and molecular based strategies. In: 5th European Conference on Grain Legumes, Djion, France, 7–11 June, 2004. AEP, 147–14
Dixon RA (2001) Natural products and plant disease resistance. Nature 411: 843-847
Dixon RA, Sumner LW (2003) Legume natural products: understanding and manipulating complex pathways for human and animal health. Plant Physiol 131:878–885
Dudal R (1976) Inventory of major soils of the world with special reference to mineral stress. In: Wright MJ (ed) Plant adaption to mineral stress in problem soils. Cornell University Agricultural Experiement Station, Ithaca, pp 3–23
Dunwell JM (2000) Crop genomics: progress and prospects. J Chem Technol Biotechnol 75:913–918
Duque AS, Almeida A, Silva AB, Silva JM, Farinha AP, Santos D, Fevereiro P, Araujo SS (2013) Abiotic stress responses in plants: unraveling the complexity of genes and networks to survive. In: Vahdati K, Leslie C (eds) Abiotic stress: plant responses and applications in agriculture. InTech, New York
Edwards O, Singh KB (2006) Resistance to insect pests: what do legumes have to offer? Euphytica 147:273–385
Eichenlaub R, Gartemann KH (2011) The Clavibacter michiganensis subspecies: molecular investigation of gram-positive bacterial plant pathogens. Ann Rev Phytopathol 49:445–464
Elmaghrabi AM, Ochatt S, Rogers HJ, Francis D (2013) Enhanced tolerance to salinity following cellular acclimation to increasing NaCl levels in Medicago truncatula. Plant Cell Tissue Organ Cult 114:61–70
Emeran AA, Sillero JC, Niks RE, Rubiales D (2005) Infection structures of host-specialized isolates of Uromyces viciae-fabae and of other species of Uromyces infecting leguminous crops. Plant Dis 89:17–22
Emeran AA, Román B, Sillero JC, Satovic Z, Rubiales D (2008) Genetic variation among and within Uromyces species infecting legumes. J Phytopathol 156:419–424
Eulgem T (2005) Regulation of the Arabidopsis defense transcriptome. Trends Plant Sci 10:71–78

Falloon RE, Viljanen-Rollinson SLH (2001) Powdery mildew. In: Kraft JM, Plfleger FL (eds) Compendium of pea diseases and pests. American Phytopathological Society, St. Paul, pp 28–29

Fan WH, Dong XN (2002) In vivo interaction between NPR1 and transcription factor TGA2 leads to salicylic acid-mediated gene activation in Arabidopsis. Plant Cell 14:1377–1389

Fang Y, Tyler BM (2016) Efficient disruption and replacement of an effector gene in the oomycete Phytophthora sojae using CRISPR/Cas9. Mol Plant Pathol 17:127–139

Fininsa C, Tefera T (2001) Effect of primary inoculum sources of bean common bacterial blight on early epidemics, seed yield and quality aspects. Int J Pest Manag 47:221–225

Flowers TJ (2004) Improving crop salt tolerance. J Exp Bot 55:307–319

Flowers TJ, Yeo AR (1989) Effects of salinity on plant growth and crop yields. In: Cherry JH (ed) Environmental stress in plants, North Atlantic Treaty Organization Advanced Science Institutes Series vol G19. pp 101–119

Flowers TJ, Yeo AR (1995) Breeding for salinity resistance in crop plants – where next. Aust J Plant Physiol 22:875–884

Foolad MR (2004) Recent advances in genetics of salt tolerance in tomato. Plant Cell Tissue Organ Cult 76:101–119

Foster-Hartnett D, Danesh D, Penuela S, Sharopova N, Endre G, Vanden BKA, Young ND, Samac DA (2007) Molecular and cytological responses of Medicago truncatula to Erysiphe pisi. Mol Plant Pathol 8:307–319

France RA, Abawi GS (1994) Interaction between Meloidogyne incognita and Fusarium oxysporum f. sp. phaseoli on selected bean genotypes. J Nematol 26:467–474

Fu Q, Yu D (2010) Expression profiles of AtWRKY25, AtWRKY26 and AtWRKY33 under abiotic stresses. Yi chuan Hereditas 32:848–856

Fuller MP, Eed MHI (2003) The development of multiple stress-resistant cauliflower using mutagenesis in conjunction with a micro shoot tissue culture technique. In: XXVI international horticultural congress: environmental stress and horticulture crops, Toronto, Canada, 2003. Ed I.A. Horticulturae

Gao H, Narayanan NN, Ellison L, Bhattacharyya MK (2005) Two classes of highly similar coiled coil-nucleotide binding-leucine rich repeat genes isolated from the Rps1-k locus encode Phytophthora resistance in soybean. Mol Plant-Microbe Interact 18:1035–1045

Garza R, Cardona C, Singh SP (1996) Inheritance of resistance to the bean-pod weevil (Apion godmani Wagner) in common beans from Mexico. Theor Appl Genet 92:357–362

Gibberd M, Gray JD, Cocks PS, Colmer TD (2001) Waterlogging tolerance among a diverse range of Trifolium accessions is related to root porosity, lateral root formation and 'aerotropic rooting'. Ann Bot 88:579–589

Gidoni D, Brosio P, Bond-Nutter D, Bedbrook J, Dunsmuir P (1989) Novel *cis*-acting elements in petunia *Cab* gene promoters. Mol Gen Genet 215:337–344

Gillard CL, Conner RL, Howard RJ, Pauls KP, Shaw L, Taran B (2009) The performance of dry bean cultivars with and without common bacterial blight resistance in field studies across Canada. Can J Plant Sci 89:405–410

Githiri SM, Watanabe S, Harada K, Takahashi R (2006) QTL analysis of flooding tolerance in soybean at anearly vegetative growth stage. Plant Breed 125:613–618

Giuliano G, Pichersky E, Malik VS, Timko MP, Scolnic PA, Cashmore AR (1988) An evolutionarily conserved protein binding sequence upstream of a plant light-regulated gene. Proc Natl Acad Sci U S A 85:7089–7093

Glazebrook J (2005) Contrasting mechanisms of defense against biotrophic and necrotrophic pathogens. Annu Rev Phytopathol 43:205–227

Graham MA, Ramırez M, Valdes-Lopez O, Lara M, Tesfaye M, Vance CP, Hernandez G (2006) Identification of candidate phosphorus stress induced genes in Phaseolus vulgaris L. through clustering analysis across several plant species. Funct Plant Biol 33:789–797

Griga M, Stejskal J, Beber K (1995) Analysis of tissue culture-derived variation in pea (*Pisum sativum* L). Preliminary results. Euphytica 85:335–339

Guenther JF, Chanmanivone N, Galetovic MP, Wallace IS, Cobb JA, Roberts DM (2003) Phosphorylation of soybean nodulin 26 on serine 262 enhances water permeability and is regulated developmentally and by osmotic signals. Plant Cell 15:981–991

Gulati A, Schryer P, McHughen A (2002) Production of fertile transgenic lentil (Lens culinaris Medik) plants using particle bombardment. In Vitro Cell Dev Biol Plant 38:316–324

Guptha AK, Hasan W, Singh D (2014) Integrated management of major pests and diseases in pigeonpea. Krishi Sewa e Publication, 20 Sept 2014

Gygi SP, Aebersold R (2000) Mass spectrometry and proteomics. Curr Opin Chem Biol:489–494

Hammer GL, Dong ZS, McLean G, Doherty A, Messina C, Schusler J, Zinselmeier C, Paszkiewicz S, Cooper M (2009) Can changes in canopy and/or root system architecture explain historical maize yield trends in the US corn belt? Crop Sci 49:299–312

Hammett KRW, Murray BG, Markham KR, Hallett IC (1994) Interspecific hybridization between Lathyrus odoratus and L. belinensis. Int J Plant Sci 155:763–771

Hamwieh A, Udupa SM, Choumane W, Sarker A, Dreyer F, Jung C, Baum M (2005) A genetic linkage map of Lens sp based on microsatellite and AFLP markers and the localization of fusarium vascular wilt resistance. Theor Appl Genet 110:669–677

Handberg K, Stiller J, Thykjær T, Stougaard J (1994) Transgenic plants: Agrobacterium mediated transformation of the diploid model legume Lotus japonicus. In: Cells JE (eds) Cell biology: a laboratory handbook. Academic Press

Hannon G (2002) Mechanisms and application of RNA interference. Biol Reprod 66:70

Hao X, Li P, Feng Y, Han X, Gao J, Lin E, Han Y (2013) Effects of fully open-air (CO_2) elevation on leaf photosynthesis and ultrastructure of *Isatis indigotica* Fort. PLoS ONE 8:e74600

He XZ, Dixon RA (2000) Genetic manipulation of isoflavone 7-O-methyltransferase enhances biosynthesis of 4′-O-methylated iso flavonoid phytoalexins and disease resistance in alfalfa. Plant Cell 12:1689–1702

He SY, Jin Q (2003) The Hrp pilus: learning from flagella. Curr Opin Microbiol 6:15–19

Henshaw TL, Gilbert RA, Scholberg JMS, Sinclair TR (2007) Soya bean (*Glycine max* L. Merr.) genotype response to early-season flooding: I. Root and nodule development. J Agron Crop Sci 193:177–188

Higgins TJV, Gollasch S, Molvig L, Moore A, Popelka C, Armstrong J, Mahon R, Ehlers J, Huesing J, Margam V, Shade R, Murdock L (2012) Insect-protected cowpeas using gene technology. In: Boukar OC, Fatokun CA, Lopez K, Tamo M (eds) Innovative research along the cowpea value chain. Proceedings of the fifth world cowpea conference, Saly, Senegal, pp 131–137

Hiremath PJ, Farmer A, Cannon SB, Woodward J, Kudapa H, Tuteja R, Kumar A, Bhanuprakash A, Mulaosmanovic B, Gujaria N (2011) Large-scale transcriptome analysis in chickpea (*Cicer arietinum* L.), an orphan legume crop of the semi-arid tropics of Asia and Africa. Plant Biotechnol J 9:922–931

Hogenhout SA, Ammar ED, Whitfield AE, Redinbaugh MG (2008) Insect vector interactions with persistently transmitted viruses. Annu Rev Phytopathol 46:327–359

Hu HH, Dai MQ, Yao JL, Xiao BZ, Li XH, Zhang QF (2006) Overexpressing *aNAM, ATAE* and *CUC (NAC)* transcription factor enhances drought resistance and salt tolerance in rice. Proc Natl Acad Sci U S A 103:12987–12992

Infantino A, Kharrat M, Riccioni L, Coyne CJ, McPhee KE, Grunwald NJ (2006) Screening techniques and sources of resistance to root diseases in cool season food legumes. Euphytica 147:201–221

International Committee on Taxonomy of Viruses (ICTV) (2012) USA 109: 1011–1018

Iqbal MJ, Yaegashi S, Ahsan R, Shopinski KL, Lightfoot DA (2005) Root response to Fusarium solani f. sp. glycines: temporal accumulation of transcripts in partially resistant and susceptible soybean. Theor Appl Genet 110:1429–1438

Jackson MB, Ram PC (2003) Physiological and molecular basis of susceptibility and tolerance of rice plants to complete submergence. Ann Bot 91:227–241

Jagtap GP, Dhopte SB, Dey U (2012) Bio-efficacy of different antibacterial antibiotic, plant extracts and bioagents against bacterial blight of soybean caused by Pseudomonas syringae pv. glycinea. Sci J Microbiol 1:1–9

Jiang H, Liao B, Ren X, Lei Y, Mace E, Fu T, Crouch JH (2007) Comparative assessment of genetic diversity of peanut (Arachis hypogaea L.) genotypes with various levels of resistance to bacterial wilt through SSR and AFLP analyses. J Genet Genomics 34:544–554

Jiang W, Zhou H, Bi H, Fromm M, Yang B, Weeks DP (2013) Demonstration of CRISPR/Cas9/sgRNA-mediated targeted gene modification in Arabidopsis, tobacco, sorghum and rice. Nucleic Acids Res 41:e188

Jiao Y, Lau OS, Wang DX (2007) Light-regulated transcriptional networks in higher plants. Nat Rev Genet 8:217

Jitsuyama Y (2015) Morphological root responses of soybean to rhizosphere hypoxia reflect waterlogging tolerance. Can J Plant Sci 95:999–1005

Kamphuis LG, Williams AH, Küster H, Trengove RD, Singh KB, Oliver RP, Ellwood SR (2012) Phoma medicaginis stimulates the induction of the octadecanoid and phenylpropanoid pathways in Medicago truncatula. Mol Plant Pathol 13:593–603

Kang J, Xie W, Sun Y, Yang Q, Wu M (2010) Identification of genes induced by salt stress from Medicago truncatula L. seedlings. Afric. J Biotechnol 9:7589–7594

Kareem KT, Taiwo MA (2007) Interaction of viruses in cowpea: effect on growth and yield parameters. Virol J 4:15

Karpinski S, Escobar C, Karpinska B, Creissen G, Mullineaux PM (1997) Photosynthetic electron transport regulates the expression of cytosolic ascorbate peroxidase genes in Arabidopsis during excess light stress. Plant Cell 9:627–640

Kassem MA, Meksem K, Kang CH, Njiti VN, Kilo V, Wood AJ, Lightfoot DA (2004) Loci underlying resistance to manganese toxicity mapped in a soybean recombinant inbred line population of 'Essex' x 'Forrest'. Plant Soil 260:197–204

Kato T, Sato S, Nakamura Y, Kaneko T, Asamizu E, Tabata S (2003) Structural analysis of a Lotus japonicus genome. V. Sequence features and mapping of sixty-four TAC clones which cover the 6.4 Mb regions of the genome. DNA Res 10:277–285

Keneni G, Bekele E, Getu E, Imtiaz M, Damte BM, Dagne K (2011) Breeding food legumes for resistance to storage insect pests: potential and limitations. Sustainability 3:1399–1415

Khan AJ, Hassan S, Tariq M, Khan T (2001) Haploidy breeding and mutagenesis for drought tolerance in wheat. Euphytica 120:409–414

Kim GB, Nam YW (2013) A novel _1-pyrroline-5-carboxylate synthetase gene of Medicago truncatula plays a predominant role in stressinduced proline accumulation during symbiotic nitrogen fixation. J Plant Physiol 170:291–302

Kim JB, Kang JY, Kim SY (2004) Over-expression of a transcription factor regulating ABA-responsive gene expression confers multiple stress tolerance. Plant Biotechnol J 2:459–466

Kim CS, Schaible GD, Garrett L, Lubowski RN, Lee DJ (2008) Economic impacts of the US soybean aphid infestation: a multi-regional competitive dynamic analysis. Agric Resour Econ Rev 37:227–242

Kim YH, Hwang S, Wagas M, Khan AL, Lee JH, Lee JD, Nguyen HT, Lee IJ (2015) Comparative analysis of endogenous hormones level in two soybean (*Glycine max* L.) lines differing in waterlogging tolerance. Front Plant Sci 6:714

Kim HJ, Cho HS, Pak JH, Kwon T, Lee J, Kim D, Lee DH, Kim C, Chung Y (2018) Molecules and cells confirmation of drought tolerance of ectopically expressed *AtABF3* gene in soybean. Mol Cell 41:413–422

Klimczak LJ, Colline MA, Farini D (1995) Reconstitution of Arabidopsis casein kinase II from recombinant subunits and phosphorylation of transcription factor GBF1. Plant Cell 7:105–115

Koenning SR, Overstreet C, Noling JW, Donald PA, Becker JO, Fortnum BA (1999) Survey of crop losses in response to phytoparasitic nematodes in the United States for 1994. J Nematol 31:587–618

Komatsu S, Deschamps T, Hiraga S, Kato M, Chiba M, Hashiguchi A, Tougou M, Shimamura S, Yasue H (2011) Characterization of a novel flooding stress-responsive alcohol dehydrogenase expressed in soybean roots. Plant Mol Biol 77:309–322

Kraft JM, Pfleger FL (2001) Compendium of pea diseases and pests. APS Press, St Paul

Kulikova O, Gualtieri G, Geurts R, Kim DJ, Cook D, Huguet T, de Jong JH, Fransz PF, Bisseling T (2001) Integration of the FISH pachytene and genetic maps of *Medicago truncatula*. Plant J 27:49–58

Kumar S, Malik J, Thakur P, Kaistha S, Sharma KD, Upadhyaya HD, Berger JD, Nayyar H (2011) Growth and metabolic responses of contrasting chickpea (*Cicer arietinum* L.) genotypes to chilling stress at reproductive phase. Acta Physiol Plant 33:779–787

Kumar S, Kaushal N, Nayyar H, Gaur P (2012) Abscisic acid induces heat tolerance in chickpea (Cicer arietinum L.) seedlings by facilitated accumulation of osmoprotectants. Acta Physiol Plant 34:1651–1658. https://doi.org/10.1007/s11738-012-0959-1

Kushwaha C, Chand R, Srivastava C (2006) Role of aeciospores in outbreaks of pea (Pisum sativum) rust (Uromyces fabae). Eur J Plant Pathol 115:323–330

Larkin PJ, Scowcroft WR (1981) Somaclonal variation – a novel source of variability from cell-cultures for plant Improvement. Theor Appl Genet 60:197–214

Lazaridou TB, Roupakias DG, Economou AS (1993) Embryo rescue in Vicia faba and Vicia narbonensis. Plant Cell Tissue Organ Cult 33:297–301

Lee GJ, Boerma HR, Villagarcia MR, Zhou XT, Carter E, Li Z, Gibbs MO (2004) A major QTL conditioning salt tolerance in S-100 soybean and descendent cultivars. Theor Appl Genet 109:1610–1619

Lee HA, Lee HY, Seo E, Lee J, Kim SB, Oh S (2017) Current understandings of plant nonhost resistance. Mol Plant-Microbe Interact 30:5–15

Levitt J (1980) Responses of plant to environmental stresses, 2nd edn. Academic Press, New York

Lewis G, Schrire B, Mackinder B, Lock M (2005) Legumes of the world. Royal Botanic Gardens, Kew

Li HY, Zhu YM, Chen Q, Conner RL, Ding XD, Li J, Zhang BB (2004) Production of transgenic soybean plants with two anti-fungal protein genes via Agrobacterium and particle bombardment. Biol Plant 48:367–374

Li WYF, Wong FL, Tsai SN, Phang TH, Shao G, Lam HM (2006) Tonoplast-located GmCLC1 and GmNHX1 from soybean enhance NaCl tolerance in transgenic bright yellow (BY)-2 cells. Plant Cell Environ 29:1122–1137

Li D, Zhang Y, Hu X, Shen X, Ma L, Su Z, Wang T, Dong J (2011) Transcriptional profiling of Medicago truncatula under salt stress identified a novel CBF transcription factor MtCBF4 that plays an important role in abiotic stress responses. BMC Plant Biol 11:109

Li H, Wang Z, Ke Q, Ji CY, Jeong JC, Lee HS, Lim YP, Xu B, Deng XP, Kwak SS (2014) Overexpression of *codA* gene confers enhanced tolerance to abiotic stresses in alfalfa. Plant Physiol Biochem 85:31–40

Li Y, Chen Q, Nan H, Li X, Lu S, Zhao X, Liu B, Guo C, Kong F, Cao D (2017) Over expression of *GmFDL19* enhances tolerance to drought and salt stresses in soybean. PLoS ONE 12:e0179554

Li DH, Li W, Li HY, Guo JJ, Chen FJ (2018a) The soybean *GmRACK1* gene plays a role in drought tolerance at vegetative stages. Russ J Plant Physiol 65:541–552

Li J, Zhang Y, Zhang Y, Yu P, Pan PL, Rollins JA (2018b) Introduction of large sequence inserts by crispr-cas9 to create pathogenicity mutants in the multinucleate filamentous pathogen *Sclerotinia sclerotiorum*. MBio 9:e00567

Liu YH, Offler CE, Ruan YL (2013) Regulation of fruit and seed response to heat and drought by sugars as nutrients and signals. Front Plant Sci 4:282

Loebenstein G, Thottapilly G (2003) Virus and virus-like diseases of major crops in developing countries. Sweet Potato. Academic Publishers pp 223–248

Lopez M, Tejera NA, Iribarne C, Lluch C, Herrera-Cervera JA (2008) Trehalose and trehalase in root nodules of Medicago truncatula and Phaseolus vulgaris in response to salt stress. Physiol Plant 134:575–582

Lozovaya VV, Lygin AV, Li S, Hartman GL, Widhohn JM (2004) Biochemical response of soybean roots to Fusarium solani f. sp glycines infection. Crop Sci 44:819–826
Luo Q, Yua B, Liua Y (2005) Differential sensitivity to chloride and sodium ions in seedlings of Glycine max and G. soja under NaCl stress. J Plant Physiol 162:1003–1012
Mackie JM, Musial JM, Armour DJ, Phan HTT, Ellwood SE, Aitken KS, Irwin JAG (2007) Identification of QTL for reaction to three races of Colletotrichum trifolii and further analysis of inheritance of resistance in autotetraploid lucerne. Theor Appl Genet 114:1417–1426
Makkouk KM, Kumari SG (2009) Epidemiology and integrated management of persistently transmitted aphid-borne viruses of legume and cereal crops in West Asia and North Africa. Virus Res 141:209–218
Makkouk KM, Kumari SG, van Leur JAG, Jones RAC (2014) Control of plant virus diseases in cool-season grain legume crops. Adv Virus Res 90:207–253
Mallikarjuna N, Moss JP (1995) Production of hybrids between Cajanus platycarpus and Cajanus cajan. Euphytica 83:43–46
Martin GB, Bogdonove AJ, Sessa G (2003) Understanding the functions of plant disease resistance proteins. Annu Rev Plant Biol 54:23–61
Massoud FI (1974) Salinity and alkalinity. In: A world assessment of soil degradation. An international program of soil conservation. Report of an expert consultation on soil degradation. FAO, UNEP, Rome, pp 16–17
Mehrtens F, Kranz H, Bednarek P, Weisshaar B (2005) The Arabidopsis transcription factor MYB12 is a flavonol-specific regulator of phenylpropanoid biosynthesis. Plant Physiol 138:1083–1096
Merchan F, de Lorenzo L, Rizzo SG, Niebel A, Manyani H, Frugier F, Sousa C, Crespi M (2007) Identification of regulatory pathways involved in the reacquisition of root growth after salt stress in Medicago truncatula. Plant J 51:11–17
Mhadhbi H, Fotopolos V, Mylona PV, Jebara M, Aouani ME, Polidoros AN (2011) Antioxidant gene-enzyme responses in Medicago truncatula genotypes with different degree of sensitivity to salinity. Physiol Plant 141:201–214
Michno JM, Wang X, Liu J, Curtin SJ, Kono TJ, Stupar RM (2015) CRISPR/Cas mutagenesis of soybean and *Medicago truncatula* using a new web-tool and a modified Cas9 enzyme. GM Crops Food 6:243–252
Miklas PN, Kelly JD, Beebe SE, Blair MW (2006) Common bean breeding for resistance against biotic and abiotic stresses: from classical to MAS breeding. Euphytica 147:105–131
Miller RW, Donahue RL (1990) Chelates and nutrient availability. In: Miller RW, Donahue RL (eds) Soils, an introduction to soils and plant growth. Prentice-Hall Inc, Englewood Cliffs, pp 295–296
Miller PR, Mc Conkey BG, Clayton GW, Brandt SA, Staricka JA, Johnston AM (2002) Pulse crop adaptation in the northern great plains. Agron J 94:261–272
Mingpeng H, Yongge G, Chengzhang W, Fangrui S, Yanhua W, Xiaoxia Z (2010) Related studies on the effects of high temperature stress on alfalfa and its heat resistance mechanism. Genom Appl Biol 29:563–569
Morphew R, Arndt R, Lauter C, Tesfaye M, Samac D, Temple G (2004) Characterization of altered metabolism in transgenic alfalfa overexpressing malate dehydrogenase to confer aluminum tolerance. Abstr Pap Am Chem Soc 227:148
Moy P, Qutob D, Chapman BP, Atkinson I, Gijzen M (2004) Patterns of gene expression upon infection of soybean plants by *Phytophthora sojae*. Mol Plant-Microbe Interact 17:1051–1062
Mutlu N, Miklas P, Reiser J, Coyne D (2005) Backcross breeding for improved resistance to common bacterial blight in pinto bean (*Phaseolus vulgaris* L.). Plant Breed 124:282–287
Nayak SN, Balaji J, Upadhyaya HD, Hash CT, Kishor PBK, Blair MW, Baum M, Chattopadhyay D, Marı L, Mcnally K (2009) Plant science isolation and sequence analysis of *DREB2A* homologues in three cereal and two legume species. Plant Sci 177:460–467
Nayyar H, Bains TS, Kumar S (2005) Chilling stressed chickpea seedlings: effect of cold acclimation, calcium and abscisic acid on cryoprotective solutes and oxidative damage. Environ Exp Bot 54:275–285. https://doi.org/10.1016/j.envexpbot.2004.09.007

Nemeskeri E (2004) Study of heat tolerance during germination in grain legumes. In: ISTA seed symposium. Abstracts. Budapest, Hungary, 17–19 May, pp 85–86
Nene YL, Reddy MV (1987) Chickpea diseases and their control. In: Saxena MC, Singh KB (eds) The chickpea, CAB International, Oxon, pp 233–270
Newton AC, Johnson SN, Gregory PJ (2011) Implications of climate change for diseases, crop yields and food security. Euphytica 179:3–18
Nguyen VT, Vuong TD, Van TT, Lee JD, Wu X, Mian MAR, Dorrance AE, Shannon JG, Nguyen HT (2012) Mapping of quantitative trait loci associated with resistance to and flooding tolerance in soybean. Crop Sci 52:2481–2493
Nguyen VL, Takahashi R, Githiri SM, Rodriguez TO, Tsutsumi N, Kajihara S, Sayama T, Ishimoto M, Harada K, Suematsu K (2017) Mapping quantitative trait loci for root development under hypoxia conditions in soybean (Glycine max L. Merr.). Theor Appl Genet 130:743–755
Nigam SN, Prasada Rao RDVJ, Bhatnagar-Mathur P, Sharma KK (2012) Genetic management of virus diseases in peanut. Plant Breeding Rev 36:293–356
Onate SL, Singh KB (2002) Identification of Arabidopsis ethylene-responsive element binding factors with distinct induction kinetics after pathogen infection. Plant Physiol 128:1313–1322
Ouedraogo JT, Tignegre JB, Timko P, Belzile FJ (2002) AFLP markers linked to resistance against *Striga gesnerioides* race 1 in cowpea (*Vigna unguiculata*). Genome 45(3):787–793
Ozawa R, Arimura G, Takabayashi J, Shimoda T, Nishioka T (2000) Involvement of jasmonate- and salicylate-related signalling pathways for the production of specific herbivore-induced volatiles in plants. Plant Cell Physiol 41:391–398
Palta JA, Ganjeali A, Turner NC, Siddique KHM (2010) Effects of transient subsurface waterlogging on root growth, plant biomass and yield of chickpea. Agric Water Manag 97:1469–1476
Pande S, Stevenson PC, Rao JN, Neupane RK, Grzywacz D, Bourai VA, Krishna-Kishnore G (2005) Riviving chickpea production in Napel through integrated crop management, with emphasis on botrytis gray mold. Plant Dis 89:1252–1262
Pande S, Sharma M, Kumari S, Gaur PM, Chen W, Kaur L, MacLeod W, Basandrai A, Basandrai D, Bakr A, Sandhu JS, Tripathi HS, Gowda CLL (2009) Integrated foliar diseases management of legumes. In: Proceedings of international conference on grain legumes: quality improvement, value addition and trade, 14–16 February 2009. Indian Institute of Pulses Research, Kanpur, pp 143–167
Parker C (2009) Observations on the current status of Orobanche and Striga problems worldwide. Pest Manag Sci 65:453–459
Pedrosa A, Sandal N, Stougaard J, Schweizer D, Bachmair A (2002) Chromosomal map of the model legume Lotus japonicus. Genetics 161:1661–1672
Pessarakli M (1999) Response of green beans (*Phaseolus vulgaris* L.) to salt stress. In: Pessarakli M (ed) Handbook of plant and crop stress, 2nd edn. MARCED LEKKERIN, C.E.E.U.U, New York, pp 827–842
Pfannschmidt T, Nilsson A, Allen JF (1999) Photosynthetic control of chloroplast gene expression. Nature 397:625–628
Piramila BHM, Prabha AL, Nandagopalan V, Stanley AL (2012) Effect of heat treatment on germination, seedling growth and some biochemical parameters of dry seeds of black gram. Int J Pharm Phytopharmacol Res 1:194–202
Poehlman JM (1991) The mungbean. IBH Publishing Co. Pvt. Ltd., New Delhi/Oxford, pp 27–30
Pooniya V, Choudhary AK, Dass A, Bana RS, Rana KS, Rana DS, Tyagi VK, Puniya MM (2015) Improved crop management practices for sustainable pulse production: an Indian perspective. Indian J Agric Sci 85(6):747–758
Popelka JC, Terryn N, Higgins TJV (2004) Gene technology for grain legumes: can it contribute to the food challenge in developing countries? Plant Sci 167:195–206
Prince SJ, Mutava RN, Pegoraro C, Oliveira ACD, Nguyen HT (2013) Root characters. In: Kole C (ed) Genomics and breeding for climate resilient crops. Springer, Berlin, pp 67–131

Pruthvi V, Narasimhan R, Nataraja KN (2014) Simultaneous expression of abiotic stress responsive transcription factors, *AtDREB2A, AtHB7* and *AtABF3* improves salinity and drought tolerance in peanut (*Arachis hypogaea* L.). PLoS One 9(12):e111152

Ramón S, Cecilia C, Gabriel I (2009) Enhanced tolerance to multiple abiotic stresses in transgenic alfalfa accumulating trehalose. Crop Sci 49:1791–1799

Reddy AA (2006) Impact assessment of pulses production technology, Research Report No 3. Indian Institute of Pulses Research, Kanpur

Reddy AR, Chaitanya KV, Vivekanandan M (2004) Drought-induced responses of photosynthesis and antioxidant metabolism in higher plants. J Plant Physiol 161:1189–1202

Riches CR, Hamilton K, Parker C (1992) Parasitism of grain legumes by Alectra species (Scrophulariaceae). Ann Appl Biol 121:361–370

Rispail N (2005) Molecular and metabolic characterisation of symbiotic interactions in Lotus japonicus. PhD thesis, Institute of Grassland and Environmental Research (IGER), University of Wales, Aberystwyth

Robatzek S, Somssich IE (2002) Targets of AtWRKY6 regulation during plant senescence and pathogen defense. Genes Dev 16:1139–1149

Robinson MF, Very AA, Sanders D, Mansfield TA (1997) How can stomata contribute to salt tolerance? Ann Bot 80:387–393

Rockstrom J, Falkenmark M (2000) Semiarid crop production from a hydrological perspective: gap between potential and actual yields. Crit Rev Plant Sci 19:319–346

Rodrigues LR, Oliveira JMS, Mariath JEA, Bodanese-Zanettini MH (2005) Histology of embryogenic responses in soybean anther culture. Plant Cell Tissue Organ Cult 80:129–137

Rodriguez LD, Lemmon ZH, Man J, Bartlett ME, Lippman ZB (2017) Engineering quantitative trait variation for crop improvement by genome editing. Cell 171:470–480

Roman B, Torres AM, Rubiales D, Cubero JI, Satovic Z (2002) Mapping of quantitative trait loci controlling broomrape (*Orobanche crenata* Forsk.) resistance in faba bean (*Vicia faba* L.). Genome 45:1057–1063

Rubiales D, Fernández-Aparicio M (2012) Innovations in parasitic weeds management in legume crops. A review. Agron Sustain Dev 32:433–449

Rubiales D, Emeran AAJ, Sillero C (2002) Rusts on legumes in Europe and North 22 Africa. Grain Legumes 37:8–9

Rubiales D, Pérez-de-Luque A, Sillero JC, Román B, Kharrat M, Khalil S, Joel DM, Riches CR (2006) Screening techniques and sources of resistance against parasitic weeds in grain legumes. Euphytica 147:187–199

Rubiales D, Fondevilla S, Chen W, Gentzbittel L, Higgins TJV, Castillejo MA (2015) Achievements and challenges in legume breeding for pest and disease resistance. Crit Rev Plant Sci 34:195–236

Rubio MC, Bustos-Sanmamed P, Clemente MR, Becana M (2009) Effects of salt stress on the expression of antioxidant genes and proteins in the model legume Lotus japonicus. New Phytol 181:851–859

Sagan M, Huguet T, Duc G (1994) Phenotypic characterization and classification of nodulation mutants of pea (*Pisum sativum* L.). Plant Sci 100:59–70

Sakazono S, Nagata T, Matsuo R, Kajihara S, Watanabe M, Ishimoto M, Shimamura S, Harada K, Takahashi R, Mochizuki T (2014) Variation in root development response to flooding among 92 Soybean lines during early growth stages. Plant Prod Sci 17:228–236

Saleh AAH, Abdel-Kader DZ, El Elish AM (2007) Role of heat shock and salicylic acid in antioxidant homeostasis in Mung bean (*Vigna radiata* L.) plant subjected to heat stress. Am J Plant Physiol 2:344–355

Salles II, Blount JW, Dixon RA, Schubert K (2002) Phytoalexin induction and beta-1,3- glucanase activities in *Colletotrichum trifolii* infected leaves of alfalfa (*Medicago sativa* L.). Physiol Mol Plant Pathol 61:89–101

Samac D, Nix RJ, Oleson AE (1998) Transmission frequency of *Clavibacter michiganensis* subsp. insidiosus to alfalfa seeds and identification of the bacterium by PCR. Plant Disease 82:1362–1367

Samac DA, Tesfaye M, Dornbusch M, Saruul PSJ, Temple (2004) A comparison of constitutive promoters for expression of transgenes in alfalfa (*Medicago sativa*). Transgenic Res 13:349–361

Samantaray S, Rout GR, Das P (1999) In vitro selection and regeneration of zinc tolerant calli from Setaria italica L. Plant Sci 143:201–209

Sanchez DH, Lippold F, Redestig H, Hannah MA, Erban A, Kramer U, Kopka J, Udvardi MK (2008) Integrative functional genomics of salt acclimatization in the model legume Lotus japonicus. Plant J 53:973–987

Santos JRP, Ndeve AD, Huynh BL, Matthews WC, Roberts PA (2018) QTL mapping and transcriptome analysis of cowpea reveals candidate genes for root-knot nematode resistance. PLoS One 13:e0189185

Sasidharan R, Bailey Serres J, Ashikari M, Atwell BJ, Colmer TD, Fagerstedt K et al (2017a) Community recommendations on terminology and procedures used in flooding and low oxygen stress research. New Phytol 214:1403–1407

Sasidharan R, Hartman S, Liu Z, Martopawiro S, Sajeev N, van Veen H, Yeung E, Voesenek LACJ (2017b) Signal dynamics and interactions during flooding stress. Plant Physiol 176:1106

Sasser JN, Eisenback JD, Carter CC, Triantaphyllou AC (1983) The International Meloidogyne project – its goals and accomplishments. Annu Rev Phytopathol 21:271–288

Sastry KS, Zitter TA (2014) Plant virus and viroid diseases in the tropics. Epidemiol Manag 2:489

Satyagopal K, Sushil SN, Jeyakumar P, Shankar G, Sharma OP, Boina DR, Sain SK, Lavanya N, Sunanda BS, Asre R, Kapoor KS, Arya S, Kumar S, Patni CS, Jacob TK, Santhosh J, Eapen CN, Biju K, Dhanapal H, Ravindra BC, Hanumanthaswamy Raju LS, Babu R, Sathyanarayana L, Latha S (2014a) AESA based IPM Package for Redgram. National Institute of Plant Health Management, Hyderabad, p 42

Satyagopal K, Sushil SN, Jeyakumar P, Shankar G, Sharma OP, Sain SK, Boina DR, Lavanya N, Sunanda BS, Asre R, Kapoor KS, Arya S, Kumar S, Patni CS, Dhanapal AN, Sabalpara SK, Beura CN, Biju BG, Naik RK, Mesta TK, Jacob N, Sathyanarayana L, Latha S (2014b) AESA based IPM package for Black gram and Green gram. National Institute of Plant Health Management, Hyderabad, p 43

Saunders J, O'Neill N (2004) The characterization of defense responses to fungal infection in alfalfa. Biol Control 49:715–728

Savitri ES, Fauziah SM (2018) Characterization of drought tolerance of *GmDREB2* soybean mutants (*Glycine max* L.) by ethyl methane sulfonate induction. AIP Conf Proc 2019:020017

Schafleitner R, Huang SM, Chu SH, Yen JY, Lin CY, Yan MR, et al (2016) Identification of single nucleotide polymorphism markers associated with resistance to bruchids (*Callosobruchus* spp.) in wild mungbean (*Vigna radiata* var. sublobata) and cultivated V. radiata through genotyping by sequencing and quantitative trait locus analysis. BMC Plant Biol 16:159

Scheelbeek PFD (2018) Effect of environmental changes on vegetable and legume yields 435 and nutritional quality. Proc Natl Acad Sci U S A 115:6804–6809

Schena M, Shalon D, Davis RW, Brown PO (1995) Quantitative monitoring of gene2 expression patterns with a complementary-DNA microarray. Science 270:467–470

Schneider KA, Brothers ME, Kelly JD (1997) Marker-assisted selection to improve drought resistance in common bean. Crop Sci 37:51–60

Schneider KA, Grafton KF, Kelly JD (2001) QTL analysis of resistance to fusarium root rot in bean. Crop Sci 41:535–542

Seki M, Kamei A, Yamaguchi-Shinozaki K, Shinozaki K (2003) Molecular responses to drought, salinity and frost: common and different paths for plant protection. Curr Opin Biotechnol 14:194–199

Shao HB, Chu LY, Jaleel CA, Zhao CX (2008) Water-deficit stress induced anatomical changes in higher plants. Comptes Rendus Biologies 331:215–225

Sharma KK, Lavanya M (2002) Recent developments in transgenics for abiotic stress in legumes of the semi-arid tropics. In: Ivanaga M (ed) Genetic engineering of crop plants for abiotic stress. JIRCAS Working Report No. 23, JIRCAS: Tsukuba, Japan, pp 61–73

Sharma KD, Nayyar H (2014) Cold stress alters transcription in meiotic anthers of cold tolerant chickpea (*Cicer arietinum* L.). BMC Res Notes 7:717. https://doi.org/10.1186/1756-0500-7-717

Shimada N, Akashi T, Aoki T, Ayabe S (2000) Induction of iso flavonoid pathway in the model legume Lotus japonicus: molecular characterization of enzymes involved in phytoalexin biosynthesis. Plant Sci 160:37–47

Siddique KHM, Sykes J (1997) Pulse production in Australia past, present and future. Aust J Exp Agric 37:103–111

Sillero JCS, Fondevilla J, Davidson MC, Vaz Patto TD, Warkentin J, Thomas J, Rubiales D (2005) Screening techniques and sources of resistance to rusts and mildews in grain legumes. Euphytica 147:255–272

Sillero JC, Fondevilla S, Davidson J, Vaz Patto MC, Warkentin TD, Thomas J, Rubiales D (2006) Screening techniques and sources of resistance to rusts and mildews in grain legumes. Euphytica 147:255–272

Sillero C, Villegas-Fernández AM, Thomas J, Rojas-Molina MM, Emeran AA, Fernández-Aparicio M, Rubiales D (2010) Faba bean breeding for disease resistance. Field Crop Res 115:297–307

Sillero JC, Moreno-Alía I, Rubiales D (2012) Identification and characterization of resistance to rust (Uromyces ciceris-arietini (Grognot) Jacz. & Boyd) in a germplasm collection of Cicer spp. Euphytica 188:229–238

Singh NH, Dhaliwal GS (1972) Effect of soil temperature on seedling emergence in different crops. Plant Soil 37:441–444

Singh DP, Singh BB (2011) Breeding for tolerance to abiotic stresses in mung bean. J Food Legumes 24:83–90

Singh KB, Foley RC, Onate-Sanchez L (2002) Transcription factors in plant defense and stress responses. Curr Opin Plant Biol 5:430–436

Solaiman Z, Colmer TD, Loss SP, Thomson BD, Siddique KHM (2007) Growth responses of cool-season grain legumes to transient waterlogging. Aust J Agric Res 58:406–412

Somers DA, Samac DA, Olhoft PM (2003) Recent advances in legume transformation. Plant Physiol 131:892–899

Somta P, Ammaranan C, Ooi PAC, Srinives P (2007) Inheritance of seed resistance to bruchids in cultivated mung bean (*Vigna radiata* L. Wilczek). Euphytica 155:47–55

Southgate BJ (1979) Biology of the bruchids. Annu Rev Entomol 24:449–473

Srinivasan A, Takeda H, Senboku T (1996) Heat tolerance in food legumes as evaluated by cell membrane thermostability and chlorophyll fluorescence techniques. Euphytica 88:35–45

Srivastava R, Kumar S, Kobayashi Y, Kusunoki K, Tripathi P, Kobayashi Y, Koyama H, Sahoo L (2018) Comparative genome-wide analysis of WRKY transcription factors in two Asian legume crops: Adzuki bean and Mung bean. Sci Rep 8:16971

Stoddard FL, Nicholas AH, Rubiales D, Thomas J, Villegas AM (2010) Integrated pest management in faba bean. Field Crop Res 115:308–318

Subramanian S, Graham MY, Yu O, Graham TL (2005) RNA interference of soybean isoflavone synthase genes leads to silencing in tissues distal to the transformation site and to enhanced susceptibility to *Phytophthora sojae*. Plant Physiol 137:1345–1353

Sumner LW, Mendes P, Dixon RA (2003) Plant metabolomics: large-scale phytochemistry in the functional genomics era. Phytochemistry 62:817–836

Taïbi K, Taïbi F, Abderrahim LA, Ennajah A, Belkhodja M, Mulet JM (2016) Effect of salt stress on growth, chlorophyll content, lipid peroxidation and antioxidant defence systems in *Phaseolus vulgaris* L. S Afr J Bot 105:306–312

Tang L, Cai H, Ji W, Luo X, Wang Z, Wu J, Wang X, Cui L, Wang Y, Zhu Y et al (2013) Overexpression of *GsZFP1* enhances salt and drought tolerance in transgenic alfalfa (*Medicago sativa* L.). Plant Physiol Biochem 71:22–30

Taylor JD, Teverson DM, Allen DJ, Pastor-Corrales MA (1996) Identification and origin of races of Pseudomonas syringae pv. phaseolicola from Africa and other bean growing areas. Plant Pathol 45:469–478

Teakle NL, Amtmann A, Real D, Colmer TD (2010) Lotus tenuis tolerates combined salinity and waterlogging: maintaining O_2 transport to roots and expression of an NHX1-like gene contribute to regulation of Na+ transport. Physiol Plant 139:358–374

Tegli S, Sereni A, Surico G (2002) PCR-based assay for the detection of Curtobacterium flaccumfaciens pv. flaccumfaciens in bean seeds. Lett Appl Microbiol 35:331–337

Thomashow MF (1999) Plant cold acclimation: freezing tolerance genes and regulatory mechanisms. Annu Rev Plant Physiol Plant Mol Biol 50:571–599

Thomashow MF (2001) So what's new in the field of plant cold acclimation? Lots!: Fig. 1. Plant Physiology 125(1):89–93

Thoquet P, Gherardi M, Journet EP, Kereszt A, Ane JM, Prosperi JM, Huguet T (2002) The molecular genetic linkage map of the model legume Medicago truncatula: an essential tool for comparative legume genomics and the isolation of agronomically important genes. BMC Plant Biol 2:1

Timmerman VGM, Moya L, Frew TJ, Murray SR, Crowhurst R (2016) *Ascochyta blight* disease of pea (*Pisum sativum* L.): defence-related candidate genes associated with QTL regions and identification of epistatic QTL. Theor Appl Genet 129:879–896

Tivoli B, Baranger A, Avila CM, Banniza S, Barbetti M, Chen W, Davidson J, Lindeck K, Kharrat M, Rubiales D, Sadiki M, Sillero JC, Sweetingham M, Muehlbauer FJ (2006) Screening techniques and sources of resistance to foliar diseases caused by major necrotrophic fungi in grain legumes. Euphytica 147:223–253

Toker C, Yadav SS (2010) Legumes cultivars for stress environments. In: Yadav SS, McNeil DL, Redden R, Patil SA (eds) Climate change and management of cool season grain legume crops. Springer, Dordrecht

Trabanco N, Pérez-Vega E, Campa A, Rubiales D, Ferreira JJ (2012) Genetic resistance to powdery mildew in common bean. Euphytica 186:875–882

Tran L, Nakashima K, Sakuma Y, Simpson SD, Fujita Y, Maruyama K, Fujita M, Seki M, Shinozaki K (2004) Isolation and functional analysis of *Arabidopsis* stress-inducible NAC transcription factors that bind to a drought responsive *cis*-element in the early responsive to dehydration stress promoter. Plant Cell 6:2481–2498

Tullu A, Banniza S, Taran B, Warkentin T, Vandenberg A (2010) Sources of resistance to ascochyta blight in wild species of lentil (Lens culinaris Medik.). Genet Resour Crop Evol 57:1053–1063

Turner M, Jauneau A, Genin S, Tavella MJ, Vailleau F, Gentzbittel L, Jardinaud MF (2009) Dissection of bacterial wilt on Medicago truncatula revealed two type III secretion system effectors acting on root infection process and disease development. Plant Physiol 150:1713–1722

Uemura M, Joseph RA, Steponkus PL (1995) Cold acclimation of *Arabidopsis thaliana*: effect on plasma membrane lipid composition and freeze-induced lesions. Plant Physiol 109:15–30

Valliyodan B, Van Toai TT, Alves JD, Goulart PDP, Lee JD, Fritschi FB, Rahman MA, Islam R, Shannon JG, Nguyen HT (2014) Expression of root-related transcription factors associated with flooding tolerance of soybean (*Glycine max*). Int J Mol Sci 15:17622–17643

Valliyodan B, Ye H, Song L, Murphy M, Shannon JG, Nguyen HT (2017) Genetic diversity and genomic strategies for improving drought and waterlogging tolerance in soybeans. J Exp Bot 68:1835–1849

Van den Boogaart T, Maule AJ, Davies JW, Lomonossoff GP (2004) Sources of target specificity associated with the recovery against Pea Seed-borne Mosaic Virus infection mediated by RNA silencing in pea. Mol Plant Pathol 5:37–43

Van Toai TT, Martin SK, Chase K, Boru G, Schnipke V, Schmitthenner AF, Lark KG (2001) Identification of a QTL associated with tolerance of soybean to soil waterlogging. Crop Sci 41:1247–1252

Vandemark GJ, Larsen RC, Hughes TJ (2006) Heritability of resistance to Verticillium wilt in alfalfa. Plant Dis 90:314–318

Varshney RK, Dubey A (2009) Novel genomic tools and modern genetic and breeding approaches for crop improvement. J Plant Biochem Biotechnol 18:127–138
Velculescu VE, Zhang L, Vogelstein B, Kinzler KW (1995) Serial analysis of gene expression. Science 270:484–487
Wahid A, Gelani S, Ashraf M, Foolad M (2007) Heat tolerance in plants: an overview. Environ Exp Bot 61:199–223
Walker DR, Narvel JM, Boerma HR, All JN, Parrott WA (2004) A QTL that enhances and broadens Bt insect resistance in soybean. Theor Appl Genet 109:1051–1057
Wang WX, Vinocur B, Altman A (2003) Plant responses to drought, salinity and extreme temperatures: towards genetic engineering for stress tolerance. Planta 218:1–14
Wang L, Zhu J, Li X, Wang S, Wu J (2018a) Salt and drought stress and ABA responses related to *bZIP* genes from *V. radiata* and *V. angularis*. Gene 651:152–160
Wang LS, Chen QS, Xin DW, Qi ZM, Zhang C, Li SN, Jin Y, Li M, Mei HY, Su AY et al (2018b) Overexpression of *GmBIN2*, a soybean glycogen synthase kinase 3 gene, enhances tolerance to salt and drought in transgenic *Arabidopsis* and soybean hairy roots. J Integr Agric 17:1959–1971
Wang W, Wang C, Pan D, Zhang Y, Luo B, Ji J (2018c) Effects of drought stress on photosynthesis and chlorophyll fluorescence images of soybean (*Glycine max* L.) seedlings. Int J Agric Biol Eng 11:196–201
War AR, Paulraj MG, Hussain B, Buhroo AA, Ignacimuthu S, Sharma HC (2013) Effect of plant secondary metabolites on legume pod borer, Helicoperpa armigera. J Pest Sci 86:399–408
Wrather JA, Anderson TR, Arzsad DM, Gai J, Ploper LD, Porta-Puglia A, Ram HH, Yorinori YT (1997) Soybean disease loss estimates for the top 10 soybean-producing countries in 1995. Plant Dis 81:107–110
Wu QD, Van Etten HD (2004) Introduction of plant and fungal genes into pea (*Pisum sativum* L.) hairy roots reduces their ability to produce pisatin and affects their response to a fungal pathogen. Mol Plant-Microbe Interact 17:798–804
Xiaoshuang L, Daoyuan Z, Haiyan L, Yucheng W, Yuanming Z, Andrew J, Wood (2014) EsDREB2B, a novel truncated DREB2-type transcription factor in the desert legume *Eremosparton songoricum*, enhances tolerance to multiple abiotic stresses in yeast and transgenic tobacco. BMC Plant Biol 14(1):44
Xu ZZ, Zhou GS (2006) Combined effects of water stress and high temperature on photosynthesis, nitrogen metabolism and lipid peroxidation of a perennial grass *Leymus chinensis. Planta* 224:1080–1090
Yamaguchi SK, Shinozaki K (2005) Organization of cis-acting regulatory elements in osmotic- and cold-stress-responsive promoters. Trends Plant Sci 10:88–94
Yamamoto E, Karakaya HC, Knap HT (2000) Molecular characterization of two soybean homologs of *Arabidopsis thaliana CLAVATA1* from the wild type and fasciation. Biochim Biophys Acta 1491:333–340
Ye H, Song L, Chen HT, Valliyodan B, Cheng P, Ali L, Vuong T, Wu CJ, Orlowski J, Buckley B et al (2018) A major natural genetic variation associated with root system architecture and plasticity improves waterlogging tolerance and yield in soybean. Plant Cell Environ 41:2169–2182
Yeung E, van Veen H, Vashisht D, Paiva ALS, Hummel M, Rankenberg T et al (2018) A stress recovery signaling network for enhanced flooding tolerance in Arabidopsis thaliana. Proc Natl Acad Sci U S A 115:E6085–E6094. https://doi.org/10.1073/pnas.1803841115
Yoo JH, Park CY, Cheol J, Do Heo W, Cheong MS, Park HC, Kim MC, Moon BC, Choi MS, Kang YH et al (2005) Direct interaction of a divergent *CaM* isoform and the transcription factor, MYB2, enhances salt tolerance in *Arabidopsis*. J Biol Chem 280:3697–3706
Yoshida M, Cowgill SE, Wightman JA (1997) Roles of oxalic and malic acids in chickpea trichome exudate in host-plant resistant to *Helicoverpa armigera*. J Chem Ecol 23:1195–1210
Yu Q, Rengel Z (1999) Waterlogging influences plant growth and activities of superoxide dismutases in narrow-leafed lupin and transgenic tobacco plants. J Plant Physiol 155:431–438

Yu Y, Buss G, Saghai Maroof M (1996) Isolation of a superfamily of candidate disease-resistance genes in soybean based on a conserved nucleotide-binding site. Proc Natl Acad Sci U S A 93:11751–11756

Yu KF, Park SJ, Zhang BL, Haffner M, Poysa V (2004) An SSR marker in the nitrate reductase gene of common bean is tightly linked to a major gene conferring resistance to common bacterial blight. Euphytica 138:89–95

Yvonne M, Jideani VA (2017) The role of legumes in human, chapter 6. In: Hueda MC (ed) Functional food improve health through adequate food. InTech, Rijeka

Zabala G, Zou J, Tuteja J, Gonzales DO, Clough SJ, Vodkin LO (2006) Transcriptome changes in the phenylpropanoid pathway of Glycine max in response to Pseudomonas syringae infection. BMC Plant Biol 6:26

Zagorska N, Dimitrov B, Gadeva P, Robeva P (1997) Regeneration and characterization of plants obtained from anther cultures in Medicago sativa L. In Vitro Cell Dev Biol Plant 33:107–110

Zahaf O, Blanchet S, de Z'elicourt A, Alunni B, Plet J, Laffont C, de Lorenzo L, Imbeaud S, Ichante JL, Diet A, Badri M, Zabalza A, Gonzalez EM, Delacroix H, Gruber V, Frugier F, Crespi M (2012) Comparative transcriptomic analysis of salt adaptation in roots of contrasting Medicago truncatula genotypes. Mol Plant 5:1068–1081

Zahran HH, Manzano MCM, Sanchez-Raya AJ, Bedmar EJ, Venema K, Rodriguez-Rosales MP (2007) Effect of salt stress on the expression of NHX-type ion transporters in Medicago intertexta and Melilotus indicus plants. Physiol Plant 131:122–130

Zair I, Chlyah A, Sabounji K, Tittahsen M, Chlyah H (2003) Salt tolerance improvement in some wheat cultivars after application of in vitro selection pressure. Plant Cell Tissue Organ Cult 73:237–244

Zhang XH, Zhong WQ, Widholm JM (2005) Expression of a fungal cyanamide hydratase in transgenic soybean detoxifies cyanamide in tissue culture and in planta to provide cyanamide resistance. J Plant Physiol 162:1064–1073

Zhang G, Chen M, Chen X, Xu Z, Li L, Guo J, Ma Y (2010) Isolation and characterization of a novel EAR-motif-containing gene GmERF4 from soybean (Glycine max L.). Mol Biol Rep 37:809–818

Zhang LL, Zhao MG, Tian QY, Zhang WH (2011) Comparative studies on tolerance of Medicago truncatula and Medicago falcata to freezing. Planta 234:445–457

Zhang T, Yu L, Mccord X, Miller P, Bhamidimarri DS, Johnson D et al (2014) Identification of molecular markers associated with *Verticillium* wilt resistance in alfalfa (*Medicago sativa* L.) using high-resolution melting. PLoS One 9:e115953

Zhang J, Duan Z, Zhang D, Zhang J, Di H, Wu F, Wang Y (2016) Co-transforming *bar* and *CsLEA* enhanced tolerance to drought and salt stress in transgenic alfalfa (*Medicago sativa* L.) Biochem. Biophys Res Commun 472:75–82

Zheng G, Fan C, Di S, Wang X, Xiang C, Pang Y (2017) Over-expression of *Arabidopsis EDT1* gene confers drought tolerance in alfalfa (*Medicago sativa* L.). Front Plant Sci 8:2125

Zhu H, Cannon SB, Young ND, Cook DR (2002) Phylogeny and genomic organization of the TIR and non-TIR NBS-LRR resistance gene family in *Medicago truncatula*. Mol Plant-Microbe Interact 15:529–539

Zhu H, Choi HK, Cook DR, Shoemaker RC (2005) Bridging model and crop legumes through comparative genomics. Plant Physiol 137:1189–1196

Zhu B, Ye C, Lu H, Chen X, Chai G, Chen J et al (2006) Identification and characterization of a novel heat shock transcription factor gene, GmHsfA1, in oybeans (Glycine max). J Plant Res 119:247–256. https://doi.org/10.1007/s10265-006-0267-1

Zou J, Rodriguez-Zas S, Aldea M, Li M, Zhu J, Gonzalez DO, Vodkin LO, DeLucia E, Clough SJ (2005) Expression profiling soybean response to *Pseudomonas syringae* reveals new defense-related genes and rapid HR specific downregulation of photosynthesis. Mol Plant-Microbe Interact 18:1161–1174

Chapter 10
Deciphering the Molecular Mechanisms of Biotic Stress Tolerance Unravels the Mystery of Plant-Pathogen Interaction

Nibedita Chakraborty, Priyanka Chakraborty, Rajib Bandopadhyay, and Jolly Basak

Abstract Plants are always exposed to the wide variety of pathogens. In compatible interaction, pathogen can successfully invade into the host plant by manipulating host immune system and causes plant disease. On the other hand, during incompatible interaction plant recognizes variable pathogen elicitor molecules and induces its defense response. The molecular mechanism of plant-pathogen interaction is very specific, complex and dynamic. A plethora of research has explored the multifaceted defense mechanisms of plant against pathogen infection which has definite implication in the identification of superior plant genotype with durable resistance to variable pathogens. Thus this treatise is an attempt to summarize the key approaches about the plant-pathogen interaction. Collectively, such information indeed will provide the insight on the molecular mechanism of plant immune response against pathogen infection and the possible protective role of different phytohormones in plant defense system.

Keywords Plant-pathogen interaction · Disease cycle · Plant immune response · Pathogen associated molecular pattern (PAMP) · Triggered immunity (PTI) · Effector triggered immunity (ETI) · Avirulance (avr) gene · Reactive oxygen species · Host and non-host resistance · Pathogenesis related proteins · Phytohormones

---

N. Chakraborty
Department of Biotechnology, National Institute of Technology, Durgapur, West Bengal, India

P. Chakraborty · R. Bandopadhyay
UGC-Center of Advanced study, Department of Botany, The University of Burdwan, Bardhaman, West Bengal, India

J. Basak (✉)
Department of Biotechnology, Sikhsha Bhavan, Visva-Bharati, Santiniketan, West Bengal, India
e-mail: jolly.basak@visva-bharati.ac.in

P. Guleria et al. (eds.), *Sustainable Agriculture Reviews 51*, Sustainable Agriculture Reviews 51, https://doi.org/10.1007/978-3-030-68828-8_10

10.1 Introduction

In recent years food security has become a global alarming matter due to increasing price of crops worldwide. The reason behind the price hike is partly because of the impact of plant diseases and one of the major causes behind these plant diseases is pathogenesis. Eukaryotic host cells are frequently subjected to infection by a wide variety of pathogens that leads to altered gene expressions in the host as well as in the pathogen. This event leads to any one of the two differential cascade mechanisms- either in the favorable environment, the pathogen invades into the host cell and causes infection, the phenomenon is known as compatible interaction; or the incompatible interaction leads to the prevention of pathogen invasion by host immune system and inhibit the pathogen infection (Glazebrook 2005). The mechanism of the plant immune system is being revealed by researchers and parallel research on pathogen biology are disclosing the pathway by which pathogen can manipulate the host immune system to cause infection and the subsequent plant defence responses are also being revealed (Dodds and Rathjen 2010).

10.2 Plant-Pathogen Interaction: An Overview

Plants are exposed to different stress factors in the environment in which they reside. The fundamental requirements for a plant to become stressed are as followings- (i) a vulnerable plant, (ii) a virulent pathogen & (iii) a favourable environment (Fig. 10.1).

Once the pathogens spread in a plant, there can be several outcomes based upon the life history of the invading pathogen. Plant pathogens can be broadly divided into three groups based upon their mode of pathogenicity, viz. Biotrophs, Nectrophs and Hemibiotrophs (Schenk et al. 2009). Biotrophs are pathogens which require

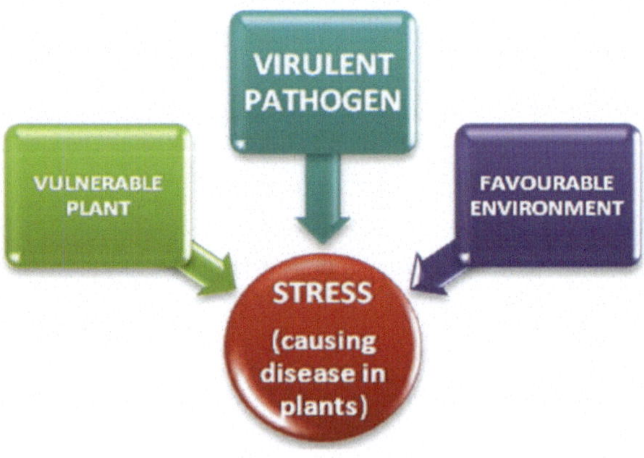

Fig. 10.1 The factors involve in diseased plant

living tissues for growth and reproduction. In case of nectrophs, at the very beginning of the infection necrotrophic pathogens kill the host tissue and feed on the dead tissue. Hemibiotrophic pathogens are hybrid in nature, i.e., they grow as biotrophs but afterwards aggressively become necrotophic in nature (Vargas et al. 2012).

10.2.1 Survival of Plant Pathogen

The disease cycle is defined as the stages of events that occurred during the development of the disease (Fig. 10.2). Initially, it was termed as "infection chain" (Gaumann 1950); later this term has been changed to "infection cycle" or "disease cycle" (Kranz 2003; Maloy 1993). A plant disease cycle involves the development of the disease caused by a pathogen and effects of the disease into the host plant. The survival of the pathogen depends either on the disease cycle sustained during the intercropping season and or on the dispersion of the pathogen to uninfected plants. A successful infection process helps in the pathogen survival. The infection process can be continuous or discontinuous (Ravichandra 2013). The continuous infection process involves a susceptible host plant through which pathogen can survive by multiplying its life cycle. In between cropping season's, pathogens can infect the alternative host and thereby use them as a carrier to continue their existence, for example, black rot disease of brassica caused by *Xanthomonas campestrus* pv. *campestris*, can easily grown on wild radish during the absence of its host (Ravichandra 2013). The discontinuous infection process includes the epiphytic and

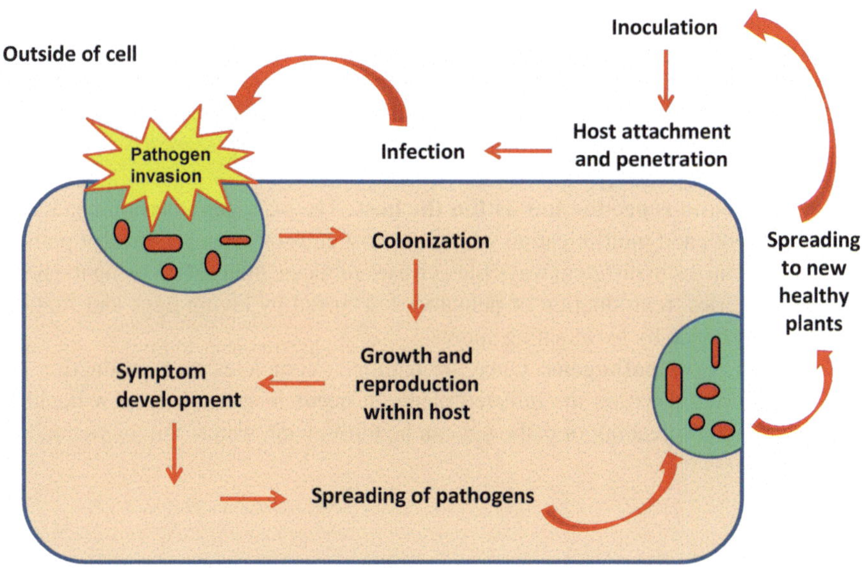

Fig. 10.2 The stages of plant disease development

saprophytic pathogens, which interrupted the disease cycle. Epiphytes are non-parasitic pathogens grown on the surface of the host or other plants (Melotto et al. 2008; Tjamos et al. 2013) whereas saprophytic pathogens unable to infect living plants and grows on the waste materials of diseased plants (Prell and Day 2013).

10.2.2 The Stages of Plant Disease Development

The development of the disease depends on the series of events (Guest and Brown 1997; Agrios 2012) that includes (i) inoculation, (ii) host attachment and penetration, (iii) infection, (iv) growth and reproduction within the host and (v) spreading of pathogens (Fig. 10.2).

(i) **Inoculation:** To establish the infection in a plant, a pathogen must contact with the potential host. The pathogen or infectious part of any pathogen is the inoculam that arrived to the host to initiate infection. Inoculam can be of two types, primary and secondary. Primary inoculam causes the first infection to start the disease cycle during the cropping season while secondary inoculam is being generated from this primary infection and causes the dispersal of the disease.
(ii) **Host attachment and penetration:** Once the suitable host becomes available, the pathogen gets attached to the surface of the host plant by either hatching of eggs, producing biofilms to adhere the host surface or by germinating spore. After adherence to the host surface, pathogen needs to get penetrate by direct entry or by wounds or natural openings.
(iii) **Infection:** Successful invasion of the pathogen to cause the disease occurs by obtaining the nutrients from the host plant and producing the disease symptoms to the plants. The process of infection generally involves incubation period and latent period. The time between the penetration of a pathogen and appearance of the disease symptoms is known as the incubation period while the phase in which infected host plant does not show any symptoms is known as the latent period.
(iv) **Growth and reproduction within the host:** The pathogen invasion results in the growth and multiplication within the host cells or tissues. Different pathogen colonizes in different ways like, viruses replicate themselves using the host machineries, reproduction of nematodes occurred by laying eggs and fungus can reproduce by germinating spores.
(v) **Spreading of pathogens:** Once the pathogen completes its reproduction or multiplication within the infected plants, it needs to spread into new healthy plants. The spreading of pathogen can be through air, water, soil, or vector.

10.2.2.1 Penetration and Spreading Mechanism of Pathogens

Fungi, bacteria, viruses, oomycetes and nematodes (Buonaurio 2008; Abramovitch et al. 2006; Faulkner and Robatzek 2012; Williamson and Gleason 2003; Vargas et al. 2012) are the different plant pathogens that are present in the environment. They invade plants in different ways by feeding and reproducing in the plant. After detecting the specific host, the pathogen faces the physical barriers of the host plant during its penetration into it, which involves cell wall and the extracellular waxy cuticle layer in aerial plant tissues (Faulkner and Robatzek 2012). Generally, pathogens mostly overcome these barriers through several specialized invading mechanisms as discussed below. Certain pathogens like bacteria adhere themselves to the plant surface through specialized structures called Pilli (Buonaurio 2008). The penetration of bacteria (*Agrobacterium*, *Erwinia*, *Pseudomonas*, *Ralstonia* and *Xanthomonas*) occurs through natural plant openings, such as stomata, hydathodes, lenticels, nectarthodes, stigma and also through abrasions or wounds on leaves, stems or roots or even through specific feeding insects. They spread through the intercellular spaces (apoplast) of various plant tissues or the xylem (Buonaurio 2008; Abramovitch et al. 2006). Viruses are disseminated by insect or vector and penetrated through wound or by insect feeding. For example, white fly (*Bemisia tabaci*) transmits the yellow mosaic virus into the vascular tissue of the host plant (Faulkner and Robatzek 2012). Nematode penetrates through cell wall and forms a feeding structure (stylet), which in turn creates a feeding site in vascular cylinder by injecting stylet secretions (Williamson and Gleason 2003). Oomycetes (*Hyaloperonospora arabidopsidis*, Hpa) makes boundaries between the pavement cells and breaks through the leaf epidemis and Spreads in to mesophyll cells through a specialized feeding structure (haustoria), protruded through hyphae that in turn grows in extracellular space (Faulkner and Robatzek 2012). Fungus invades directly in epidermal cells through appressorium, a thin penetration peg, which later develops into an enlarged, irregular primary hypha. This primary hypha forms one or more branches and spreads from one cell to another, growing between plant plasma membrane and plant cell wall (Vargas et al. 2012).

10.2.2.2 Assessment of Plant Disease and Crop Loss

To measure the incidence and severity of plant disease during the growing season the assessment of the plant disease and crop losses has been performed (Campbell and Buchwalter 2012; Jones 2013). The term disease incidence can be described as the measurement of the proportion of plant populations that shows disease symptoms and can be estimated using the formula, disease incidence $I = \Sigma\ x/N$, where x represents the number of diseased plants and N is the total number of plants evaluated (Cardoso et al. 2004). Disease severity is the measurement of the proportion of the area of diseased plants and can be evaluated by the following equation, disease severity $S = \Sigma(x_i n_i)/n$, where x is the status of disease, n_i stands for the number of infected plants corresponding to i^{th} grade of disease scale and n represents the total numbers of plants estimated (Cardoso et al. 2004). A disease index (DI) is the product of disease incidence and severity and expressed as $DI = \Sigma(x_i n_i)/N$ (Groth et al. 1999). There are mainly five steps involved in the assessment of the plant disease and crop lossess in productivity (Campbell and Buchwalter 2012): (i) assessment of growth and development of a healthy plant during cropping season, (ii) development of the methods that can accurately measure the disease incidence and severity, (iii) development of the statistical sampling method to measure the amount of the disease, (iv) experimental or statistical assessment of the infected plants in which the particular disease level can be controlled and (v) development of different methods used in disease reduction for economically important crops.

10.2.2.3 Disease Management

Once a pathogen already establishes a successful infection into a plant, severity of infection can only be minimized by manipulating or reducing the amount of inoculum. There are various methods available that can reduce the amount of inoculum, such as crop rotation programs and various chemical or physical treatments.

(i) **Quarantine:** The term quarantine means prevention of the pathogen entry into new region. The main aim of quarantine is to establish a barrier against the dispersal of the infectious pathogen to protect the environment and agriculture (Rodoni 2009). The disease management can be achieved by exclusion, eradication and elimination of the disease (Horsfall 2012). The term exclusion is defined as the prevention of pathogen entry into the infection free area accomplished by quarantine or by treating with propagating material. Propagating material such as seeds, cuttings, root-stocks, buds, tubers, rhizomes, bulbs and corms should be free of pathogen when entering into new area. Therefore, elimination of disease can be achieved by using the pathogen free propagating materials (Horsfall 2012). Eradication is the effective quarantine practices to destroy the infected plant materials from an area. It can be achieved by irradiation with UV light, chlorination, bromination or ozonation.

(ii) **Chemical control:** Plant disease can be controlled by chemicals that are either used as disinfectant or as a toxic chemical to target a particular pathogen. Fungicides are the toxic chemicals that kill or destroy fungus. Fungicides are classified in two types, protectant and systemic fungicides (Hirooka and Ishii 2013). Protectants can be used against broad range of fungus and applicable for the fungal infection on the surface of the host plants whereas systemic fungicides can be easily absorbed by the infected plants and transported to the site of fungal infection (Hirooka and Ishii 2013). Bacteriocides are usually the antibiotics generated from microorganisms to destroy the particular bacteria (Lewis 2001). The application of nematicides is restricted to horticultural crops as these are very expensive (Ortiz et al. 2012). No chemical substances so far have been found to destroy viral pathogens.

(iii) **Biological control:** Biological control can be defined as the use of beneficial living organism that destroy plant pathogen and induce defense response in plants (Junaid et al. 2013). Biological control can be achieved by various processes including antagonists, plant breeding or modifications of microorganisms to destroy infectious pathogens (Alabouvette et al. 2006). The mechanisms of action behind the biological control agent against plant pathogen are competition, parasitism, induced resistance and the generation of antimicrobial compounds (Alabouvette et al. 2006). Competition exists between organisms sharing similar resource or nutrient for growth and survival. Therefore, the treatment of an infected plant with non-pathogenic organisms of a related species can be useful to compete with the pathogenic organism sharing the same resource. For example, fungi is generally used as an antagonist against nematode (Sarhy-Bagnon et al. 2000; Brand et al. 2004) as they share same infection process: penetration through cuticle, immobilization, invasion and degradation of the nematode (Huang et al. 2004). Parasitism is another mode of action of biological control where a parasitic fungus can be used to destroy other pathogens present on the plant surface by either formation of hyphae of parasitic fungi as a physical barrier to their victim or by producing cellular degrading enzymes to the plant surface. For instance, *Trichoderma harzianum* act as an active bio-control agent against nematode (Harman 2000). The term induced resistance can be described as inoculation with inactive or low doses of pathogens as well as pathogen derived chemicals to induce immune response to plants. In this process when the plant becomes infected, it can induce systemic defense response by activating antioxidative enzymes, pathogenesis related (PR) proteins or salicylic acid (SA) mediated signalling pathway. For example, *Bacillus amyloliquefaciens* LJ02FB can induce systemic resistance against the powdery mildew fungal disease (Li et al. 2015). Antimicrobial compounds are the secondary metabolites produced by one microorganism which are toxic to another pathogenic organism (Alabouvette et al. 2006).

10.3 Defensive Mechanisms of Plants

Plants are constantly interacting with a wide variety of potential pathogens within their environment. Pathogen can invade and manipulate the host machinery to sustain their growth and complete their life cycle which results in the development of diseased plant (Fig. 10.2). On the contrary, the ability of the plant's recognition of the pathogen results in activation of series of defensive events which ultimately causes prevention of the pathogen entry or restriction of its progression and multiplication. Based on pathogen confrontation, the plant defense mechanism can be classified in two category, passive defense and active defense (Guest and Brown 1997), which has been depicted in Fig. 10.3.

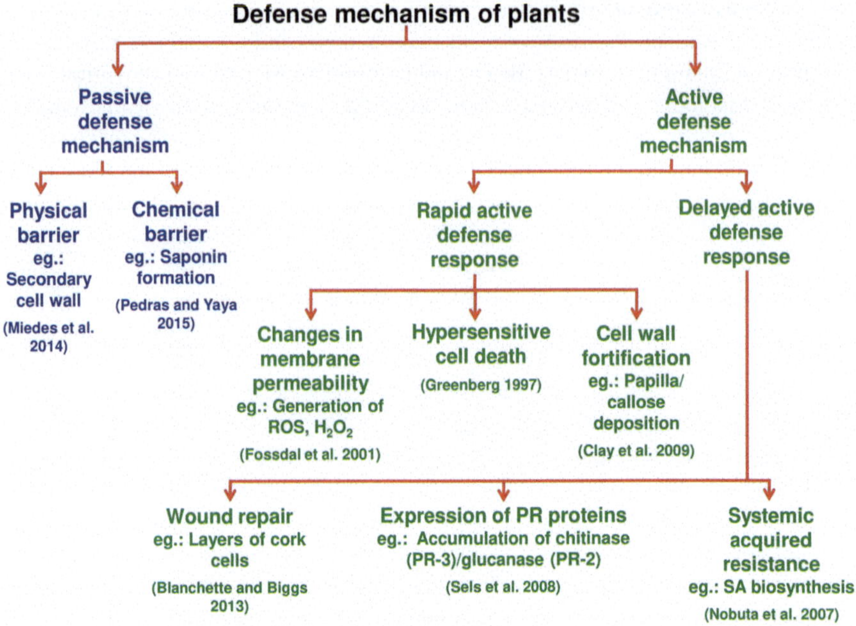

Fig. 10.3 The categorization of plant defense mechanism

10.3.1 Passive Defense Mechanism

The mechanism of passive defense depends on the defensive properties that are already present in plants before pathogen recognition (Guest and Brown 1997). To gain access into the host plant machinery, a pathogen must overcome the natural barrier present in the uninfected healthy plants. These barriers can be physical or chemical barrier.

(i) **Physical barrier:** The secondary cell wall, cuticle, stomata, lenticels etc. presents on the plant surface act as physical barriers against pathogen. The thickness of cuticle layers, sizes of stomatal pores and the presence of secondary cell wall can prevent the pathogen invasion (Miedes et al. 2014). The vertical orientation of leaves can also affect the pathogen attachment to the leaf surface. The time of stomata opening and size of stomatal pores also act as physical barrier and make the plants tolerant to the pathogen attack.
(ii) **Chemical barrier:** There are some antimicrobial compounds present in plant vacuoles in inactive forms that get activated during pathogen attack, such as, Lactones, cyanogenic glucosides, saponins, terpenoids etc. These antimicrobial compounds are known as "phytoanticipins" (Pedras and Yaya 2015) which prevent the pathogen attack, specially the vectors of plant viruses. During wounding or infection in plants, the releases of hydrolase enzymes activate saponin to destroy the membrane integrity and function of pathogens (Pedras and Yaya

2015). Phenol is another antimicrobial compound which has antiviral, antifungal and antibacterial activity (Sivaprakasan and Vidhyasekaran 1993). Pathogen attack induces intra- and inter- molecular cross linking of phenolic compounds in plant cell wall, which in turn increases the plant cell wall rigidity and inhibits the pathogen invasion (Matern and Kneusal 1988). Defensins are the plant peptides that act as an inhibitor by interfering pathogen nutrition and hinder their development (Stotz et al. 2009).

10.3.2 Active Defense Mechanism

The defensive mechanism activated by the recognition of pathogen is known as active defense. This recognition depends on the elicitor released from the pathogens that induces plant defense response (Wood 2012). The elicitors are of two types, non-specific and specific. Non-specific elicitors are the general indicators like, peptides, hydrolytic enzymes, glycoprotein releases from host's own cell wall or pathogen origin. All these elicitors induces plant defense response. Specific elicitors activate defense response against a particular pathogen. These types of elicitors are the pathogen avirulence (avr) genes which can be recognized by plant's corresponding resistant (R) genes (Jia et al. 2000) and activate a series of signalling pathways to induce several plant defense activities. The defensive ability of the plant can be rapid or delayed (Guest and Brown 1997).

10.3.2.1 Rapid Active Defense Response

The rapid active defense depends on changes in membrane permeability, hypersensitivity and cell wall reinforcement.

(i) **Changes in membrane permeability:** The initial stage of pathogen recognition results in changes in membrane permeability by leaking the electrolytes, such as loss of K^+ and uptake of H^+ within the cellular environment (Guest and Brown 1997). The influx of Ca^{2+} results in the activation of different signalling pathways which in turn results in different defense gene activation. The plasma membrane "oxidative burst" is another active defense response which generates reactive oxygen species such as H_2O_2 (Mehdy 1994; Baker and Orlandi 1995). H_2O_2 act as a signalling molecule by activating plant defense related genes (Apel and Hirt 2004) and also helps in cross linking of plant cell wall which improves the plant resistance (Fossdal et al. 2001).

(ii) **Hypersensitive cell death:** Hypersensitive cell death is another well-known active defense response against the pathogen (Greenberg 1997). In this event, the infected host cell become over sensitive due to pathogen infection and destroys host's own cell to prevent further pathogen spreading (Greenberg

1997). Hypersensitivity is also associated with other defense response like lignifications and synthesis of antimicrobial compounds (Guest and Brown 1997).

(iii) **Cell wall fortification:** The host cell resistance is also associated with cell wall reinforcement. This is the first defense response of pathogen penetration. This defense response is characterized by the accumulation of cytoplasmic aggregates containing cellular apparatus surrounding the pathogen penetration site. Several types of cell wall fortification occur during pathogen penetration. Papilla deposition (Guest and Brown 1997; Clay et al. 2009) is one type of cell wall reinforcement which is composed of branched P-1, 3 glucan, callose, silicon, lignin and proteins (Guest and Brown 1997). During powdery mildew fungus (*Blumeria graminis*) penetration, a rapid deposition of papillae forms at the inner side of outer epidermal cell wall, directly under the penetration peg (Aist and Bushnell 1991; Thordal-Christensen et al. 1997). Lignitubers are the lignified callose deposition formed to prevent the invading hyphal tips. Secondary cell wall thickening is another type of cell wall fortification, formed by hydroxyproline-rich glycoproteins (HRGP) (Malinovsky et al. 2014). Pathogen attack results in host cell oxidative burst which release H_2O_2 and helps in crosslinking of HRGPs molecule. Pathogen penetration induces the expression of the genes encoding HRGPs, making the plant more resistant to pathogen invasion.

10.3.2.2 Delayed Active Defense Response

Once pathogen infection is established in host plant, delayed defense response is activated to restrict the spread of pathogen. These defense responses include wound repair, expression of pathogenesis related (PR) proteins and induction of systemic acquired resistance.

(i) **Wound repair:** The ability to repair the primary wound minimizes the chances of secondary infection by opportunistic pathogens (Guest and Brown 1997). The infection area can be repaired by the formation of cork cell with thick, suberized wall. The cork cells are made up of secondary meristem in fleshy tissues, fruits, roots and bark and the cork cambium (Blanchette and Biggs 2013). These cells make layers around the infection site and separate the infected cells from the healthy ones. The layers of cork cells also act as barrier and prevent the pathogen from further colonization to the host cell (Guest and Brown 1997). Some wooden tree trunk seal their wound by secreting gums which also helps in infection prevention (Blanchette and Biggs 2013).

(ii) **Expression of pathogenesis related (PR) proteins:** Pathogen recognition initiates synthesis of different defense related proteins such as PR proteins (Kitajima and Sato 1999; van Loon et al. 2006). Most of the PR proteins have antimicrobial activity (Veronese et al. 2003) and some have b-glucanase, chitinase or lysozyme activity. Many PR proteins are located in extracellular region and some of them accumulates in plant cell vacuole such as chitinase

(PR-3, PR-4, PR-8, and PR-11) and glucanase (PR-2) and degrade fungal cell wall components (Sels et al. 2008). PR-6 act as proteinase inhibitor and prevents the pathogen to complete its replication cycle and reduces the generation of secondary elicitors (Sels et al. 2008). According to the primary structure, enzymatic and biological functions, PR proteins are broadly classified into 17

Table 10.1 The classification of PR proteins and their identified targets

Family	Sources	Protein	Identified targets
PR-1	*Nicotiana tabacum*, and many other species	Unknown (antifungal)	*Phytophthora infestans*
PR-2	*Hordeum vulgare*, *N. tabacum*, *Oryza sativa*	β-1,3-Glucanase (antifungal)	*Colletotrichum lagenarium*, *Fusarium solani*, *Rhizoctonia solani*
PR-3	*N. tabacum*, *Phaseolus vulgaris*, *Triticum aestivum*	Chitinase (antifungal)	*Fusarium oxysporum*, *Physalospora piricola*
PR-4	*H. vulgare*, *N. tabacum*, *Solanum tuberosum*, *Solanum lycopersicum*, *Zea mays*	Chitinase (antifungal)	*Alternaria radicina*, *Fusarium moniliforme*, *F. solani*, *Trichoderma viride*
PR-5	*Cucurbita maxima*, *H. vulgare*, *N. tabacum*, *Zea mays*	Thaumatin-like (antifungal)	*Candida albicans*, *F. oxysporum*, *Neurospora crassa*, *P. infestans*, *Trichoderma reesei*, *Trich. viride*
PR-6	*S. lycopersicum*	Proteinase inhibitor	Unknown
PR-7	*S. lycopersicum*	Endoproteinase	Unknown
PR-8	*Cucumis sativus*	Chitinase (antifungal)	*C. lagenarium*
PR-9	*N. tabacum*	Lignin-forming peroxidase	Unknown
PR-10	*Capsicum annuum*, *Lupinus albus*, *Oxalis tuberosa*, *Petroselinum crispum*	Ribonuclease-like (antifungal)	*Colletotrichum gloeosporioides*, *F. oxysporum*, *Phytophthora capsici*, *R. Solani*
PR-11	*N. tabacum*	Chitinase (antifungal)	*A. radicina*, *Trichoderma viride*
PR-12	*Raphanus sativus*	Defensin (antifungal)	*Aspergillus flavus*, *F. moniliforme*
PR-13	*Arabidopsis thaliana*	Thionin (antifungal)	*F. solani*
PR-14	*H. vulgare*	Lipid-transfer protein	Unknown
PR-15	*H. vulgare*	Oxalate oxidase (antifungal)	*Blumeria graminis*
PR-16	*H. vulgare*	Oxalate-oxidase like	*B. graminis*
PR-17	*H. vulgare*, *N. tabacum*, *T. aestivum*	Unknown (antifungal)	*B. graminis*

Adapted from van Loon et al. (2006)

families. Table 10.1 depicts the activities of the PR proteins and their identified targets (van Loon et al. 2006).

(iii) **Systemic acquired resistance:** The accumulations of PR proteins activate multiple signalling pathways which induce defense reponse not only at the site of infection; the activated signals restrict the pathogen entry in distal tissues and make the host plant resistant to the pathogen. This enhanced state of resistance phenomenon is known as systemic acquired resistance (Vlot et al. 2008). In this defense response, a number of defense related genes along with chitinase and glucanse (PR proteins) are activated in local and systemic tissues (Murphy et al. 1999; Durrant and Dong 2004). Salicylic acid (SA) is an important key regulator of this defense pathway (Alvarez 2000). Virus invasion results in the development of hypersensitive response (HR) lesion which in turn activates SA biosynthesis at the site of lesion and later SA mediated signals get transmit throughout the entire plant (Nobuta et al. 2007; Strawn et al. 2007).

10.4 Plant Immune Response

A plant defence response involves a number of genes and signalling pathways. Two types of immune reactions are being followed by plants during such defense reaction; pathogen associated molecular pattern (PAMP) – triggered immunity (PTI) and effector triggered immunity (ETI) (Jones and Dangl 2006; Fig. 10.4a, b). The phenomenon of plants' initial recognition and response to pathogen is known as PTI (Fig. 10.4a). In this approach PAMPs, conserved microbial elicitors and essential components of a whole class of pathogens, are recognised by plant receptor proteins called pattern recognition receptors (PRRs) (Fig. 10.4a) (Jones and Dangl 2006). PRR consists of mainly extracellular leucine-rich repeat (LRR) domain (Fig. 10.4a, b; blue coiled region) (Jones and Jones 1996) and intracellular kinase domain (Fig. 10.4a, b; yellow oval shaped). Examples of some well known PRRs are flagellin sensing 2 (FLS2) (Boller and Felix 2009) – recognising the bacterial flagellin protein; EF-TU Receptor (Zipfel et al. 2006) – recognising bacterial elongation factor and LysM (chitin binding) – recognising chitins. PAMPs are bound by the extracellular domain of the PRRs on the outside of the plant cell. PAMP recognition initiates a signal transduction process which reprograms gene transcription with the increase of cytosolic Ca^{2+}, activation of mitogen activated protein kinase (MAPKs) and reactive oxygen species (ROS) production. Pathogens that produce effectors (Fig. 10.4b) to boost their own pathogenicity are not affected by PTI and thus induces effector triggered susceptibility (ETS) (Fig. 10.4b; yellow rectangular box). The modes of pathogenicity by effectors are usually species-specific or race-specific. The second line of defense response of plant immune system is the ETI (Fig. 10.4b) where plants recognise the specific pathogen effector through intracellular resistance protein (R protein) (Torres et al. 2006). R proteins consist of three domains (Fig. 10.4b); a variable N terminus containing either a toll/interleukin-1

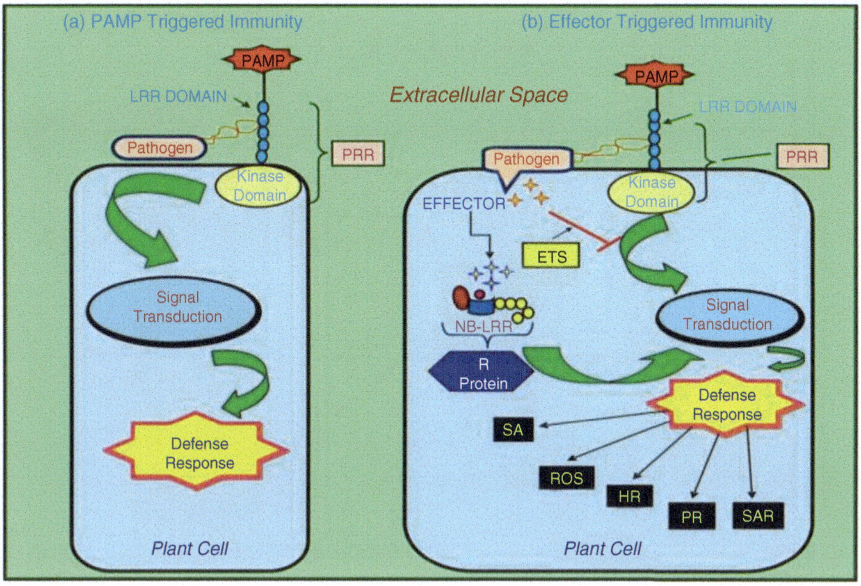

Fig. 10.4 (**a**) PAMP Triggered Immunity. Pathogen releases molecules known as pathogen associated molecular patterns (PAMPs) into the extracellular spaces that are recognised by cell surface PRRs which consist of extracellular leucine-rich repeat (LRR) domain and intracellular kinase domain and induce PAMP Triggered Immunity (PTI) by subsequent signalling pathway and induce Defence Response. (**b**) Effector Triggered Immunity. Pathogens (bacteria, fungi oomycetes) release effectors that often repress PTI response. This effector induced repression is known as effector triggered susceptibility (ETS). When PTI response is repressed, the plant builds another immune response with the help of R protein which consists of intracellular nucleotide- binding (NB)-LRR receptor. NB-LRR consist of a carboxyl terminal LRR domain (yellow), a central NB domain (blue) that binds ATP or ADP (pink) and an amino-terminal Toll, interleukin-1 receptor, resistance protein (TIR) or coiled-coil (CC) domain (red). Activated R protein enhances a heightened defence response which includes salicylic acid (SA), Reactive Oxygen Species (ROS), Hypersensitive Response (HR), Pathogenesis Related (PR) proteins and Systematic Acquired Resistant (SAR) and these defence response is known as Effector Triggered Immunity (ETI)

receptor (TIR) domain (Fig. 10.4b; red oval shaped) or a coiled coil domain (CC), a central NB-ARC (nucleotide binding, human apaf-1, resistance *Caenorhabditis elegans*) domain (blue small shaped) that binds ATP or ADP (pink) (Peso et al. 2000; van der Biezen and Jones 1998) and a C terminal LRR domain (yellow coiled structure). The N-terminal domain helps in different downstream signalling processes (Aarts et al. 1998), the NB-ARC domain helps in nucleotide binding and exchange/hydrolysis (Collier and Moffett 2009; Eitas and Dangl 2010), while the LRR domain is responsible primarily for effector recognition and defense signaling activation (Jones and Jones 1996; Cesari et al. 2014). According to the "gene-for-gene hypothesis" (Flor 1971), R genes recognize pathogen avirulence genes (Avr) directly or indirectly (Jia et al. 2000), and the presence of both genes are required to limit pathogen growth and subsequent defence gene expression. Jones and Dangl

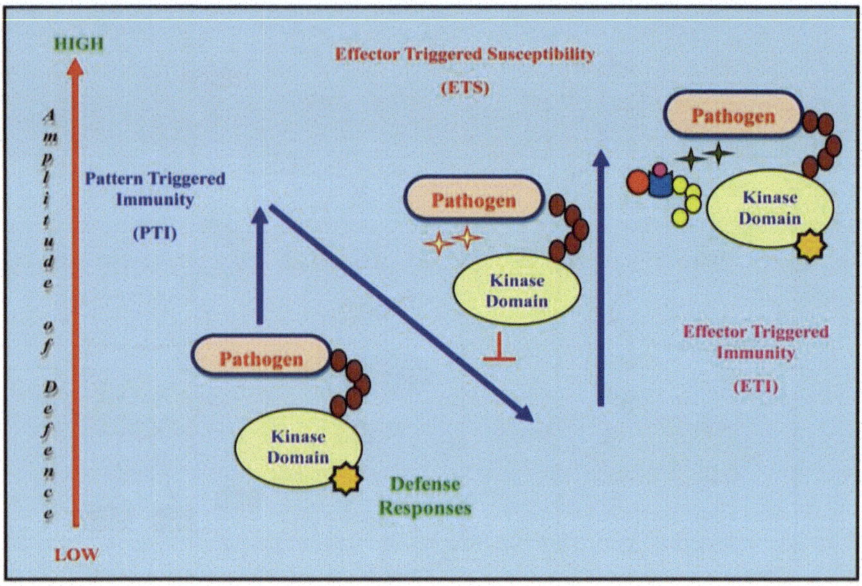

Fig. 10.5 Zig-zag model of plant-pathogen interaction

have contributed to a relatively logical scenario of the molecular interactions through which plants recognise and respond to pathogens, known as the Zig-zag model (Fig. 10.5) (Jones and Dangl 2006), provide a framework and describe this plant immune response system.

10.5 Host-Pathogen Specificity in Plant Disease Resistance: A Genetic Approach

Depending on the defense response of different races of pathogen, plants are broadly classified into two categories (Fraser 1990), (i) non-host resistance and (ii) host resistance. Both of these resistance responses are the product of plant immune response (Gill et al. 2015).

10.5.1 Non-Host Resistance

This type of defense response is mediated by all genotypes of a host plant species against all races of a particular pathogen. This resistance response involves activation of passive defense or non-specific elicitor induced active defense. This defense response is long-lasting and appropriate for the development for virus-resistant plants (Maiti et al. 2014). Depending on the presence or absence of disease symptoms, non-host resistance is divided into two types, type I and type II non-host resistance (Mysore and Ryu 2004). Type I does not show any visual symptoms whereas type II non-host resistance links with the development of HR/symptoms of cell death against non-specific pathogens (Mysore and Ryu 2004). Type I non-host resistance is associated with generally passive defense, like preformed barriers or general pathogen (PAMPs) induced active defense, such as PTI (Mysore and Ryu 2004; Jones and Dangl 2006). Type II non-host resistance resemble with pathogen effector recognition and shows ETI response against particular pathogens that mostly infect non-host species (Schulze-Lefert and Panstruga 2011).

10.5.2 Host-Resistance

Host specific resistance is associated with single or multiple resistant (R) genes introduced against infected cultivars (Gill et al. 2015). This is less durable resistance response and subdivided into three major categories on the basis of particular pathogen race, host cultivar specific and host-pathogen specific response (Mysore and Ryu 2004).

(i) **Race specific resistance:** A particular race of pathogen induces this defense response to all cultivars of the host plant (Mysore and Ryu 2004). This type of resistance is associated with genetic variation within the particular race of pathogens.
(ii) **Host cultivar specific resistance:** In this resistance response, a specific host cultivar activates the defense response against all races of a pathogen species (Mysore and Ryu 2004). This type of resistance is associated with genetic variation within the host plant species.
(iii) **Host-pathogen specific response:** This type of resistant response depends on the interaction between the specific race of pathogen and specific cultivar of host species (Mysore and Ryu 2004). This defense response is associated with the recognition of pathogen specific Avr gene by host plant specific R gene and generally the event is termed as 'gene for gene resistance' (Flor 1971).

10.6 The Role of Phytohormones in Plant Defense

Plants have evolved highly sophisticated mechanism mediated by the regulatory network of phytohormones to combat the biotic stresses. Phytohormones are endogenous, low molecular weight molecules which are known to play a significant role in plant growth and development. During stress, the production, transportation and the signaling network of the phytohormones are affected resulting in the morphological, physiological and molecular changes through which plants can resist against stresses. There are several phytohormones, for example salicylic acid (SA), jasmonic acid (JA), ethylene (ET), abscisic acid (ABA), auxins, cytokinins (CKs), gibberellins (GAs), and brassinosteroids (BRs) respond against stress through synergistic and antagonistic way (Mauch-Mani and Mauch 2005). Among these, SA, JA and ET are generally known to be involved in plant defense against pathogens or insects (Bari and Jones 2009). SA is a phenolic compound, derived from shikimic acid pathway showing defensive response against hemibiotrophic and biotrophic pathogens (Sticher et al. 1997). The biosynthesis of SA has been reported as two distinct pathways, phenylalanine ammonia-lyase (PAL) pathway occuring in cytosol (Verberne et al. 1999) and isochorismate synthase (ICS) pathway occuring in chloroplast. Pathogen infection initiates the accumulation of SA through isochorismate synthases one and two (ICS1/2) and promotes pathogen resistance (Fragnière et al. 2011). Moreover, to induce the localized defense response, SA activates systemic acquired resistance to prevent the entry of the pathogens. The induction of SAR promotes the induced expression of PR proteins, phytoalexin, and reinforcement of plant cell wall which prevents the spread of invading pathogens and restricts the pathogen colonization (Lu et al. 2016). Unlike, hemibiotrophic and biotrophic pathogen specific resistance, SA induction is not effective against necrotrophic pathogens since these pathogens are not hindered by HR or cell death. JA and ET signaling networks are more efficient against necrotrophic pathogens (Glazebrook 2005). Pathogen infection elicits a signal that induces the cascades of phosphorylation to initiate JA biosynthesis. The perception of the signal by JA receptors activates different defense related genes which results in the formation of pathogen barrier (Kazan and Manners 2008). In addition, jasmonates and its derivates, methyl jasmonates (MeJA) has been reported as intra- and interplant signaling molecules as it communicates with neighboring plants during pathogen attack and prevents the spread of infections to the nearby healthy plants (Wasternack 2007; Seo et al. 2011). This is because MeJA is a volatile compound and it can diffuse through membrane within the plants. Earlier reports showed that exogenous application of MeJA lead to the expression of different defense related genes during necrotrophic fungus, insect or virus attacks (Mei et al. 2006; Chakraborty and Basak 2018). In plants, ET plays important regulatory role in plant growth, development and biotic stress tolerance. The ET signaling network and biosynthesis are conserved among plant species. S-adenosylmethionine (SAM) is the main precursor of ET synthesis pathway, which is converted into 1-aminocyclopropane-1-carboxylic acid (ACC) and 5′-methylthioadenosine (MTA) by ACC synthase. The ET signaling is mainly regulated by mitogen-activated protein kinase (MAPK) phosphorylation pathway (Takahashi et al. 2007). There are five ET receptors divided into two subfamilies known as ethylene response 1 (ETR1) and ethylene response sensor 1 (ERS1) I subfamily 1 and subfamily 2 includes ETR2, ERS2, and ethylene insensitive 4 (EIN4). It is reported that ET signaling is governed by ethylene response factors (ERFs), which works downstream of EIN3 to activate defense response (Solano et al. 1998). SA and JA/ET works as antagonistic way. Van der Does et al. (2013) reported that the transcription regulator of JA/ET-induced defense genes, (OCTADECANOID-RESPONSIVE ARABIDOPSIS AP2/ERF-domain protein 59,

ORA59), are activated by JA or ET signaling whereas negatively regulated by SA signaling network. On the other hand, ET signaling network is positively regulated by JA pathway. When JA is accumulated, it degrades the JASMONATE ZIM-DOMAIN (JAZ) protein, and permits the binding of EIN3 to the ERF1 promoter, which results in the transcription of ERF1. Moreover, EIN3 activates ORA59 which leads to the activation of defense pathway (Zhu et al. 2011; Xu and Brosche 2014).

10.7 RNAi Mediated Plant Defense Mechanism

An emergent strategy to improve biotic stress tolerance is the RNA interference or RNA silencing process, which are regulated by small non-coding RNAs (ncRNAs). The ncRNAs are mainly categorized into two groups, small interfering RNA (siRNA) and microRNA (miRNA). Although the biogenesis of these ncRNAs differs from one another, yet the regulatory mechanism of target gene suppression mediates through a common protein, ribonucleoprotein silencing complex called as RNA-induced silencing complex (RISC) (Kamthan et al. 2015).

siRNA is a small double stranded RNA (dsRNA) molecule (18–25 ntds), produced by the cleavage of dsRNA by an enzyme known as, dicer. The siRNAs have two nucleotide overhangs at the 3′end. The ribonucleoprotein from RISC unwinds the siRNA and binds with mRNA strand with a sequence specific manner, which can be recognized by the RISC complex and cleaves in the middle of the binding region. This cleaved mRNA can be recognized by the cell as abnormal and eventually destroyed. On the other hand, miRNAs are the small hairpin-shaped precursor molecules which are cleaved by the dicer enzyme. The mature miRNA forms miRNA-induced silencing complex (miRISC) that binds to the targeted mRNA molecule and retards or blocks translation (Wilson and Doudna 2013; Kamthan et al. 2015).

There are several RNAi strategies have been successfully applied in the crop improvement against biotic stresses. The first application of ncRNA in *Arabidopsis* was performed against *Pst* infection. The accumulation of *miR393* induced resistance by suppression of auxin signaling (Navarro et al. 2016). The *miR393* negatively regulates F-box auxin receptors and prevents *Pst* infection. Moreover, plants overexpressing *miR393* showed more resistance against *Pst*. RNAi mediated defense has also been shown against fungus infection. The suppression of Ω-3 fatty acid desaturase encoded by *OsFAD7* and *OsFAD8* enhanced resistance against *Magnaporthe grisea* (Yara et al. 2007). In soybean, RNAi targeting the genes involved in lignin pathway results in reduced lignin content which has been shown to induce resistance against *Sclerotinia sclerotiorum* (Peltier et al. 2009). Patwa et al. (2018) identified 422 differetially expressed miRNAs in *Phaseolus vulgaris* against mungbean yellow mosaic India virus infection. Among these 14 were validated through RT-qPCR and targets of the seven miRNAs were predicted, that are involved in virus resistance.

10.8 Future Perspective

The mechanism of plant-pathogen interaction is very specific, complex and dynamic. For several years, the plant breeders tried to identify the resistance genes or defense related genes in the wild type species of field crops and their subsequent integration into the commercial cultivars for the crop improvement. Recently, several techniques are successfully applied in the identification, characterization of several plant genes involved in biotic stress tolerance and their functional analyzation in pathogen resistance. For instance, RNA sequencing, suppression subtractive hybridization, cDNA-AFLP and microarrays, all are used to identify the defense related genes during plant-pathogen interaction. Moreover, the use of recombinant DNA technology to identify the superior genotype for the beginning phase of cultivar selection is also the emergent tools in plant science. Finally, the identification of the novel genes which are involved in recognition, signaling, and preventing pathogen invasion will provide the information for the generation of superior genotypes with durable resistance that will eventually stabilize the level of crop yields for the sustainable agriculture.

References

Aarts N, Metz M, Holub E, Staskawicz BJ, Daniels MJ, Parker JE (1998) Different requirements for EDS1 and NDR1 by disease resistance genes define at least two R gene-mediated signaling pathways in *Arabidopsis*. Proc Natl Acad Sci U S A 95:10306–10311

Abramovitch RB, Anderson JC, Martin GB (2006) Bacterial elicitation and evasion of plant innate immunity. Nat Rev Mol Cell Biol 7:601–611

Agrios G (2012) Plant pathology. Elsevier Science, Burlington

Aist JR, Bushnell WR (1991) Invasion of plants by powdery mildew fungi, and cellular mechanisms of the resistance. Plenum Press, New York

Alabouvette C, Olivain C, Steinberg C (2006) Biological control of plant diseases: the European situation. Eur J Plant Pathol 114:329–341

Alvarez ME (2000) Salicylic acid in the machinery of hypersensitive cell death and disease resistance. Plant Mol Biol 44:429–442

Apel K, Hirt H (2004) Reactive oxygen species: metabolism, oxidative stress, and signal transduction. Annu Rev Plant Biol 55:373–399

Baker CJ, Orlandi EW (1995) Active oxygen in plant/pathogen interactions. Annu Rev Phytopathol 33:299–321

Bari R, Jones JD (2009) Role of plant hormones in plant defence responses. Plant Mol Biol 69:473–488

Blanchette R, Biggs A (2013) Defense mechanisms of woody plants against fungi. Springer, Berlin/Heidelberg

Boller T, Felix G (2009) A renaissance of elicitors: perception of microbe-associated molecular patterns and danger signals by pattern-recognition receptors. Annu Rev Plant Biol 60:379–406

Brand D, Oishi BO, Roussos S, Soccol CR (2004) Spore production of *Paecilomyces lilacinus* by solid state culture and application in pot experiments to control *Meloidogyne incognita*. Appl Biochem Biotechnol 118:81–88

Buonaurio R (2008) Infection and plant defense responses during plant-bacterial interaction. Plant-Microbe Interact 8:169–197

Campbell CL, Buchwalter D (2012) Epidemiology and management of root diseases. Springer, Berlin/Heidelberg

Cardoso JE, Santos AA, Rossetti AG, Vidal JC (2004) Relationship between incidence and severity of cashew gummosis in semiarid north-eastern Brazil. Plant Pathol 53:363–367

Cesari S, Maud B, Philippe M, Thomas K, Peter ND (2014) A novel conserved mechanism for plant NLR protein pairs: the "integrated decoy" hypothesis. Front Plant Sci 5:606

Chakraborty N, Basak J (2018) Exogenous application of methyl jasmonate induces defense response and develops tolerance against mungbean yellow mosaic India virus in *Vigna mungo*. Funct Plant Biol 46(1):69–81

Clay NK, Adio AM, Denoux C, Jan-der G, Ausubel FM (2009) Glucosinolate metabolites required for an Arabidopsis innate immune response. Science 323:95–100

Collier SM, Moffett P (2009) NB-LRRs work a "bait and switch" on pathogens. Trends Plant Sci 14:521–529

Dodds PN, Rathjen JP (2010) Plant immunity: towards an integrated view of plant-pathogen interactions. Nature 11:1–10

Durrant WE, Dong X (2004) Systemic acquired resistance. Annu Rev Phytopathol 42:185–209

Eitas TK, Dangl JL (2010) NB-LRR proteins: pairs, pieces, perception, partners, and pathways. Curr Opin Plant Biol 13:472–477

Faulkner C, Robatzek S (2012) Plants and pathogens: putting infection strategies and defence mechanisms on the map. Curr Opin Plant Biol 15:699–707

Flor HH (1971) Current status of the gene for gene concept. Annu Rev Phytopathol 9:275–296

Fossdal CG, Sharma P, Lonneborg A (2001) Isolation of the first putative peroxidase cDNA from a conifer and local and systemic accumulation of related proteins upon pathogen infection. Plant Mol Biol 47:423–435

Fragnière C, Serrano M, Abou-Mansour E, Métraux JP, L'Haridon F (2011) Salicylic acid and its location in response to biotic and abiotic stress. FEBS Lett 585:1847–1852

Fraser RSS (1990) The genetics of resistance to plant viruses. Annu Rev Phytopathol 28:179–200

Gaumann E (1950) Principles of plant disease infection. Hafner, New York

Gill US, Lee S, Mysore KS (2015) Host versus nonhost resistance: distinct wars with similar arsenals. Phytopathology 105:580–587

Glazebrook J (2005) Contrasting mechanisms of defense against biotrophic and necrotrophic pathogens. Annu Rev Phytopathol 43:205–227

Greenberg JT (1997) Programmed cell death in plant-pathogen interactions. Annu Rev Plant Physiol Plant Mol Biol 48:525–545

Groth JV, Ozmon EA, Busch RH (1999) Repeatability and relationship of incidence and severity measures of scab of wheat caused by *Fusarium graminearum* in inoculated nurseries. Plant Dis 83:1033–1038

Guest DI, Brown JF (1997) Plant pathogens and plant diseases. Rockvale Publications, Armidale

Harman GE (2000) Changes in perceptions derived from research on *Trichoderma harzianum* T22. Plant Dis 84:377–393

Hirooka T, Ishii H (2013) Chemical control of plant diseases. J Gen Plant Pathol 79:390–401

Horsfall JG (2012) Plant disease: an advanced treatise: how disease is managed. Elsevier Science, Burlington

Huang XW, Zhao NH, Zhang KQ (2004) Extracellular enzymes serving as virulence factors in nematophagous fungi involved in infection of the host. Res Microbiol 155:811–816

Jia Y, Sean AM, Gregory TB, Howard PH, Barbara V (2000) Direct interaction of resistance gene and avirulence gene products confers rice blast resistance. EMBO J 19:4004–4014

Jones DG (2013) The epidemiology of plant diseases. Springer, Dordrecht

Jones JDG, Dangl JL (2006) The plant immune system. Nature 444:323–329

Jones JD, Jones JDG (1996) The roles of leucine-rich repeats in plant defences. Adv Bot Res 24:89–167

Junaid JM, Dar NA, Bhat TA, Bhat AH, Bhat MA (2013) Commercial biocontrol agents and their mechanism of action in the management of plant pathogens. Int J Modern Plant Anim Sci 1:39–57
Kamthan A, Chaudhuri A, Kamthan M, Datta A (2015) Small RNAs in plants: recent development and application for crop improvement. Front Plant Sci 6:208
Kazan K, Manners JM (2008) Jasmonate signaling: toward an integrated view. Plant Physiol 146:1459–1468
Kitajima S, Sato F (1999) Plant pathogenesis-related proteins: molecular mechanisms of gene expression and protein function. J Biochem 125:1–8
Kranz J (2003) Comparative epidemiology of plant diseases. Springer, New York
Lewis RA (2001) CRC dictionary of agricultural sciences. Taylor & Francis, Boca Raton
Li Y, Gu Y, Li J, Xu M, Wei Q, Wang Y (2015) Biocontrol agent *Bacillus amyloliquefaciens* LJ02 induces systemic resistance against cucurbits powdery mildew. Front Microbiol 6:883
Lu H, Greenberg JT, Holuigue L (2016) Editorial: salicylic acid signaling networks. Front Plant Sci 7:238
Maiti S, Basak J, Pal A (2014) Current understanding on plant r-genes/proteins and mechanisms of defense responses against biotic stresses. Rev Plant Pathol 6:93–126
Malinovsky FG, Fangel JU, Willats WGT (2014) The role of the cell wall in plant immunity. Front Plant Sci 5:178
Maloy OC (1993) Plant disease control. Wiley, New York
Matern U, Kneusal RE (1988) Phenolic compounds in plant disease resistance. Phytopathology 16:153–170
Mauch-Mani B, Mauch F (2005) The role of abscisic acid in plant-pathogen interactions. Curr Opin Plant Biol 8:409–414
Mehdy MC (1994) Involvement of active oxygen species in plant defense against pathogens. Plant Physiol 105:467–472
Mei C, Qi M, Sheng G, Yang Y (2006) Inducible overexpression of a rice allene oxide synthase gene increases the endogenous jasmonic acid level, PR gene expression, and host resistance to fungal infection. Mol Plant-Microbe Interact 19:1127–1137
Melotto M, Underwood W, Yang He S (2008) Role of stomata in plant innate immunity and foliar bacterial diseases. Annu Rev Phytopathol 46:101–122
Miedes E, Vanholme R, Boerjan W, Molina A (2014) The role of the secondary cell wall in plant resistance to pathogens. Front Plant Sci 5:358
Murphy AM, Chivasa S, Singh DP, Carr JP (1999) Salicylic acid induced resistance to viruses and other pathogens: a parting of the ways? Trends Plant Sci 4:155–160
Mysore KS, Ryu CM (2004) Nonhost resistance: how much do we know? Trends Plant Sci 9:97–104
Navarro L, Dunoyer P, Jay F, Arnold B, Dharmasiri N, Estelle M, Voinnet O, Jones JD (2016) A plant miRNA contributes to antibacterial resistance by repressing auxin signalling. Science 312:436–439
Nobuta K, Okrent RA, Stoutemyer M, Rodibaugh N, Kempema L, Wildermuth MC, Innes RW (2007) The GH3 acyl adenylase family member PBS3 regulates salicylic acid-dependent defense responses in *Arabidopsis*. Plant Physiol 144:1144–1156
Ortiz BV, Perry C, Sullivan D, Lu P, Kemerait R, Davis RF, Smith A, Vellidis G, Nichols R (2012) Variable rate application of nematicides on cotton fields: a promising site-specific management strategy. J Nematol 44:31–39
Patwa N, Nithin C, Bahadur RP, Basak J (2018) Identification and characterization of differentially expressed *Phaseolus vulgaris* miRNAs and their targets during mungbean yellow mosaic India virus infection reveals new insight into *Phaseolus*-MYMIV interaction. Genomics 111:1333–1342
Pedras MS, Yaya EE (2015) Plant chemical defenses: are all constitutive antimicrobial metabolites phytoanticipins? Nat Prod Commun 10:209–218

Peltier AJ, Hatfield RD, Grau CR (2009) Soybean stem lignin concentration relates to resistance to *Sclerotinia sclerotiorum*. Plant Dis 93:149–154
Peso L, Gonzalez VM, Inoharat N, Ellis RE, Nunez G (2000) Disruption of the Ced-9.Ced 4 complex by EGL-1 is a critical step for programmed cell death in *Caenorhabditis elegans*. J Biol Chem 275:27205–27211
Prell HH, Day P (2013) Plant-fungal pathogen interaction: a classical and molecular view. Springer, Berlin/Heidelberg
Ravichandra NG (2013) Fundamentals of plant pathology. PHI Learning, New Delhi
Rodoni B (2009) The role of plant biosecurity in preventing and controlling emergingplant virus disease epidemics. Virus Res 141:150–157
Sarhy-Bagnon V, Lozano P, Saucedo-Castaneda G, Roussos S (2000) Production of 6-pentyl-alpha-pyrone by *Trichoderma harzianum* in liquid and solid state cultures. Process Biochem 36:103–109
Schenk PM, Choo JH, Wong CL (2009) Microarray analyses to study plant defense and rhizosphere microbe interaction. CAB Rev Perspect Agric Vet Sci Nutr Nat Resour 4:45–46
Schulze-Lefert P, Panstruga R (2011) A molecular evolutionary concept connecting nonhost resistance, pathogen host range, and pathogen speciation. Trends Plant Sci 16:117–125
Sels J, Mathys J, De Coninck BM, Cammue BP, De Bolle MF (2008) Plant pathogenesis-related (PR) proteins: a focus on PR peptides. Plant Physiol Biochem 46:941–950
Seo S, Mitsuhara I, Feng J, Iwai T, Hasegawa M, Ohashi Y (2011) Cyanide, a coproduct of plant hormone ethylene biosynthesis, contributes to the resistance of rice to blast fungus. Plant Physiol 155:502–514
Sivaprakasan K, Vidhyasekaran P (1993) Phenylalanine ammonia lyase gene for crop disease management. In: Vidhyasekaran P (ed) Genetic engineering, molecular biology and tissue culture for crop pest and disease management. Daya Publishing House, New Delhi, pp 113–122
Solano R, Stepanova A, Chao Q, Ecker JR (1998) Nuclear events in ethylene signaling: a transcriptional cascade mediated by ETHYLENE-INSENSITIVE3 and ETHYLENE-RESPONSE-FACTOR1. Genes Dev 12:3703–3714
Sticher L, MauchMani B, Metraux JP (1997) Systemic acquired resistance. Annu Rev Phytopathol 35:235–270
Stotz HU, Thomson JG, Wang Y (2009) Plant defensins. Plant Signal Behav 4:1010–1012
Strawn MA, Marr SK, Inoue K, Inada N, Zubieta C, Wildermuth MC (2007) *Arabidopsis* isochorismate synthase functional in pathogen-induced salicylate biosynthesis exhibits properties consistent with a role in diverse stress responses. J Biol Chem 282:5919–5933
Takahashi F, Yoshida R, Ichimura K, Mizoguchi T, Seo S, Yonezawa M, Maruyama K, Yamaguchi-Shinozaki K, Shinozaki K (2007) The mitogen-activated protein kinase cascade MKK3-MPK6 is an important part of the jasmonate signal transduction pathway in Arabidopsis. Plant Cell 19:805–818
Thordal-Christensen H, Zhang Z, Wei Y, Collinge DB (1997) Subcellular localization of H2O2 in plants: H2O2 accumulation in papillae and hypersensitive response during the barley-powdery mildew interaction. Plant J 11:1187–1194
Tjamos EC, Papavizas GC, Cook RJ (2013) Biological control of plant diseases: progress and challenges for the future. Springer, Boston
Torres MA, Jones JD, Dangl JL (2006) Reactive oxygen species signaling in response to pathogens. Plant Physiol 141:373–378
van der Biezen EA, Jones JD (1998) The NB-ARC domain: a novel signalling motif shared by plant resistance gene products and regulators of cell death in animals. Curr Biol 8:226–227
Van der Does D, Leon-Reyes A, Koornneef A, Van Verk MC, Rodenburg N, Pauwels L, Goossens A, Körbes AP, Memelink J, Ritsema T, Van Wees SC, Pieterse CM (2013) Salicylic acid suppresses jasmonic acid signaling downstream of SCFCOI1-JAZ by targeting GCC promoter motifs via transcription factor ORA59. Plant Cell 25:744–761
van Loon LC, Rep M, Pieterse CMJ (2006) Significance of inducible defense-related proteins in infected plants. Annu Rev Phytopathol 44:135–162

Vargas WA, Sanz Martín JM, Rech GE, Rivera LP, Benito EP, Díaz-Mínguez JM, Thon MR, Sukno SA (2012) Plant defense mechanisms are activated during biotrophic and necrotrophic development of *Colletotricum graminicola* in Maize1[W][OA]. Plant Physiol 158:1342–1358

Verberne MC, Budi Muljono AB, Verpoorte R (1999) Salicylic acid biosynthesis. In: Libbenga K, Hall M, Hooykaas PJJ (eds) Biochemistry and molecular biology of plant hormones, vol 33. Elsevier, London, pp 295–312

Veronese P, Ruiz MT, Coca MA, Hernandez-Lopez A, Lee H (2003) In defense against pathogens. Both plant sentinels and foot soldiers need to know the enemy. Plant Physiol 131:1580–1590

Vlot AC, Klessig DF, Park SW (2008) Systemic acquired resistance: the elusive signal(s). Curr Opin Plant Biol 11:436–442

Wasternack C (2007) Jasmonates: an update on biosynthesis, signal transduction and action in plant stress response, growth and development. Ann Bot 100:681–697

Williamson VM, Gleason CA (2003) Plant-nematode interactions. Curr Opin Plant Biol 6:1–7

Wilson RC, Doudna JA (2013) Molecular mechanisms of RNA interference. Annu Rev Biophys 42:217–239

Wood R (2012) Active defense mechanisms in plants. Springer, Boston

Xu E, Brosche M (2014) Salicylic acid signaling inhibits apoplastic reactive oxygen species signaling. BMC Plant Biol 14:155

Yara A, Yaeno T, Hasegawa M, Seto H, Montillet JL, Kusumi K (2007) Disease resistance against *Magnaporthe grisea* is enhanced in transgenic rice with suppression of o-3 fatty acid desaturases. Plant Cell Physiol 48:1263–1274

Zhu Z, An F, Feng Y, Li P, Xue L, Mu A, Jiang Z, Kim JM, To TK, Li W, Zhang X, Yu Q, Dong Z, Chen WQ, Seki M, Zhou JM, Guo H (2011) Derepression of ethylene-stabilized transcription factors (EIN3/EIL1) mediates jasmonate and ethylene signaling synergy in Arabidopsis. Proc Natl Acad Sci U S A 108:12539–12544

Zipfel C, Kunze G, Chinchilla D, Caniard A, Jones JDG, Boller T, Felix G (2006) Perception of the bacterial PAMP EF-Tu by the receptor EFR restricts *Agrobacterium*-mediated transformation. Cell 125:749–760

Index

P. Guleria et al. (eds.), *Sustainable Agriculture Reviews 51*, Sustainable Agriculture Reviews 51, https://doi.org/10.1007/978-3-030-68828-8

The manufacturer's authorised representative in the EU is Springer Nature Customer Service Centre GmbH, Europaplatz 3, 69115 Heidelberg, Germany. If you have any concerns regarding our products, please contact ProductSafety@springernature.com

Printed and bound by CPI Group (UK) Ltd, Croydon, CR0 4YY
30/09/2026
02225705-0005